AYUSH Sponsored Two Day Conference cum Workshop on 'Advancement in Animal Handling and Generative AI for Pre-clinical Studies'

This Conference on Generative AI in Animal Experiments in Pharmacology represents a groundbreaking convergence of two distinct fields, artificial intelligence (AI) and pharmacological research involving animal models. This unique conference brings together experts from both domains to explore the potential of generative AI techniques in revolutionizing the way pharmaceutical compounds are tested and evaluated in preclinical studies. By leveraging generative AI algorithms, researchers aim to simulate and predict the effects of drugs on biological systems, reducing the reliance on animal experimentation while maintaining scientific rigor and accuracy. Attendees' present cutting-edge research on AI-driven drug discovery, computational modelling of drug interactions and virtual screening of pharmaceutical compounds, showcasing the transformative potential of AI in advancing pharmacological research practices. Furthermore, the conference serves as a catalyst for interdisciplinary collaboration, fostering dialogue between AI developers, pharmacologists, and regulatory authorities to address ethical considerations, regulatory challenges, and the responsible integration of AI technologies in animal experiments within the field of drug development.

Edited by

Dr. Gurudutta Pattnaik
Dean, School of Pharmacy
Centurion University of Technology and Management
Bhubaneswar, Odisha, India

Dr. Soumya Jal
Dean, School of Paramedics and Allied Health Sciences
Centurion University of Technology and Management
Bhubaneswar, Odisha, India

Advancement in Animal Handling and Generative AI for Pre-clinical Studies

Animal Handling and Generative AI for Pre-clinical Studies

Edited by

Gurudutta Pattnaik
Soumya Jal

CRC Press
Taylor & Francis Group
Boca Raton London New York

CRC Press is an imprint of the
Taylor & Francis Group, an **informa** business

First edition published 2026
by CRC Press
4 Park Square, Milton Park, Abingdon, Oxon, OX14 4RN

and by CRC Press
2385 NW Executive Center Drive, Suite 320, Boca Raton FL 33431

CRC Press is an imprint of Informa UK Limited

British Library Cataloguing-in-Publication Data
A catalogue record for this book is available from the British Library

ISBN: 9781041140795 (hbk)
ISBN: 9781041140818 (pbk)
ISBN: 9781003672869 (ebk)

DOI: 10.1201/9781003672869

Typeset in Times New Roman
by HBK Digital

Contents

Lists of figures

Lists of tables

List of contributors

Abhijith S.
Medical Radiology and Diagnostic
Imaging
School of Allied Health Sciences
REVA University
Bangalore, Karnataka, India
Medical Imaging Technology
Department of Radiology, K S Hegde
Medical Academy
NITTE University
Mangalore, Karnataka, India

Babita Adhikary
Department of Pharmaceutical Analysis
and Quality Assurance
Centurion University of Technology and
Management
Odisha, India

Sk Hasibur Ali
School of Pharmacy and Life Sciences
Centurion University of Technology and
Management
Odisha, India

Aswathi P.
Department of Radiology
School of Paramedics and Allied Health
Science
Centurion University of Technology and
Management
Bhubaneswar, Odisha, India
Medical Radiology and Diagnostic
Imaging
School of Allied Health Sciences
REVA University
Bangalore, Karnataka, India

Sucharita Babu
Department of Pharmacology
School of Pharmacy and Life Sciences
Centurion University of Technologies
and Management
Odisha, India

Amulyaratna Behera
School of Pharmacy and Life Sciences
Centurion University of Technology and
Management
Odisha, India

Preetha Bhadra
Department of Biotechnology
Centurion University of Technology and
Management
Odisha, India

Kalpita Bhatta
Department of Botany
School of Applied Sciences
Centurion University of Technology and
Management
Bhubaneswar, Odisha, India

Sonali Bhujabal
School of Pharmacy and Life Sciences
Centurion University of Technology and
Management
Odisha, India

Chandan Kumar Brahma
School of Pharmacy and Life Sciences
Centurion University of Technology and
Management
Jatni, Khurda, Odisha, India

Sangeeta Chhotaray
*School of Paramedics and Allied Health
Sciences*
*Centurion University of Technology and
Management*
Odisha, India

Jayeeta Dandapat
Department of Optometry
*School of Paramedics and Allied Health
Science*
*Centurion University of Technology and
Management*
Bhubaneswar, Odisha, India

Sanjoy Das
Department of Pharmaceutical Sciences
Dibrugarh University
Dibrugarh, Assam, India

Subhashree Das
School of Pharmacy and Life Sciences
*Centurion University of Technology and
Management*
Odisha, India

Anup Kumar Dash
*Shri Rawatpura Sarkar College of
Pharmacy*
*Shri Rawatpura Sarkar University
Raipur*
Chattishgarh, India

Jyoshna Rani Dash
School of Pharmacy and Life Sciences
*Centurion University of Technology and
Management*
Odisha, India

Sujit Dash
Institute of Pharmacy and Technology
Salipur, Cuttack, Odisha, India

Medisetty Gayatri Devi
Department of Pharmaceutics
*Viswanadha Institute of Pharmaceutical
Sciences*
Visakhapatnam, Andhra Pradesh, India
Department of Pharmaceutics
*GITAM School of Pharmacy, GITAM
(Deemed to be University), Rushikonda*
Visakhapatnam, Andhra Pradesh, India

Kumar Dhiraj
Department of Pharmacy
*Institute of Technology and
Management, GIDA*
Gorakhpur, Uttar Pradesh, India

Anushka Valeska Fernandes
Department of Radiology
*School of Paramedics and Allied Health
Science*
*Centurion University of Technology and
Management*
Bhubaneswar, Odisha, India

Yashwant Giri
School of Pharmacy and Life Sciences
*Centurion University of Technology and
Management*
Odisha, India

SriKavya Grandhi
*Department of Physiotherapy, School of
Paramedics and Allied Health Science*
*Centurion University of Technology and
Management*
Bhubaneswar, Odisha, India

Soumya Jal
*School of Paramedics and Allied Health
Sciences*
*Centurion University of Technology and
Management*
Odisha, India

Bikash Ranjan Jena
School of Pharmacy and Life Sciences
Centurion University of Technology and
Management
Bhubaneswar, Odisha, India

Sameer Jena
Department of Botany
School of Applied Sciences
Centurion University of Technology and
Management
Odisha, India

B. Jyotirmayee
Department of Botany
School of Applied Sciences
Centurion University of Technology and
Management
Odisha, India

Ravi Kumar Kalari
Sree Rama Educational Trust College of
Physiotherapy
MIMS Hospital
Vizianagaram, Andhra Pradesh, India

Pritish Kanungo
School of Pharmacy and Life Sciences
Centurion University of Technology and
Management
Odisha, India

Biswakanth Kar
School of Pharmaceutical Sciences
Siksha O Anusandhan Deemed to be
University
Bhubaneswar, India

Ladi Alik Kumar
Department of Pharmaceutics
Centurion University of Technology and
Management
Odisha, India

Preety Kumari
Department of Physiotherapy
CT University
Ludhiana, Punjab, India

Anusha Kusuma
Department of Pharmaceutics
Balaji Institute of pharmaceutical
Sciences
Warangal, Telangana, India

Gyanranjan Mahalik
Department of Botany
School of Applied Sciences
Centurion University of Technology and
Management
Odisha, India

Sonalika Mahapatra
School of Paramedic and Allied Health
Sciences
Centurion University of Technology and
Management
Odisha, India

Kirtimaya Mishra
School of Pharmacy and Life Sciences
Centurion University of Technology and
Management
Bhubaneswar, Odisha, India

Monali Priyadarshini Mishra
School of Paramedic and Allied Health
Sciences
Centurion University of Technology and
Management
Odisha, India

Biswaranjan Mohanty
Institute of Pharmacy and Technology
Salipur, Odisha, India

Pratikshya Mohanty
Department of Botany
School of Applied Sciences
Centurion University of Technology and
Management
Bhubaneswar, Odisha, India

Rozalika Mohanty
School of Pharmacy and Life Sciences
Centurion University of Technology and
Management
Odisha, India

Tamosa Mukherjee
School of Forensic Sciences
Centurion University of Technology and
Management
Odisha, India

Nagarjuna Narayansetti
Sai Sri College of Physiotherapy
Eluru, Andhra Pradesh, India

Gopal Krishna Padhy
Department of Pharmaceutical
Chemistry
Centurion University of Technology and
Management
Odisha, India

DattaSai Pamidimarri
Department of Physiotherapy, School of
Paramedics and Allied Health Science
Centurion University of Technology and
Management
Bhubaneswar, Odisha, India

Debabrata Panda
School of Pharmacy and Life Sciences
Centurion University of Technology and
Management
Bhubaneswar, Odisha, India

Jagadeesh Panda
Department of Pharmaceutical
Chemistry
Raghu College of Pharmacy
Dakamarri, Visakhapatnam, India

Jnanranjan Panda
Department of Science
Sri Sri University
Cuttack, Odisha, India

Sagorika Panda
Department of Botany
Dhenkanal Mahila Mahavidyalaya
Dhenkanal, Odisha, India

Aradhana Panigrahi
School of Pharmacy and Life Sciences
Centurion University of Technology and
Management
Odisha, India

Soumya Saswati Panigrahi
Department of Physiotherapy, School of
Paramedics and Allied Health Science
Centurion University of Technology and
Management
Bhubaneswar, Odisha, India

Prasanna Parida
Research Scholar (Pharmacy)
Biju Patnaik University of Technology
Rourkela, Odisha, India
Institute of Pharmacy and Technology
Salipur, Cuttack, Odisha, India

Gurudutta Pattnaik
School of Pharmacy and Life Sciences
Centurion University of Technology and
Management
Odisha, India

Omprakash Pradhan
School of Pharmacy and Life Sciences
Centurion University of Technology and
Management
Odisha, India

Rajesh Kumar Pradhan
School of Pharmacy and Life Sciences
Centurion University of Technology and
Management
Odisha, India

Victor Pradhan
Department of Biotechnology
Centurion University of Technology and
Management
Odisha, India

Amiya Kumar Prusty
Faculty of Pharmacy
C.V. Raman Global University
Bhubaneswar, Odisha, India

Rajat Kumar Prusty
School of Pharmacy and Life Sciences
Centurion University of Technology and
Management
Bhubaneswar, Odisha, India

Biswajeet Puhan
School of Pharmacy and Life Sciences
Centurion University of Technology and
Management
Odisha, India
Sagar Rout
School of Pharmaceutical Sciences
Siksha O Anusandhan University
Bhubaneswar, Odisha, India

Debaprasad Routray
School of Pharmacy and Life Sciences
Centurion University of Technology and
Management
Odisha, India

Santosh Kumar R.
Department of Pharmaceutics
GITAM School of Pharmacy, GITAM
(Deemed to be University), Rushikonda
Visakhapatnam, Andhra Pradesh, India

Biswa Mohan Sahoo
School of Pharmacy and Life Sciences
Centurion University of Technology and
Management
Jatni, Khurda, Odisha, India

Chinmaya Sahoo
School of Pharmacy and Life Sciences
Centurion University of Technology and
Management
Bhubaneswar, Odisha, India

Mamalisa Sahoo
School of Paramedic and Allied Health
Sciences
Centurion University of Technology and
Management
Odisha, India

Pralaya Kumar Sahoo
School of Paramedics and Allied Health
Sciences
Centurion University of Technology and
Management
Odisha, India

Guptanjali Sahu
School of Pharmacy and Life Sciences
Centurion University of Technology and
Management
Odisha, India

Dipika Rani Sahu
Pandavesar School of Pharmacy
Pandaveswar, Paschim Bardhaman
West Bengal, India

Himansu Bhusan Samal
Department of Pharmaceutics
School of Pharmacy and Life Sciences
Centurion University of Technology and
Management
Bhubaneswar, Odisha, India

Ipsita Priyadarsini Samal
Department of Botany
School of Applied Sciences
Centurion University of Technology and
Management
Odisha, India

Bhabani Sankar Satapathy
GITAM School of Pharmacy
GITAM Deemed to be University
Hyderabad Campus, Telangana, India

Kuldip Singh
School of Pharmacy and Life Sciences
Centurion University of Technology and
Management
Odisha, India

Satya Narayan Tripathy
School of Pharmacy and Life Sciences
Centurion University of Technology and
Management
Jatni, Khurda, Odisha, India

Kanika Verma
Natural Products for Neuroprotection
and Anti-ageing Research Unit
Chulalongkorn University
Bangkok, Thailand

Foreword or Series Editor Introduction

Lead Editors:

Dr. Gurudutta Pattnaik perused his BPharm from PES College of Pharmacy, Bangalore in 1996, MPharm from University Department of Pharmaceutical Sciences, Utkal University, Bhubaneswar, in 2001 and Doctoral degree from Department of Pharmaceutical Technology, Jadavpur University, Kolkata, West Bengal, India. Having over 25 years of UG and PG teaching experience he served various faculty positions at various Pharmacy Institutions of high repute in India and abroad, like National Institute of Pharmaceutical Education & Research (NIPER), Raebareli, India and College of Health Science, Mekelle University, Ethiopia. His area of Research Interest includes Nanoparticle, Liposomal drug delivery, Pharmacokinetics etc. He has supervised 9 PhD scholars under CUTM and 01 under BPUT, Odisha, Convenor in DST-SERB, ICMR sponsored National Seminars, Evaluator, Question paper setter at NIPER, Raebareli, India. He also served as Resource persons in various National and International Conference, QIPs and published papers, Book chapters in reputed indexing high impact factor journals.

Dr. Soumya Jal is currently serving as the Associate Professor and Dean of the School of Paramedics and Allied Health Sciences, Centurion University of Technology and Management (Odisha). With over a decade of experience in healthcare education and research, Dr. Jal has made significant contributions to the field, particularly in healthcare management and biotechnology. She holds a PhD in Biotechnology from Vellore Institute of Technology University, Vellore, with her research focusing on the tetrodotoxin profile of puffer fish and its bacterial origins in eastern India. She has published extensively in reputed journals and contributed chapters to significant academic books. Her expertise includes clinical biochemistry, immunology, microbiology, and molecular biology techniques. Throughout her career, Dr. Jal has been recognized with several awards and achievements, including a Research Award from VIT University and a Merit Scholarship Award for outstanding academic performance. She is also actively involved in interdisciplinary collaborations and initiatives aimed at improving healthcare delivery and outcomes. She has chaired a session at the 3rd GSN Conference and acted as a Jury member at ICMSE 2024 international conference. Dr. Soumya Jal has held several prestigious academic positions in her career. She has served as HoD (SoPAHS); Disciplinary Committee member; Nodal Officer from CUTM at Board of Practical Training (Ministry of Higher Education, GOI).

Series Editors:

Dr. Himansu Bhusan Samal

Dr. Himansu Bhusan Samal, MPharm, PhD, FIC is working as Associate Professor & Associate Dean, in School of Pharmacy and Life Sciences, Centurion University of Technology and Management, Bhubaneswar, Odisha. Prior to working at Centurion University, he was working as Assistant Professor at Guru Nanak Institutions Technical Campus-School of Pharmacy (An Autonomous Institution under UGC, NBA, NAAC A+ Accredited), Ibrahimpatnam, Hyderabad. He is having 16 years of experience in Teaching and Research. Currently he is working as. He has published 40 scientific research papers in international and national peer reviewed journals. He has also published 6 book chapter and 7 patents. He is a life member of APTI and has guided more than 30 M. Pharm students. Currently guiding 5 PhD research scholars. His primary research interest are in the field of Periodontal drug delivery, wound healing, solubility and dissolution rate enhancement, nanoparticulate drug delivery, and mucoadhesive drug delivery.

Dr. Kirtimay Mishra

Prof. (Dr.) Kirtimaya Mishra is a young and dynamic professor cum researcher. He completed his PhD, from Annamalai University, Chidambaram, India. He completed his Master of Pharmacy from Berhampur University, Berhampur, Odisha, India; and Bachelor of Pharmacy as well as Diploma in Pharmacy from Roland Institute of Pharmaceutical Sciences, Berhampur, Odisha, India. He is meritorious student during his studies. Currently he is working as Professor in Centurion University, Department of Pharmacy, Bhubaneswar, Odisha, India.

Prof. (Dr.) Kirtimaya Mishra has 14 years of research and teaching experience. He has published many articles at national/international journals of repute. He has presented his research work at national/international conferences and attended more than 250 conferences/seminars/workshops. He has guided 14 MPharm students to carry out their research work and 4 PhD students continuing research under his guidance. He has developed some new methods for analytical validation using Quality by Design technology. He is expertise in analytical instruments handling like HPLC, LC-MS/MS, HPTLC, UV-Visible Spectrophotometer etc.

Dr. Bikash Ranjan Jena

Dr. Bikash Ranjan Jena is Associate Professor at Centurion University, Bhubaneswar, Odisha, India. He has a total 13 years of academic and research experience. He has published 85+ research papers, edited 4 international textbooks, 35+ book chapters, 37 in Scopus, 18+ WoS, articles. He holds H index 10 in Google Scholar, 8 in Scopus. He received the Young Achiever Award in 2021 from the Institute of Scholars (InSc), India, Received 'Faculty Performer Award-2023' and 'Research Incentive Award'- 2022 from Centurion University for his article contributions. He is a life member of the Association of Pharmaceutical Teachers of India, the Indian Association of Nuclear Chemists and Allied Scientists, and the Asian Council of Science

Editors (ACSE). He is currently guiding 3 PhD scholars, and guided 19 MPharm, 25+ BPharm students and currently taken additional responsibility as Placement Co-ordinator for Training & Placement, Career counselling activities, and working as Central Academic Coordinator for managing Academic and Administrative activities of School of Pharmacy, Centurion University, Odisha since 2021 onwards.

Preface

AYUSH encompasses traditional Indian medical systems like Ayurveda, Yoga, Naturopathy, Unani, Siddha, and Homeopathy. The CCRAS, funded by AYUSH, supports research programs to scientifically validate traditional medicine's efficacy. India's Ministry of AYUSH promotes and regulates these practices, aiming for their integration into modern healthcare while preserving their cultural significance. Centurion University of Technology and Management (CUTM), established in 2010, offers quality education across various fields. Noteworthy for its holistic approach, CUTM emphasizes practical skills, industry collaboration, and societal contributions. It's School of Pharmacy and Life Sciences, along with the School of Paramedics and Allied Health Sciences, lead in providing quality healthcare education, maintaining robust ecosystems to bolster healthcare facilities.

This Conference on Generative AI in Animal Experiments in Pharmacology represents a groundbreaking convergence of two distinct fields, artificial intelligence (AI) and pharmacological research involving animal models. This unique conference brings together experts from both domains to explore the potential of generative AI techniques in revolutionizing the way pharmaceutical compounds are tested and evaluated in preclinical studies. By leveraging generative AI algorithms, researchers aim to simulate and predict the effects of drugs on biological systems, reducing the reliance on animal experimentation while maintaining scientific rigor and accuracy. Attendees' present cutting-edge research on AI-driven drug discovery, computational modelling of drug interactions and virtual screening of pharmaceutical compounds, showcasing the transformative potential of AI in advancing pharmacological research practices. Furthermore, the conference serves as a catalyst for interdisciplinary collaboration, fostering dialogue between AI developers, pharmacologists, and regulatory authorities to address ethical considerations, regulatory challenges, and the responsible integration of AI technologies in animal experiments within the field of drug development.

Goals:
Develop and implement advanced animal handling techniques and protocols.

Utilize Generative AI to minimize the use of experimental animals and generate precise in vivo data.

Provide a platform to exchange information and ideas on recent trends in Pharmaceutical and Biomedical research.

Key Feature: By leveraging generative AI algorithms, researchers aim to simulate and predict the effects of drugs on biological systems, reducing the reliance on animal experimentation while maintaining scientific rigor and accuracy.

Intended audience: Students, Research Scholars, Faculty Members, Pharmacists, Industry Personnel and Health Care Professionals.

Conference highlights: This groundbreaking conference unites AI and pharmacological research to explore generative AI's potential in transforming preclinical drug testing, aiming to reduce reliance on animal experimentation. Key focus on simulating drug effects with precision. Intended for students, researchers, faculty, pharmacists, industry personnel, and healthcare professionals seeking innovative advancements.

Acknowledgements or Credits List

The Lead Editors, Dr. Gurudutta Pattnaik and Dr. Soumya Jal are grateful to the Centurion University of Technology and Management, Odisha and list of contributors for their dedication supports. The Lead editors also grateful to the Internal Peer-Reviewers, Dr. Himansu Bhusan Samal, Dr. Kirtimay Mishra, Dr. Bikash Ranjan Jena, Mr. Debaprasad Routray (Ph.D.), and Dr. Yashwant Giri for in-time review, compilation and processing of all submitted manuscripts to External Peer Review members. This book begins by unraveling the nuances of drug discovery, detailing innovative strategies and methodologies used to identify potential therapeutic compounds. Emphasis is placed on pharmacokinetics, elucidating how drugs move through the body, ensuring optimal dosing and efficacy. Analytical methods are thoroughly examined for their role in drug formulation and quality control, ensuring the safety and potency of pharmaceutical products. The book explores the integration of natural products and phytochemistry into drug development, highlighting their potential as sources of novel therapeutics. Quality control processes are scrutinized to meet rigorous standards, essential for safeguarding public health. It also delves into the intersections of biotechnology, microbiology, and public health, addressing their pivotal roles in combating infectious diseases and advancing medical treatments. Furthermore, the book discusses different interventions and medical laboratory techniques, essential in managing chronic health conditions and optimizing patient care. Each chapter offers a blend of knowledge insights with applied aspects, making it an indispensable resource for researchers, healthcare professionals, and students alike in the fields of biomedical sciences and drug development.

1 AI-driven mathematical modelling of perturbated cancer metabolic pathways: Predicting doxorubicin-induced PHGDH inhibition through the PI3K/Akt/mTOR pathway

Victor Pradhan[a] and Preetha Bhadra[b]

Department of Biotechnology, Centurion University of Technology and Management, Odisha, India

Abstract: Signal transduction and gene law are pivotal in reorganizing metabolic sports to assist cellular proliferation. The PI3K/Akt/mTOR pathway, a master regulator of aerobic glycolysis and cellular biosynthesis, is regularly hyperactive in most cancer cells, facilitating their metabolic adaptations. We developed a computational model using JWS (Java Web Analysis) to analyze the effect of PHGDH inhibition on most cancer metabolism to simulate and predict the metabolic effects of doxorubicin-induced PHGDH inhibition in cancer cells through the PI3K/Akt/mTOR pathway using computational modeling and in silico analysis. The model simulates metabolic states: one representing a proliferating cancer cell with active PI3K/Akt/mTOR signaling and aerobic glycolysis and the other with decreased glycolytic flux and enhanced mitochondrial metabolism. Doxorubicin, a PHGDH inhibitor, was delivered to assess changes in metabolic pathways, including glycolysis, the pentose phosphate pathway (PPP) and nucleotide biosynthesis. Our simulations indicate that doxorubicin-induced inhibition of PHGDH results in 3-phosphoglycerate (PG3) accumulation and reduced glycolytic flux. This reorganization reduces lactic acid production and disrupts serine synthesis, impairing nucleotide biosynthesis. Consequently, DNA replication and repair processes are compromized, slowing most cancer cell proliferation. This work highlights the capacity to leverage cancer cells' metabolic vulnerabilities via targeted drug interventions. By inhibiting PHGDH, doxorubicin reduces glycolysis and impacts critical biosynthetic pathways, supplying a therapeutic advantage. Our computational model gives insights into the metabolic dynamics of cancer cells and underscores the application of generative AI algorithms in predicting drug effects on organic systems.

Keywords: PI3K/Akt/mTOR pathway, glycolysis, PHGDH inhibition, doxorubicin, cancer metabolism, computational modelling, artificial intelligence, machine learning, deep learning

1. Introduction

1.1. Background on cancer metabolism

The PI3K/Akt/mTOR signaling pathway is an essential cascade that significantly influences cellular functions, including growth, metabolism, survival and proliferation. This pathway is frequently hyperactivated in numerous cancer types, facilitating the metabolic reprogramming that underpins accelerated cellular growth and division, commonly called the **Warburg effect**. Significant metabolic reprogramming occurs in

[a]victor.pradhan@cutm.ac.in, [b]preetha.bhadra@cutm.ac.in

DOI: 10.1201/9781003672869-1

cancer cells to facilitate their unchecked growth and multiplication. The PI3K/ Akt/mTOR pathway, which controls food absorption, biosynthesis and cellular metabolism and allows cancer cells to proliferate in harsh environments, is essential to this adaptation. The Warburg effect, in which cells preferentially favor aerobic glycolysis over oxidative phosphorylation even in the presence of oxygen, is one of the hallmark metabolic alterations in cancer cells [1–5].

Microbes and other unicellular creatures are under evolutionary pressure to multiply as fast as possible when nutrients are available. Their metabolic regulation systems have developed to detect when enough nutrition is available and direct the necessary amounts of carbon, nitrogen and free energy into creating the building blocks required to create a new cell [6–8]. When there is a shortage of nutrients, cells stop producing biomass and modify their metabolism to take as much free energy as possible from the available resources to survive the starving phase. Different regulatory mechanisms have evolved to control cellular metabolism in proliferating vs non-proliferating cells, reflecting these fundamental distinctions in metabolic demands [9–13].

The majority of cells in multicellular organisms are always in contact with an abundance of nutrients. Control systems that stop abnormal individual cell proliferation when nutrition availability surpasses the levels required to support cell division are essential for the organism's survival. Because mammalian cells generally do not take up nutrients from their surroundings unless induced to do so by growth hormones, uncontrolled proliferation is prevented [14–16]. Through the acquisition of genetic mutations that functionally modify

receptor-initiated signaling pathways, cancer cells can circumvent this growth factor reliance. There is mounting evidence that several of these pathways are essential for triggering food absorption and metabolism, which supports cell growth and survival [17].

Oncogenic mutations can lead to an increased uptake of nutrients, especially glucose, which can satisfy or surpass the bioenergetic needs associated with cellular growth and division. This understanding has reignited interest in Otto Warburg's 1924 finding that cancer cells process glucose differently than normal tissue cells. By exploring the implications of Louis Pasteur's findings on the fermentation of glucose to ethanol in relation to mammalian tissues [18, 19]. Warburg discovered that, in contrast to most normal tissues, cancer cells often convert glucose into lactate even when there is ample oxygen available for mitochondrial oxidative phosphorylation. A conclusive explanation for Warburg's findings has proven difficult to ascertain, partly because the energy demands of cell proliferation initially seem to be more effectively fulfilled through the complete breakdown of glucose via mitochondrial oxidative phosphorylation, which optimizes adenosine 5′-triphosphate (ATP) production [2023].

The present comprehension of metabolic pathways is predominantly derived from investigations of nonproliferating cells within differentiated tissues. In aerobic conditions, most differentiated cells primarily convert glucose into carbon dioxide through the oxidation of glycolytic pyruvate within the mitochondrial tricarboxylic acid (TCA) cycle [23, 24]. This process generates NADH (nicotinamide adenine dinucleotide, reduced), which subsequently drives oxidative phosphorylation to optimize ATP synthesis, resulting in minimal lactate production. It is only in anaerobic environments that differentiated cells generate significant quantities of lactate. Conversely, most cancer cells produce substantial amounts of lactate irrespective of oxygen availability, which characterizes their metabolism as 'aerobic glycolysis' [25–29]. Warburg initially proposed that cancer cells exhibit mitochondrial defects that hinder aerobic respiration, increasing their dependence on glycolytic pathways. However, later research has indicated that mitochondrial function is generally intact in most cancer cells, implying an alternative rationale for the phenomenon of aerobic glycolysis in these cells [30–34].

For Warburg, numerous inquiries persisted without resolution, particularly regarding the rationale behind cancer cells' inefficient diversion of glucose-derived pyruvate towards lactate synthesis rather than directing it into the TCA cycle, which would yield considerably greater ATP output. Warburg proposed that the lactate production observed in cancer cells stemmed from a disruption in oxidative phosphorylation resulting from mitochondrial damage [35–37].

1.2. PI3K/Akt/mTOR pathway in cancer metabolism

In cancer cells, the PI3K/Akt/mTOR signaling pathway is often found to be hyperactivated due to mutations in upstream regulators, notably PTEN. PTEN functions as a tumor suppressor by dephosphorylating PIP3 back to PIP2, negatively regulating PI3K (Figure 1.2). When PTEN function is lost or activating mutations occur in PI3K or Akt, the pathway becomes excessively activated,

promoting cancer cell proliferation and metabolic reprogramming [38–40].

The hyperactivation of the PI3K/Akt/mTOR pathway facilitates the Warburg effect, wherein cancer cells predominantly rely on aerobic glycolysis for energy production, even in the presence of oxygen, rather than utilizing oxidative phosphorylation [41]. This metabolic alteration enables cancer cells to swiftly generate ATP and produce essential biosynthetic intermediates, including nucleotides, amino acids and lipids. Akt enhances glucose uptake by increasing the expression and translocation of GLUT1 transporters to the cell membrane, thereby elevating glycolysis rates in these cells. Furthermore, Akt phosphorylates and activates hexokinase 2 (HK2), the enzyme that catalyzes the initial glycolysis step, further boosting glycolytic activity [42].

The activation of mTOR is crucial for synthesizing macromolecules required for cellular growth. mTOR promotes protein synthesis through its downstream effectors, S6K and 4E-BP1 and stimulates lipid biosynthesis by activating sterol regulatory element-binding proteins (SREBPs), transcription factors that govern lipid metabolism. The increase in lipid biosynthesis is vital for forming cellular membranes in rapidly proliferating cancer cells [43].

1.3. Mathematical approach to the cancer metabolism

1. Many unanswered cancer metabolism questions are still impossible in vivo. So, in this work, we have proposed a mathematical approach to cancer metabolism.

2. Mathematical modeling is essential for comprehending the intricate

metabolic networks in cancer cells and forecasting their reactions to different disturbances, including drug treatments. In the context of cancer, metabolic reprogramming is a defining characteristic that promotes swift growth and proliferation [44]. Gaining insights into these metabolic alterations from a systems perspective can aid in developing targeted therapeutic strategies.

Cancer cells demonstrate modified metabolic pathways to facilitate their accelerated growth and division. A key characteristic of cancer cell metabolism is the Warburg effect, wherein cells preferentially utilize aerobic glycolysis despite oxygen availability. This metabolic adaptation enables cancer cells to fulfill the energy and biosynthetic precursor requirements for rapid proliferation [45]. Additional metabolic alterations encompass changes in the pentose phosphate pathway (PPP) for nucleotide synthesis, increased glutaminolysis and augmented lipid synthesis [46]. Modeling these metabolic pathways presents challenges, as cancer metabolism is governed by numerous feedback mechanisms and is influenced by signaling pathways, including the PI3K/Akt/mTOR cascade. Mathematical models provide a structured approach to simulate and forecast metabolic behavior under varying conditions, offering insights into potential therapeutic targets [47].

1.4. Development and application of mathematical model

Mathematical representations of cancer metabolism generally utilize ordinary differential equations (ODEs) to depict the kinetics of various biochemical reactions within metabolic pathways [48].

These models are grounded in mass-action kinetics or **Michaelis-Menten kinetics** to characterize enzyme-catalyzed reactions, enabling the simulation of both steady-state and dynamic responses to perturbations [49].

The primary steps in developing a mathematical model of cancer metabolism include:

1. **Identifying the Metabolic Pathways:** This step involves selecting pertinent metabolic pathways, such as glycolysis, the PPP and the tricarboxylic acid (TCA) cycle, modified in cancer.

2. **Defining Kinetic Parameters:** Experimental data estimates parameters like reaction rates, enzyme affinities and metabolite concentrations. These parameters are vital for accurately simulating metabolic fluxes.

3. **Integrating Signaling Pathways**: Given that cancer metabolism is intricately regulated by signaling pathways such as PI3K/Akt/mTOR, these pathways are frequently incorporated into the model to capture the feedback mechanisms involved [50].

Incorporating artificial intelligence (AI) and machine learning (ML) has markedly enhanced the mathematical modeling of cancer metabolism. AI-based models can scrutinize extensive datasets, including metabolomics and transcriptomics, thereby improving model parameters and enabling more precise predictions of drug responses [51]. For instance, deep learning algorithms are adept at recognizing patterns in metabolic fluxes, facilitating the optimization of drug combinations by forecasting how cancer cells may respond to various treatments. These models are especially valuable for investigating the metabolic plasticity of cancer cells, which can shift between metabolic pathways to endure challenging conditions such as nutrient scarcity or therapeutic interventions [4, 6].

1.5. Computational tools

Various computational tools and platforms have been established to facilitate the mathematical modeling of cancer metabolism. These include:

- **JWS Online:** This web-based application enables users to construct, simulate and analyze kinetic models of metabolic networks. JWS accommodates the use of ODEs to simulate both steady-state and dynamic behaviors of metabolic pathways.
- **COBRA Toolbox:** This software, based on MATLAB, is designed for constraint-based modeling of metabolic networks and is extensively utilized for simulating metabolic fluxes in cancer cells.
- **Metabo Tools:** An AI-driven platform that synthesizes multi-omics data, including metabolomics and transcriptomics, to create predictive models of metabolic behavior in cancer, etc.

2. Methodology

In this work, we constructed a mathematical model to examine the impact of doxorubicin-induced inhibition of PHGDH on cancer metabolism. PHGDH, or phosphoglycerate dehydrogenase, is the enzyme that redirects 3-phosphoglycerate (3-PG) from glycolysis into the serine biosynthesis pathway (Figure 1.1). The inhibition of PHGDH interferes with serine production, which is essential for nucleotide synthesis and cellular growth [52].

2.1. Model fitting

We have used JWS to perform the kinetic behavior of the proposed metabolism. Our models are presently housed in model repositories, including JWS Online (http://jjj.biochem.sun.ac.za) [52]. Our ultimate objective is to integrate models of various cellular system components to develop intricate dynamic models at the cellular level. Furthermore, this approach is beneficial

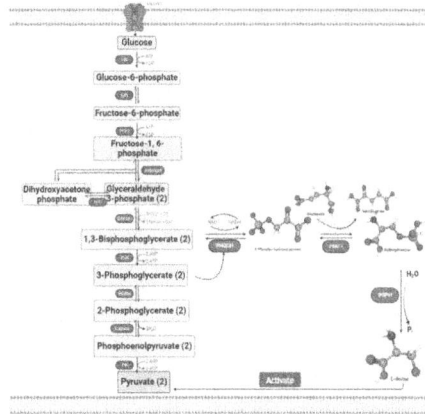

Figure 1.1. Glycolysis linked with serine synthesis showing PHGDH activation [5, 6].

Source: Author.

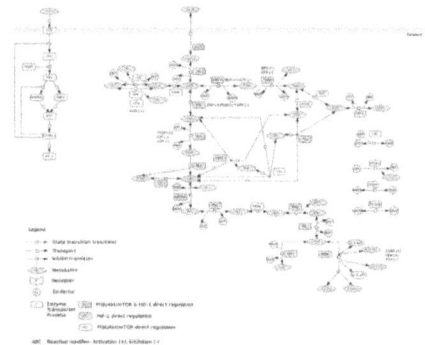

Figure 1.2. Glycolysis linked with PI3K-Akt-mTor pathway [5].

Source: Adapted from Mosca et.al., 2012.

for assessing research papers that utilize kinetic models. JWS Online adheres to this methodology and comprises the following components: **Species** refer to chemical compounds and reactants identified by their names and initial quantities. **Reactions** encompass mathematical representations of the transformations and corresponding rate laws among specific species. Quantities that symbolize local or global characteristics of a system are referred to as **parameters** (Table 1.1). **Units** are employed to articulate quantities within the model. **Rules** consist of mathematical expressions that establish parameters or impose constraints on the model, which are not classified under Reactions [52].

A response to these challenges was the development of the Systems Biology Markup Language (SBML), a set of electronic document standards aimed at Internet publications. SBML is founded on XML (eXtensible Markup Language) and offers a standardized 'language' for exchanging models among different software applications.

2.2. Equations

In order to decrease the concentration of PG3 through doxorubicine induced inhibition, we have taken v_{PGM23} and v_{PGP23} into consideration. Followings are the equation for the same:

$$v_{PGM23} = \frac{k_{PGM23}\left([PG13] - \frac{1}{K_{eq.}^{PGM23}}[PG23]\right)}{1 + \frac{[PG23]}{K_{PG23}}} \quad (1)$$

$$v_{PGP23} = \frac{V_{max}\left([PG23] - \frac{1}{K_{eq.}^{PG23}}[PG3]\right)}{[PG23] + K_{PG23}} \quad (2)$$

Both v_{PGM23} and v_{PGP23} plays a crucial role in glycolysis and regulates serine synthesis and PI3K-Akt-mTOR metabolism. In our simulation, we have used these two

enzymes as primary for the doxorubic inhibition.

Here are the list of enzymatic expressions used in our whole simulation in ODE form [5, 6],

Hexokinase (HK):

$$v_{HK} = \frac{\frac{V_{max1}}{K_{MgATP}}\left([MgATP] + \frac{V_{max2}}{V_{max1}}\frac{[MgATP][Mg]}{K_{MgATP,Mg}} \cdot \frac{(G6P)[MgADP]}{K_{eq}^{HK}}\right)}{\left(1 + \frac{[MgATP]}{K_{MgATP}}\right)\left(1 + \frac{[Mg]}{K_{MgATP,Mg}}\right) + \frac{[Mg]}{K_{Mg}}\left(\frac{[G6P]}{K_{G6P}} + 1.55\right)}$$
$$\left(1 + \frac{[Mg]}{K_{Mg}}\right) + \frac{[PG23]}{K_{PG23}}$$
$$+ \frac{[Mg][PG23]}{K_{Mg}K_{MgPG23}}$$

$V_{max} = 89.91mM \cdot h^{-1}, K_m^{DHAP} = 0.0364mM, K_m^{GAP} = 0.1906mM, K_m^{FBP} = 0.0071mM \ K_{eq}^{ALD} = 0.114, K_i^{GAP} = 0.0572mM, K_{ii}^{GAP} = 0.176Mm$

Glucose-6-phosphate   isomerase (PGI):

$$v_{PGI} = \frac{V_{max}\left([G6P] - \frac{1}{K_{eq.}^{PGI}}[F6P]\right)}{[G6P] + K_{G6P}\left(1 + \frac{[F6P]}{K_{F6P}}\right)} V_{max}$$
$$= 935mM \cdot h^{-1}, K_{G6P} = 0.182mM \ K_{F6P} = 0.071mM, K_{eq.}^{PGI} = 0.444$$

Phosphofructokinase (PFK):

$$V_{PFK} = \frac{V_{max}\left([F6P][MgATP] - \frac{[FBP][MgADP]}{K_{eq}^{PFK}}\right)}{([F6P] + K_{F6P})([MgATP] + K_{MgATP})}$$
$$\left(1 + L_0\frac{\left(1 + \frac{[ATP]}{K_{ATP}}\right)^4\left(1 + \frac{[Mg]}{K_{Mg}}\right)^4}{\left(1 + \frac{[AMP]}{K_{AMP}}\right)^4\left(1 + \frac{[F6P]}{K_{F6P}}\right)^4}\right)$$

Aldolase (ALD):

$$v_{ALD} = \frac{\frac{V_{max}}{K_m^{FBP}}([FBP]) - \frac{[GAP][DHAP]}{K_{eq}^{ALD}}}{1 + \frac{[FBP]}{K_m^{FBP}} + \frac{[GAP]}{K_i^{GAP}} +}$$
$$\frac{[DHAP]([GAP] + K_m^{GAP})}{K_m^{DHAP} * K_i^{GAP}} + \frac{[FBP][GAP]}{K_m^{FBP}K_{ii}^{GAP}}$$

Triosephosphate isomerase (TPI):

$$v_{TPI} = \frac{V_{max}\left([DHAP] - \frac{1}{K_{eq.}^{TPI}}[GAP]\right)}{[DHAP] + K_{DHAP}\left(1 + \frac{[GAP]}{K_{GAP}}\right)} V_{max} = 5456.6mM \cdot$$

$h^{-1}, K_{eq.}^{TPI} = 0.051 \ K_{DHAP} = 0.838mM, K_{GAP} = 0.428mM$

Glyceraldehyde phosphate dehydrogenase (GPDH):

v_{GPDH}
$$= \frac{\frac{V_{max}}{K_{NAD}K_{GAP}R_{P_i}}\left([NAD][GAP][P_i] - \frac{1}{K_{eq}^{GPDH}}[PG13][NADH]\right)}{-1 + \left(1 + \frac{[NAD]}{K_{NAD}}\right)\left(1 + \frac{[GAP]}{K_{GAP}}\right)}$$
$$\left(1 + \frac{[P_i]}{K_{P_i}}\right) + \left(1 + \frac{[NADH]}{K_{NADH}}\right)\left(1 + \frac{[GAP]}{K_{PG13}}\right) V_{max}$$
$$= 4300mM \cdot h^{-1}, K_{eq}^{GPDH} = 0.00109 \ K_{GAP}$$
$$= 0.005mM, K_{P_i} = 3.9mM \ K_{NAD} = 0.05mM, K_{NADH}$$
$$= 0.0083mM, K_{PG13} = 0.0035mM$$

Phosphoglycerate kinase (PK):

$$v_{PGK} = \frac{\frac{v_{max}}{K_{MgADP}K_{PG13}}\left([MgADP][PG13]-\frac{1}{K_{eq}^{PGK}}[MgATP][PG3]\right)}{-1+\left(1+\frac{MgADP}{K_{MgADP}}\right)\left(1+\frac{[PG13]}{K_{PG13}}\right)+\left(1+\frac{[MgATP]}{K_{MgATP}}\right)\left(1+\frac{[PG3]}{K_{PG3}}\right)}$$

$V_{max} = 5000mM^{-m^{-1}}, K_{eq.}^{PGK} = 2232, K_{MgADP} = 0.35mM$ $K_{MgATP} = 0.48mM, K_{PG13} = 0.002mM, K_{PG3} = 1.2mM$

2,3 Biphosphoglycerate mutase (PGM23):

$$v_{PGM23} = \frac{k_{PGM23}\left([PG13]-\frac{1}{K_{eq}^{PGM23}}[PG23]\right)}{1+\frac{[PG23]}{K_{PG23}}}$$

$k_{PGM23} = 110000\ h^{-1}, K_{PG23} = 0.04mM, K_{eq.}^{PGM23} = 100000$

2,3 Biphosphoglycerate phosphatase (PGP23):

$$v_{PGP23} = \frac{v_{max}\left([PG23]-\frac{1}{K_{eq}^{P23}}[PG3]\right)}{[PG23]+K_{PG23}}$$

$V_{max} = 0.52mM \cdot h^{-1}, K_{eq.}^{PGP2} = 100000, K_{PG23} = 0.2mM$

3-phosphoglycerate mutase (PGLM):

$$v_{PGLM} = \frac{V_{max}\left([PG3]-\frac{1}{K_{eq.}^{PGLM}}[PG2]\right)}{[PG3]+K_{PG3}\left(1+\frac{[PG2]}{K_{PG2}}\right)}$$

$V_{max} = 2000mM^{-1}, K_{eq.}^{PGLM} = 0.05, K_{PG3} = 5.0mM, K_{PG2} = 1.0mM$

Enolase (ENO):

$$V_{ENO} = \frac{V_{max}\left([PG2]-\frac{1}{K_{eq}^{ENO}}[PEP]\right)}{[PG2]+K_{PG2}\left(1+\frac{[PEP]}{K_{PEP}}\right)}$$

$V_{max} = 1500mM^{-1}, K_{eq.}^{ENO} = 1.2\ K_{PEP} = 1.0mM, K_{PG2} = 1.0mM$

Pyruvate kinase (PK):

$$v_{PK} = \frac{v_{max}\left([PEP]\cdot[MgADP]-\frac{[PYR][MgATP]}{K_{eq.}}\right)}{([PEP]+K_{PEP})([MgADP]+K_{MgADP})}$$

$$\left(1+L_0\frac{\left(1+\frac{[ATP]}{K_{ATP}}\right)^4}{\left(1+\frac{[PEP]}{K_{PEP}}\right)^4\left(1+\frac{[FBP]}{K_{FBP}}\right)^4}\right)$$

$V_{max} = 570mM \cdot h^{-1}, K_{eq.}^{PK} = 3349\ L_0 = 19, K_{PEP} = 0.225mM, K_{MgADP} = 0.474mM$

To replicate the consequences of PHGDH inhibition, the model incorporated parameters for the accumulation of PG3 and its subsequent impacts on serine and nucleotide biosynthesis. Doxorubicin (by decreasing the concentration of PG using v_{PGM23} and v_{PGP23}) was integrated into the system as a PHGDH inhibitor, modifying the metabolic flow between glycolysis and the serine bio-synthesis pathway. The kinetic parameters for PHGDH and its involvement in serine biosynthesis were obtained which emphasizes the enzyme's significance in facilitating cancer metabolism [53].

In addition the mechanistic simulation, ML algorithms were utilized to forecast the dynamic responses of cancer cells to the inhibition of PHGDH. Deep learning models, which were developed using experimental data from prior studies, were employed to anticipate alterations in metabolic flux in response to different dosages of doxorubicin. The predictions generated by these AI-driven models were corroborated with experimental results documented in the literature, emphasizing variations in glycolytic flux, nucleotide biosynthesis and mitochondrial function.

3. Result and Discussion

Our computational model (Figure 1.3) revealed that the inhibition of PHGDH induced by doxorubicin led to a notable increase in the levels of PG3. This increase reduced the overall glycolytic flux, redirecting glucose metabolism from lactic acid production to mitochondrial oxidative phosphorylation. Consequently, there was a decrease in ATP production via glycolysis, which is essential for the rapid proliferation of cancer cells. Furthermore, the inhibition of PHGDH hindered the biosynthesis of serine, which serves as a precursor for nucleotide synthesis.

Time v/s Concentration analysis using JWS model shows the perturbation of PG3 resulting from doxorubicin-induced inhibition of PHGDH to evaluate its effects on cancer cell metabolism. PHGDH is an essential enzyme in

Table 1.1. List of species and their initial conditions [5]

Id	Name	Initial condition
species_1	GLC	0.000897
species_2	G6P	0.00109
species_3	ADP	0.0027
species_4	ATP	0.0087
species_5	F6P	0.0000362
species_6	F16P	0.000367
species_7	E4P	0.00093
species_8	PGN	0.0001
species_9	GLC_e	0.01
species_10	NADP	0.0000006118
species_11	NADPH	0.0000187082
species_12	BPG	0.0000629
species_13	RU5P	0.000143
species_14	X5P	0.000242
species_15	R5P	0.0000274
species_16	GAP	0.000153
species_17	S7P	0.0000858
species_18	NADH	0.0000050000 0000000001
species_19	NAD	0.00134
species_20	AMP	0.00311
species_21	PRPP	0.001
species_22	G1P	0.0000341
species_23	Pi	0.02
species_24	GLY	0.208403745497308
species_25	CIT	0.00108
species_26	F26P	0.00000367
species_27	DHAP	0.000553
species_28	PG3	0.0000307
species_29	PG2	0.00000498
species_30	PEP	0.0000579
species_31	PYR	0.00183
species_32	LAC	0.0155
species_33	CO2	0.0214
species_34	O2	0.000065

Source: Author.

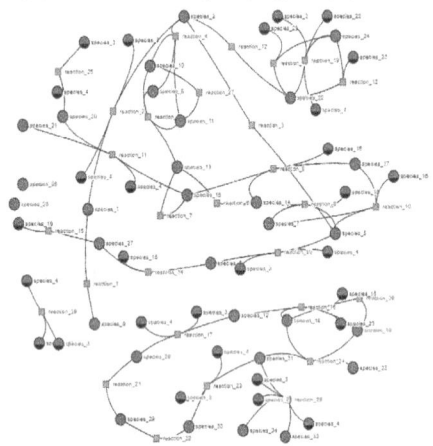

Figure 1.3. Simulated model for cancer metabolism through JWS, where Green represents the enzymes taking place, Blue represents species nodes and the other represents cofactors like NADH and ATP.

Source: Author.

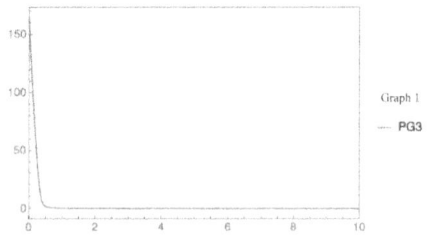

Figure 1.4. JWS generated plot showing the level of PG3 after giving doxorubicin.

Source: Author.

serine biosynthesis, connecting glycolytic intermediates to serine production. The in silico model examines two scenarios: before and following the perturbation induced by doxorubicin treatment, focusing on analyzing metabolic fluxes.

Graph 1: PG3 Accumulation The initial graph illustrates a time-dependent simulation of PG3 levels in cancer cells following doxorubicin administration. Initially, there is a rapid increase in PG3

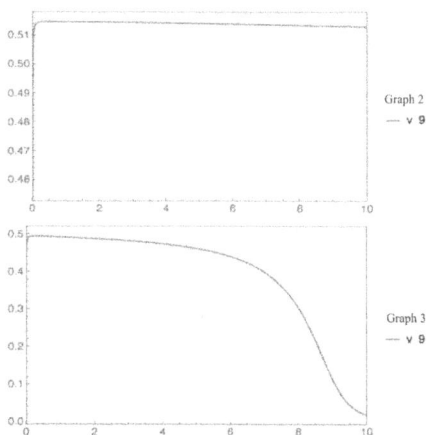

Figure 1.5. JWS generated plot showing the condition before and after Doxorubicin.

Source: Author.

concentration, indicative of the accumulation of glycolytic intermediates resulting from the inhibition of PHGDH activity (Figure 1.4). This phenomenon arises because the inhibition of PHGDH interferes with the subsequent conversion of PG3 into the serine synthesis pathway. Consequently, PG3 levels rise significantly in the initial 2 minutes, achieving a peak concentration before gradually stabilizing.

Graph 2: Fluxes before Perturbation: The second graph illustrates the condition before any perturbation, depicting the fluxes within a system where PHGDH is functioning and doxorubicin treatment is absent. The flux exhibits stability over time, suggesting that the cancer cell sustains consistent glycolysis, accompanied by the normal synthesis of serine and other downstream metabolites. This stable state signifies active aerobic glycolysis and elevated biosynthetic requirements characteristic of proliferating cancer cells (Figure 1.5).

Graph 3: Fluxes following Perturbation: The third graph illustrates

the metabolic condition after perturbation by doxorubicin, revealing a significant decrease in flux. The administration of doxorubicin inhibits PHGDH, resulting in the accumulation of PG3 and a reduction in glycolytic flux. This alteration leads to a metabolic transition in which the cell's dependence on glycolysis diminishes, potentially enhancing mitochondrial metabolism. The reduction in glycolytic flux signifies that the cancer cell's energy production and biosynthetic functions are impaired, directly impacting its ability to proliferate.

3.1. Interpretation and impact on cancer metabolism

The disruption caused by doxorubicin significantly alters the metabolic network of cancer cells by leading to the accumulation of PG3 and a decrease in glycolytic flux.

This metabolic shift carries several important consequences:

Decreased Nucleotide Biosynthesis: The reduced flow of PG3 into the synthesis of serine and glycine restricts nucleotide biosynthesis, thereby hindering DNA replication and repair processes.

Reduced Lactic Acid Production: A decline in glycolytic flux leads to lower production of lactic acid, causing cancer cells to move away from aerobic glycolysis (Warburg effect) and potentially enhancing oxidative phosphorylation.

Impaired Cell Proliferation: With nucleotide biosynthesis and energy production reduced, the overall rate of cancer cell proliferation decreases, rendering them more vulnerable to therapeutic strategies.

This analysis highlights the promise of exploiting metabolic weaknesses in cancer cells through the use of agents such as doxorubicin, which obstruct critical enzymes like PHGDH. Integrating computational modelling with drug perturbation experiments yields a significant understanding of cancer cell metabolism, establishing a basis for forecasting the effectiveness of therapies aimed at metabolic targets.

Stoichiometric visualization shows the quantitative relation between all the

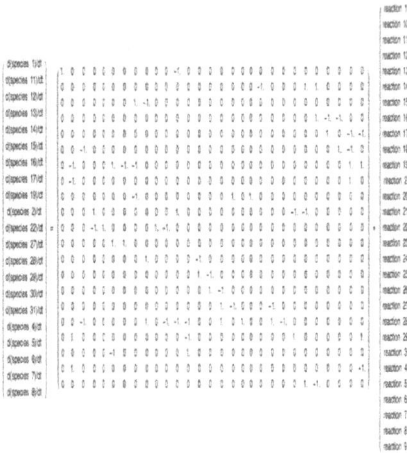

Figure 1.6. Stoichiometric analysis of the simulated model.

Source: Author.

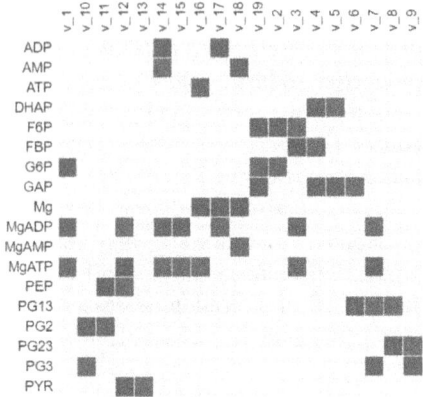

Figure 1.7. Heatmap generated diagram.

Source: Author.

Reactants and Products (Figure 1.6). The variable that comes IN will be taken as −1 and that goes OUT will be taken as +1. This matrix for the metabolic network includes glycolysis, the PPP and nucleotide biosynthesis pathways influenced by the PI3K/Akt/mTOR signaling.

The heat map presented in Figure 1.7 illustrates the complex interplay of various metabolites and cofactors across multiple experimental conditions or fluxes. This visualization provides insights into the dynamic behavior of key molecules involved in cellular metabolism.

The y-axis enumerates a diverse set of biochemical entities, including high-energy phosphate compounds (ATP, ADP, AMP), glycolytic intermediates (GAP, F6P, FBP) and important cofactors (Mg, MgADP, MgAMP, MgATP). The x-axis likely represents the studied biological process's different experimental conditions or temporal progression. Blue cells indicate positive correlations or increases in concentration/activity. Red cells signify negative correlations or decreases in concentration/activity. Gray cells suggest no significant change or correlation.

AI and Computational Modeling in Cancer Research: This research highlights the effectiveness of AI-based computational modeling in forecasting the metabolic reactions of cancer cells to targeted treatments. By combining machine learning and deep learning techniques with biochemical simulations, we can anticipate the impact of drugs such as doxorubicin on cancer metabolism from a systems perspective. This methodology presents a valuable opportunity for creating personalized cancer therapies, allowing metabolic models to be customized according to the specific tumor profiles of individual patients [54, 57].

4. Conclusion

This study concludes by introducing a computational model for the PI3K/Akt/mTOR pathway, which incorporates AI-driven predictions to investigate the impact of PHGDH inhibition through doxorubicin on cancer metabolism. The results indicate that inhibiting PHGDH interferes with glycolysis and nucleotide synthesis, thereby impeding the proliferation of cancer cells. This research highlights the promise of utilizing AI and computational modeling in cancer studies, offering valuable insights that may guide the creation of innovative therapeutic approaches.

5. Acknowledgment

I wish to express my heartfelt appreciation to my mentor, for the exceptional support and guidance during this research endeavour. Profound knowledge and direction in mathematical modelling have been instrumental in shaping this work. I am immensely grateful for the assistance in the laboratory, where their valuable insights and encouragement played a significant role in successfully executing the experiments.

References

[1] Alessi, D. R., James, S. R., Downes, C. P., Holmes, A. B., Gaffney, P. R., Reese, C. B., et al. (1997). Characterization of a 3-phosphoinositide-dependent protein kinase which phosphorylates and activates protein kinase Balpha. *Curr Biol*, *7*, 261–269.

[2] Almuhaideb, A., Papathanasiou, N., & Bomanji, J. (2011). 18F-FDG PET/CT imaging in oncology. *Ann Saudi Med*, *31*, 3–13. [PubMed: 21245592]

[3] Astuti, D., Latif, F., Dallol, A., Dahia, P. L., Douglas, F., George, E., Skoldberg, F., Husebye, E. S., Eng, C., & Maher, E. R. (2001). Gene mutations in the succinate dehydrogenase subunit SDHB cause susceptibility to familial pheochromocytoma and to familial paraganglioma. *Am J Hum Genet*, *69*, 49–54. [PubMed: 11404820]

[4] Bali, M., & Thomas, S. R. (2001). A modelling study of feedforward activation in human erythrocyte glycolysis. *C. R. Acad. Sci. III*, *324*(3), 185–199.

[5] Balss, J., Meyer, J., Mueller, W., Korshunov, A., Hartmann, C., & von Deimling, A. (2008). Analysis of the IDH1 codon 132 mutation in brain tumors. *Acta Neuropathol*, *116*, 597–602. [PubMed: 18985363]

[6] Barthel, A., Okino, S. T., Liao, J., Nakatani, K., Li, J., Whitlock, J. P. Jr, & Roth, R. A. (1999). Regulation of GLUT1 gene transcription by the serine/threonine kinase Akt1. *J Biol Chem*, *274*, 20281–20286. [PubMed: 10400647]

[7] Bauer, D. E., Hatzivassiliou, G., Zhao, F., Andreadis, C., & Thompson, C. B. (2005). ATP citrate lyase is an important component of cell growth and transformation. *Oncogene*, *24*, 6314–6322. [PubMed: 16007201]

[8] Bazil, J. N., Buzzard, G. T., & Rundell, A. E. (2010). Modeling mitochondrial bioenergetics with integrated volume dynamics. *PLoS Comput. Biol*, *6*, e1000632. doi:10.1371/journal.pcbi.1000632

[9] Ben-Sahra, I., Howell, J. J., Asara, J. M., & Manning, B. D. (2013). Stimulation of de novo pyrimidine synthesis by growth signaling through mTOR and S6K1. *Science*, *339*, 1323–1328. [PubMed: 23429703]

[10] Benz, M. R., Herrmann, K., Walter, F., Garon, E. B., Reckamp, K. L., Figlin, R., Phelps, M. E., Weber, W. A., Czernin, J., & Allen-Auerbach, M. S. (2011). (18)F-FDG PET/CT for monitoring treatment responses to the epidermal growth factor receptor inhibitor

erlotinib. *J Nucl Med*, *52*, 1684–1689. [PubMed: 22045706]

[11] Bertout, J. A., Patel, S. A., & Simon, M. C. (2008). The impact of O2 availability on human cancer. *Nat Rev Cancer*, *8*, 967–975. [PubMed: 18987634]

[12] Berwick, D. C., Hers, I., Heesom, K. J., Moule, S. K., & Tavare, J. M. (2002). The identification of ATP-citrate lyase as a protein kinase B (Akt) substrate in primary adipocytes. *J Biol Chem*, *277*, 33895–33900. [PubMed: 12107176]

[13] Birsoy, K., Wang, T., Chen, W. W., Freinkman, E., Abu-Remaileh, M., & Sabatini, D. M. (2015). An essential role of the mitochondrial electron transport chain in cell proliferation is to enable aspartate synthesis. *Cell*, *162*, 540–551. [PubMed: 26232224]

[14] Biswas, S., Troy, H., Leek, R., Chung, Y. L., Li, J. L., Raval, R. R., et al. (2010). Effects of HIF-1alpha and HIF2alpha on growth and metabolism of clear-cell renal cell carcinoma 786-0 xenografts. *J. Oncol, 2010*, 757908.

[15] Blaschke, K., Ebata, K. T., Karimi, M. M., Zepeda-Martinez, J. A., Goyal, P., Mahapatra, S., Tam, A., Laird, D. J., Hirst, M., Rao, A., et al. (2013). Vitamin C induces Tet-dependent DNA demethylation and a blastocyst-like state in ES cells. *Nature*, *500*, 222–226. [PubMed: 23812591]

[16] Borger, D. R., Tanabe, K. K., Fan, K. C., Lopez, H. U., Fantin, V. R., Straley, K. S., Schenkein, D. P., Hezel, A. F., Ancukiewicz, M., Liebman, H. M., et al. (2012). Frequent mutation of isocitrate dehydrogenase (IDH)1 and IDH2 in cholangiocarcinoma identified through broad-based tumor genotyping. *Oncologist*, *17*, 72–79. [PubMed: 22180306]

[17] Bowles, T. L., Kim, R., Galante, J., Parsons, C. M., Virudachalam, S., Kung, H. J., & Bold, R. J. (2008). Pancreatic cancer cell lines deficient in argininosuccinate synthetase are sensitive to arginine deprivation by arginine

deiminase. *Int J Cancer*, *123*, 1950–1955. [PubMed: 18661517]

[18] Boxer, R. B., Stairs, D. B., Dugan, K. D., Notarfrancesco, K. L., Portocarrero, C. P., Keister, B. A., Belka, G. K., Cho, H., Rathmell, J. C., Thompson, C. B., et al. (2006). Isoform-specific requirement for Akt1 in the developmental regulation of cellular metabolism during lactation. *Cell Metab*, *4*, 475–490. [PubMed: 17141631]

[19] Boya, P., Reggiori, F., & Codogno, P. (2013). Emerging regulation and functions of autophagy. *Nat Cell Biol*, *15*, 713–720. [PubMed: 23817233]

[20] Brady, D. C., Crowe, M. S., Turski, M. L., Hobbs, G. A., Yao, X., Chaikuad, A., Knapp, S., Xiao, K., Campbell, S. L., Thiele, D. J., et al. (2014). Copper is required for oncogenic BRAF signalling and tumorigenesis. *Nature*, *509*, 492–496. [PubMed: 24717435]

[21] Butner, J. D., Dogra, P., Chung, C., Pasqualini, R., Arap, W., Lowengrub, J., Cristini, V., & Wang, Z. (2022). Mathematical modeling of cancer immunotherapy for personalized clinical translation. *Nat Comput Sci*, *2*(12), 785–796. doi:10.1038/s43588-022-00377-z. Epub 2022 Dec 19. PMID: 38126024; PMCID: PMC10732566.

[22] Cantley, L. C. (2002). The phosphoinositide 3-kinase pathway. *Science*, *296*(5573), 1655–1657. doi:10.1126/science.296.5573.1655. PMID: 12040186.

[23] Carmeliet, P., Dor, Y., Herbert, J. M., Fukumura, D., Brusselmans, K., Dewerchin, M., et al. (1998). Role of HIF-1alpha in hypoxia-mediated apoptosis, cell proliferation and tumour angiogenesis. *Nature*, *394*, 485–490.

[24] Carnero, A. (2010). The PKB/Akt/PKB pathway in cancer. *Curr. Pharm. Des*, *16*, 34–44.

[25] Casazza, J. P., & Veech, R. L. (1986). The interdependence of glycolytic and pentose cycle intermediates in ad libitum fed rats. *J. Biol. Chem*, *26*, 690–698.

[26] Christofk, H. R., Vander Heiden, M. G., Harris, M. H., Ramanathan, A., Gerszten, R. E., Wei, R., et al. (2008). The M2 splice isoform of pyruvate kinase is important for cancer metabolism and tumour growth. *Nature, 452*, 230–233.

[27] Clem, B., Telang, S., Clem, A., Yalcin, A., Meier, J., Simmons, A., et al. (2008). Small-molecule inhibition of phosphofructo-2-kinase activity suppresses glycolytic flux and tumor growth. *Mol. Cancer Ther, 7*, 110–120.

[28] DeBerardinis, R. J., Lum, J. J., Hatzivassiliou, G., & Thompson, C. B. (2008). The biology of cancer: Metabolic reprogramming fuels cell growth and proliferation. *Cell Metab, 7*, 11–20.

[29] Deprez, J., Vertommen, D., Alessi, D. R., Hue, L., & Rider, M. H. (1997). Phosphorylation and activation of heart 6-phosphofructo-2-kinase by protein kinase B and other protein kinases of the insulin signalling cascades. *J. Biol. Chem, 272*, 17269–17275.

[30] Edinger, A. L. (2005). Growth factors regulate cell survival by controlling nutrient transporter expression. *Biochem Soc Trans, 33*, 225–227.

[31] Elstrom, R. L., Bauer, D. E., Buzzai, M., Karnauskas, R., Harris, M. H., Plas, D. R., et al. (2004). Akt stimulates aerobic glycolysis in cancer cells. *Cancer Res, 64*, 3892–3899.

[32] Evans, M. J., Saghatelian, A., Sorensen, E. J., & Cravatt, B. F. (2005). Target discovery in small-molecule cell-based screens by in situ proteome reactivity profiling. *Nat Biotechnol, 23*, 1303–1307.

[33] Fan, Y., Dickman, K. G., & Zong, W. X. (2010). Akt and c-Myc differentially activate cellular metabolic programs and prime cells to bioenergetic inhibition. *J Biol Chem, 285*, 7324–7333.

[34] Fell, D. (1997). *Understanding the Control of Metabolism*. London: Portland Press.

[35] Firth, J. D., Ebert, B. L., & Ratcliffe, P. J. (1995). Hypoxic regulation of lactate dehydrogenase A. Interaction between hypoxia-inducible factor 1 and cAMP response elements. *J Biol Chem, 270*, 21021–21027.

[36] Furuta, E., Okuda, H., Kobayashi, A., & Watabe, K. (2010). Metabolic genes in cancer: Their roles in tumor progression and clinical implications. *Biochim Biophys Acta, 1805*, 141–152.

[37] Gao, H., & Leary, J. A. (2004). Kinetic measurements of phosphoglucomutase by direct analysis of glucose-1-phosphate and glucose-6-phosphate using ion/molecule reactions and Fourier transform ion cyclotron resonance mass spectrometry. *Anal Biochem, 329*, 269–275.

[38] Gerber, G., Preissler, H., Heinrich, R., & Rapoport, S. M. (1974). Hexokinase of human erythrocytes: purification, kinetic model and its application to the conditions in the cell. *Eur J Biochem, 45*(1), 39–52.

[39] Ghosh, S., Matsuoka, Y., Asai, Y., Hsin, K. Y., & Kitano, H. (2011). Software for systems biology: From tools to integrated platforms. *Nat Rev Genet, 12*, 821–832.

[40] Hanahan, D., & Weinberg, R. A. (2011). Hallmarks of cancer: The next generation. *Cell, 144*(5), 646–674. doi:10.1016/j.cell.2011.02.013. PMID: 21376230.

[41] Joshi, A., & Palsson, B. O. (1989). Metabolic dynamics in the human red cell. Part I—A comprehensive kinetic model. *J Theor Biol, 141*(4), 515–528. doi:10.1016/s0022-5193(89)80233-4

[42] Kinoshita, A., Tsukada, K., Soga, T., et al. (2007). Roles of haemoglobin Allostery in hypoxia-induced metabolic alterations in erythrocytes: Simulation and its verification by metabolome analysis. *J Biol Chem, 282*, 10731–10741.

[43] Koh, D. M., Papanikolaou, N., Bick, U., et al. (2022). Artificial intelligence and machine learning in cancer imaging. *Commun Med, 2*, 133. https://doi.org/10.1038/s43856-022-00199-0

[44] Kuchel, P. W. (2004). Current status and challenges in connecting models of erythrocyte metabolism to experimental reality. *Prog Biophys Mol Biol*, *85*(2–3), 325–342.

[45] Lukey, M. J., Katt, W. P., & Cerione, R. A. (2017). Targeting amino acid metabolism for cancer therapy. *Drug Discov*, *22*, 796e804.

[46] Manning, B. D., & Toker, A. (2017). AKT/PKB signaling: Navigating the network. *Cell*, *169*(3), 381–405. doi:10.1016/j.cell.2017.04.001. PMID: 28431241; PMCID: PMC5546324.

[47] Mosca, E., Alfieri, R., Maj, C., Bevilacqua, A., Canti, G., & Milanesi, L. (2012). Computational modeling of the metabolic States regulated by the kinase akt. *Front Physiol*, *3*, 418. doi:10.3389/fphys.2012.00418. PMID: 23181020; PMCID: PMC3502886.

[48] Mulquiney, P. J., & Kuchel, P. W. (1999). Model of 2,3-bisphosphoglycerate metabolism in the human erythrocyte based on detailed enzyme kinetic equations: Equations and parameter refinement. *Biochem J*, *342*(Pt 3), 581–596.

[49] Jiang, P., Du, W., & Wu, M. (2014). Regulation of the pentose phosphate pathway in cancer. *Protein Cell*, *5*, 592–602.

[50] Possemato, R., Marks, K. M., Shaul, Y. D., Pacold, M. E., Kim, D., Birsoy, K., Sethumadhavan, S., Woo, H. K., Jang, H. G., Jha, A. K., & Chen, W. W. (2011). Functional genomics reveal that the serine synthesis pathway is essential in breast cancer. *Nature*, *476*(7360), 346–350. doi:10.1038/nature10350. PMID: 21760589; PMCID: PMC3353325.

[51] Dolfi, S. C., Chan, L. L. Y., Qiu, J., Tedeschi, P. M., Bertino, J. R., Hirshfield, K. M., Oltvai, Z. N., & Vazquez, A. (2013). The metabolic demands of cancer cells are coupled to their size and protein synthesis rates. *Cancer & Metabolism*, *1*, 1–13.

[52] Pradhan, V., Bhadra, P., & Pradhan, R. K. (2024). Constraint-based Modelling of Human Erythrocyte Glycolysis and Role of 3PG in Breast Cancer. *2024 3rd International Conference for Innovation in Technology (INOCON)*. Bangalore, India, pp. 1–8. doi: 10.1109/INOCON60754.2024.10512070

[53] Vander Heiden, M. G., Cantley, L. C., & Thompson, C. B. (2009). Understanding the Warburg effect: The metabolic requirements of cell proliferation. *Science*, *324*(5930), 1029–1033. doi:10.1126/science.1160809. PMID: 19460998; PMCID: PMC2849637.

[54] Warburg, O. (1956). On the origin of cancer cells. *Science*, *123*(3191), 309–314. doi:10.1126/science.123.3191.309. PMID: 13298683.

[55] Warburg, O. (1924). Über den stoffwechsel der carcinomzelle. *Naturwissenschaften, 1924*, 1131–1137.

[56] Tsun, Z. Y., & Possemato, R. (2015). Amino acid management in cancer. *Semin Cell Dev Biol*, *43*, 22–32.

[57] Zhao, J. Y., Feng, K. R., Wang, F., Zhang, J. W., Cheng, J. F., Lin, G. Q., Gao, D., & Tian, P. (2021). A retrospective overview of PHGDH and its inhibitors for regulating cancer metabolism. *Eur J Med Chem*, *217*, 113379. doi:10.1016/j.ejmech.2021.113379. Epub 2021 Mar 16. PMID: 33756126.

2 Design and synthesis of novel benzimidazole-pyrazoline hybrid molecules as anticancer agent

Gopal Krishna Padhy[1,a] and Jagadeesh Panda[2]

[1]Department of Pharmaceutical Chemistry. Centurion University of Technology and Management, Odisha, India

[2]Department of Pharmaceutical Chemistry, Raghu College of Pharmacy, Dakamarri, Visakhapatnam, Andhra Pradesh, India

Abstract: The development of resistance to existing anticancer drugs is a pressing issue, making the discovery of new effective compounds crucial. Recent studies have highlighted pyrazolines and benzimidazoles as potent agents against cancer cells. Consequently, we have adopted a hybrid approach, joining the core moiety of pyrazoline and benzimidazole into a single molecule to enhance biological activity. The hybrid compounds were designed using Autodock vina software employing a molecular docking approach against platelet-derived growth factor receptor A (PDGFRA). The targeted hybrid molecules were synthesized by cyclizing benzimidazole chalcones with hydrazine. These molecules were then identified using numerous spectroscopic methods, like Infrared (IR) Spectroscopy, to recognize functional groups. Proton Nuclear Magnetic Resonance to determine the hydrogen environment. Carbon-13 Nuclear Magnetic Resonance to analyze the carbon skeleton. Mass Spectrometry (MS): For molecular weight and structural confirmation. This method produced the desired compounds with the expected core structures. The synthesized compounds were tested for their inhibitory effects on the human breast cancer cell line MDA-MB-231 growth. The results were quantified using the LC_{50} value, that indicates the concentration needed to inhibit 50% of the cells: Compound 5b and Compound 5c exhibited significant anticancer activity with GI_{50} value of 3.3 μM and 19.1 μM, with Compound 5b showing a more potent effect. The hybrid compounds combining benzimidazole and pyrazoline structures show promising biological activities. These findings suggest that such hybrid molecules could be valuable leads in the development of new antibacterial and anticancer therapies. Future work could emphasize on optimizing these molecules for better efficacy and lower toxicity.

Keywords: Chalcone, benzimidazole, pyrazoline, anticancer, SRB assay, PDGFRA

1. Introduction

Benzimidazole is a promising candidate for the development of medicinally relevant molecules. Studies have shown that benzimidazole derivatives exhibit a variety of therapeutic activities, including anticancer [1–3], anti-inflammatory [4], antiviral [5] and antibacterial [6–7]. Additionally, pyrazolines are widely recognized for their therapeutic potential [8–10]. The platelet-derived factor receptor A (PDGFRA), is found on the surface of various cell types. Specific isoforms of platelet-derived growth factors (PDGFs) bind to this receptor, promoting cellular growth and division. PDGFRA is also essential for the embryonic development of certain tissues and mutations in PDGFRA can lead to clinically significant neoplasms. In this research endeavour,

[a]Indiagopalmedchem@gmail.com

DOI: 10.1201/9781003672869-2

novel 1H-benzimidazolamide-linked pyrazoline hybrids were created and assessed for anticancer properties [11–12].

2. Results and Discussion

2.1. Chemistry

The desired molecules were synthesized following the procedure mentioned in Figure 2.1. The desired chalcones (2a-e) were obtained by reacting 2-acetylbenzimidazole (1) with benzaldehyde derivatives in the company of a base [13]. N-substituted benzimidazole chalcone amides (4a-e) were prepared by reacting 1H-benzimidazole chalcones (2a-e) with chloroacetanilide derivatives (3) in the company of acetone as a solvent and anhydrous K_2CO_3. Finally, the benzimidazole chalcone amides (4a-e) were reacted with hydrazine hydrate in 99% ethanol to yield benzimidazole-linked pyrazolines (5a-e). In this process, the α, β-unsaturated carbonyl group underwent cyclization to create the pyrazoline ring.

2.2. Spectral studies

^{1}H NMR, ^{13}C NMR, IR and Mass spectra of all the molecules were recorded and

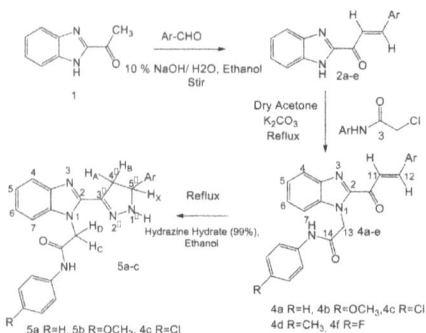

Figure 2.1. Preparation of benzimidazole-pyrazoline hybrid.

Source: Author.

found in accordance with the desired compounds. TLC is utilized to examine the purity of the molecules. In compounds 4a-e, IR spectra confirmed the stretching of the α,β-unsaturated keto group within the anticipated region of 1663–1668 cm^{-1} and (C=N) stretching bands at 1599–1611 cm^{-1} in molecules 5a-c, providing proof of the alteration of chalcone to the pyrazoline. typical absorption bands at 3306–3414 & 3204–3254 cm^{-1} due to the -NH group are observed for the compounds 5a-c.

Two doublets having coupling constant within 16–16.5 Hz In the ^{1}H NMR spectrum of compounds 4a-e indicates the unsaturated protons in E form. Peak at δ 182.43–182.46 corresponding to keto group are also observed in the ^{13}C NMR spectrum of compounds 4a-e.

In the ^{1}H NMR spectra of compounds 5a-c, an ABX pattern was observed for the protons in the pyrazoline system. The stereochemistry of protons H_A, H_B and H_X was determined through J coupling values. H_A and H_X showed cis coupling (J_{AX} = 5.5–6.5 Hz), while H_B and H_X were trans coupled (J_{BX} = 5.5 Hz). The J_{AB} coupling constant (10–11 Hz) confirmed a geminal relationship at the C-4 position. Doublets of doublets was observed for H_A and H_B protons at δ 3.03–3.05 ppm and δ 3.67 ppm, correspondingly, while H_X seemed as a triplet of doublets at δ 4.89–4.90 ppm. The acetamido methylene protons appeared as doublets at δ 5.3–6.08 ppm and δ 5.53–5.56 ppm. The protons of the amino group on the pyrazoline ring were observed as a doublet at δ 8.26–8.27 ppm and the NH protons of the amide group in 5a-d showed singlets at δ 10.25–10.54 ppm. ^{13}C NMR confirmed key signals at δ 48.18–48.37 ppm (methylene protons), δ 42.10–42.15 ppm (C4) and δ 63.18–63.21 ppm (C5).

Elemental analysis and mass spectrometry further supported the structures.

2.3. Molecular docking

These docking studies have revealed that the benzimidazole and pyrazoline ring system binds to a hydrophobic pocket in PDGFRA where the aromatic benzimidazole ring interacts with the amino acid ASP-681 via a pi anion bond and similarly, pi allyl bonding interactions observed between the aromatic rings and the amino acids ALA625, LEU599, CYS677, VAL683 and ARG822. Figure 2.2 establishes the binding modes

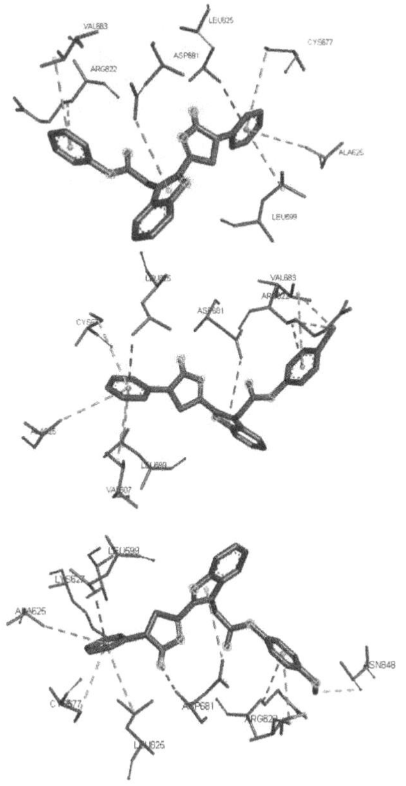

Figure 2.2. Binding modes of the molecules 5a, 5b and 5c with the PDGFRA protein.

Source: Author.

Table 2.1. In vitro cytotoxicity study of molecules 5a, 5b and 5c (GI_{50} μM)

Molecule	GI_{50} (μM)	Binding affinity (Kcal/mol)
5a	>100	−8.0
5b	3.3	−8.5
5c	19.1	−8.6

Source: Author.

of the molecules in the binding site. The binding free energy of the synthesized compounds with the PDGFRA protein is mentioned in Table 2.1.

2.4. Anticancer activity

The in vitro anticancer activity of selected molecules were determined using a cell viability assay (SRB assay) using MDA-MB-231 cell line. Compounds 5b (GI_{50} = 3.3 μM) and 5c (GI_{50} = 19.1 μM) significantly inhibited the proliferation of the cell line. All other molecules related to the standard drug adriamycin were found to be inactive. The cytotoxicity test results are presented in Table 2.1.

3. Experimental

3.1. Materials and Instrumentation

The chemicals utilized in the study were supplied by Finar (India) and Merck (India). The ^1H and ^{13}C NMR spectra were determined using a Bruker AVANCE III 500 MHz (AV 500) spectrometer. The IR spectra were measured using a Bruker ALPHA-T Fourier-transform IR spectrometer. Mass spectra were measured using either a Varian Inc. 410 Prostar Binary LC-MS. Open tube capillary procedure was utilized to measure the melting points and are uncorrected. The advancement of the reaction was checked by TLC using iodine vapor or UV light.

3.2. Synthesis

a. *Synthesis of 2-(2-cinnamoyl-benzo-imidazolyl)-N—(substituted phenyl) acetamide (4a-e):*
Benzimidazole chalcone amides **4a-e** were obtained by substitution reaction of 1H-Benzimidazole chalcone 2a-e with 2-chloro-N-(substituted phenyl)acetamide (3) in the presence of anhydrous K_2CO_3 and acetone.

1. *2-(2-cinnamoyl-1H-benzoimid-azolyl)-N-phenylacetamide (4a):*
Yield: 76%; m.p. 240–242°C; FTIR cm^{-1} 1597 (C=N), 1663 (C=O); ^{1}H-NMR: 10.52 (1H, s, N-H), 8.22 (1H, H_β), 7.94 (1H), 7.87–7.82 (4H), 7.60 (2H), 7.50–7.48 (4H), 7.43–7.40 (1H), 7.33–7.30 (2H), 7.07–7.04 (1H), 5.58 (2H); ^{13}C NMR: 182.46 (methylene), 166.06 (C14), 147.41 (C2),144.58 (C12), 141.63 (C8), 139.34, 137.75 (C9), 134.75, 131.55, 129.62, 129.38, 129.31, 126.43 (C6), 124.18 (C5), 123.86, 123.01 (C11), 121.79 (C4), 119.48, 112.13 (C7), 48.76 (C13); mass (m/z): 382.1 [M+H].

2. *2-(2-cinnamoyl-1H-benzoim-idazolyl)-N-(4-methoxyphenyl) acetamide (4b):* Yield: 64%; m.p. 254–256°C; FTIR cm^{-1} 1666 (C=O), 1603 (C=N); ^{1}H-NMR: 10.37 (1H, s, N-H), 8.22 (1H,d, J = 16.0 Hz, Hβ), 7.94–7.81 (5H), 7.52–7.49 (6H), 7.41 (1H), 6.89 (2H), 5.54 (2H), 3.71 (3H); mass (m/z): 412.1 [M+H].

3. *2-(2-cinnamoyl-1H-benzoimid-azolyl)-N-(4-chlorophenyl)acet-amide (4c):* Yield: 78%; m.p.

228–230°C, FTIR cm^{-1} 1668 (C=O), 1600 (C=N); ^{1}H-NMR: 10.67 (1H,N-H), 8.21 (1H, Hβ), 7.94 (1H), 7.88–7.83 (4H), 7.63 (2H), 7.51–7.48 (4H), 7.43–7.37 (3H), 5.57 (2H, s, –CH2); ^{13}C NMR: 182.43 (methylene), 166.29, 147.34 (C2), 144.63 (C12), 141.60 (C8), 138.26, 137.72 (C9), 134.72, 131.57, 129.6, 129.38,163, 129.24, 127.43, 126.46 (C6), 124.21(C5), 122.94 (keto), 121.79 (C4), 121.06, 112.12 (C7), 48.78 (C13); mass (m/z): 416.0 [M+H].

4. *2-(2-cinnamoyl-1H-benzoimidaz-olyl)-N-(p-tolyl)acetamide (4d):* Yield: 80%; m.p. 250–252°C; FTIR cm^{-1} 1666 (C=O), 1603 (C=N); ^{1}H-NMR: 10.44 (1H, s, N-H), 8.22 (1H,d, J = 16.5 Hz, Hβ), 7.94 (1H), 7.88–7.82 (4H), 7.50–7.47 (6H), 7.41 (1H), 7.12 (2H), 5.56 (2H, s, –CH₂); ^{13}C NMR: 182.44 (keto), 165.77 (methylene), 147.41 (C2), 144.56 (C12), 141.60 (C8), 137.74 (C9), 136.81, 134.73, 132.77, 131.56, 129.66, 129.63, 129.38, 126.42 (C6), 124.17 (C5), 123.01 (C11), 121.77 (C4), 119.49, 112.13 (C7), 48.68 (C13), 20.89 (-CH3); mass (m/z): 396.1 [M+H].

5. *2-(2-cinnamoyl-1H-benzoimidaz-olyl)-N-(4-fluorophenyl)acetamide (4e):* Yield: 76%; m.p. 218–220°C; FTIR cm^{-1} 1665 (C=O), 1604 (C=N); ^{1}H-NMR: 10.62 (1H, s, N-H), 8.22 (1H, d, J = 16 Hz, Hβ), 7.94 (1H), 7.88–7.82 (4H), 7.65–7.63 (2H), 7.51–7.48 (4H), 7.43–7.40 (1H), 7.18–7.15 (2H); 5.58 (2H, s, –CH2); ^{13}C NMR: 182.45 (keto), 166.01 (methylene), 159.47–157.56, 147.40 (C2), 144.59 (C12), 141.62 (C8), 137.73

(C9), 135.72, 134.7, 131.5 129.61, 129.37, 126.44 (C6), 124.19 (C5), 123.00 (C11), 121.78 (C4), 121.29–121.23, 115.96–115.79, 112.11(C7), 48.68 (C13).

b. *Synthesis of 2-(2-(5-phenyl-1H-pyrazoline)-benzimidazole)-N-(substituted phenyl)acetamide (5a-c):* N-(substituted phenyl)acetamide containing benzimidazole pyrazolines 5a-c are prepared by reacting N-(substituted phenyl)acetamide benzimidazole chalcones 4a-c with hydrazine hydrate in refluxing ethanol. This reaction likely played a role in the formation of hydrazones intermediate, followed by the addition of an N-H group to the olefinic bond in the propenone structure.

1. *2-(2-(5-phenyl-1H-pyrazolinyl)-benzimidazolyl)-N-phenyl acetamide (5a):* Yield: 72%; m.p. 154–156°C: FTIR cm^{-1} 3306 & 3204 (-NH), 1669 (C=O), 1599 (C=N); ^{1}H-NMR: 10.39 (1H, s, CO-NH), 8.26 (1H) 7.66 (1H), 7.61–7.57 (3H), 7.37–7.30 (6H), 7.28–7.21 (3H), 7.06 (1H), 5.56 (1H), 5.50 (1H), 4.89 (1H), 3.67 (1H) 3.03 (1H); ^{13}C NMR: 166.19 (C11), 147.06 (C2), 143.03 (C14 pyrazoline), 142.81, 142.34 (C8), 139.44, 137.48 (C9), 129.28, 128.94, 127.77, 126.99, 123.78, 123.47 (C6), 122.46 (C5), 119.52 (C4), 119.49, 110.69 (C7), 63.21 (C2'), 48.34 (C10), 42.14 (C3'); mass (m/z): 396.1905 [M+H].

2. *2-(2-(5-phenyl-1H-pyrazolinyl)-benzimidazol yl)-N-(p-tolyl) acetamide (5b):* Yield: 72%; m.p. 200–202°C; FTIR cm^{-1} 3364 & 3254 (-NH), 1668 (C=O), 1602 (C=N); ^{1}H-NMR: 10.25 (1H, s, CO-NH), 8.27 (1H) 7.66 (1H),

7.57 (1H), 7.52 (2H), 7.37–7.31 (4H), 7.29–7.20 (3H), 6.89 (2H), 5.55 (1H), 5.48 (1H), 3.72 (3H, s, -OCH$_3$), 4.90 (1H), 3.67 (1H) 3.05 (1H); ^{13}C NMR: 165.63 (C11), 155.71, 147.04 (C2), 143.03 (C14 pyrazoline), 142.82, 142.37 (C8), 137.47 (C9), 132.58, 128.93, 127.75, 126.99, 123.43 (C6), 122.41 (C5), 121.00, 119.50 (C4), 114.38, 110.68 (C7), 63.18 (C2'), 55.64 (-OCH3), 48.18 (C10), 42.15 (C3'); mass (m/z): 412.9 [M+H].

3. *2-(2-(5-phenyl-1H-pyrazolinyl)-benzoimidazo lyl)-N-(4-chlorophenyl) acetamide (5c):* Yield: 70%; m.p. 198–202 °C; FTIR cm^{-1} 3414 (br, NH), 1668 (C=O), 1611 (C=N); ^{1}H-NMR: 10.54 (1H, s, CO-NH), 8.27 (1H) 7.67–7.63 (3H), 7.59 (1H), 7.39–7.31 (6H), 7.29–7.26 (3H), 5.53 (1H), 5.50 (1H), 4.89 (1H), 3.67 (1H) 3.03 (1H);^{13}C NMR: 166.42 (C11), 147.02 (C2), 143.03 (C14 pyrazoline), 142.82, 142.32 (C8), 138.37, 137.46 (C9), 129.18, 128.93, 127.76, 127.37, 126.97, 123.48 (C6), 122.47 (C5), 121.08, 119.53 (C4), 110.67 (C7), 63.20 (C2'), 48.37 (C10), 42.10 (C3'); mass (m/z): 430.1404 [M+H].

3.3. Molecular docking

The crystal structure of PDGFRA in complex with sunitinib was obtained from the protein data bank PDB (6jok) [14]. The ligand (sunitinib) was separated from the native protein structure. The structure of the protein and native ligand was then imported to Python Molecular Viewer-1.5.7 and coordinates for the binding site were then identified (center_x = 16.879, center_y = 131.582, center_z = -6.262). The molecular docking work was accomplished using the autodock vina-1.1.2 [15] running on intel i3 quad-core processor. The interaction of protein and ligand molecules was visualized in the biovia discovery studio software [16].

3.4. Anticancer activity

The in vitro cytotoxicity of the synthesized molecules against MDA-MB-231 cell line was evaluated using the SRB assay [17]. The cell line was cultured in RPMI 1640 medium consisting of 10% fetal bovine serum containing 2 mM L-glutamine. Cells were planted at 5,000 cells per well in 90 µL of medium in 96-well microtiter plates and kept for 24 hours at 37°C under 5% carbon dioxide and 95% air. After treatments with drug (0.1–100 µM), the plates were kept for extra 48 hours. To fix the cells, 25 µL of 10% trichloroacetic acid was added and the plates were set for 60 minutes at 4°C. The supernatant was rejected and the plates were cleaned multiple times with water. For staining, 50 µL of sulforhodamine solution was added to each well. Afterward, the absorbance was measured at 515 nm after the cells were liquified in a 10 mM trim base. The percentage of cell viability at each drug concentration was calculated.

4. Conclusion

Eight novel benzimidazoles containing chalcone or pyrazoline structures were prepared, utilizing 2-acetyl benzimidazole as a precursor. These pyrazolines were tested for their anticancer properties. The pyrazoline derivative 5b notably inhibited cancer cell proliferation at a concentration of 3.3 µM.

References

[1] Zouaghi, M. O., Bensalah, D., Hassen, S., Arfaoui, Y., Mansour, L., Özdemir, N., Bülbül, H., Gurbuz, N., Özdemir, I., & Hamdi, N. (2024). Benzimidazole derivatives as a new scaffold of anticancer agents: Synthesis, optical properties, crystal structure and DFT calculations. *Heliyon.*, *10*(12), e32905

[2] Koh, B., Ryu, J. Y., Noh, J. J., Hwang, J. R., Choi, J. J., Cho, Y. J., Jang, J., Jo, J. H., Lee, K., & Lee, J. W. (2024). Anti-cancer effects of benzimidazole derivative BNZ-111 on paclitaxel-resistant ovarian cancer. *Gynecol Oncol, 188*, 60–70.

[3] Acar Çevik, U., Işik, A., Kaya, B., Kapavarapu, R., Rudrapal, M., Halimi, G., Karakaya, A., Maryam, Z., Celik, İ., Evren, A. E., & Ünver, H. (2024). Benzimidazole-Containing Compounds as Anticancer Agents. *Chemistry Select*, *9*(32), e202401566.

[4] Bano, S., Nadeem, H., Zulfiqar, I., Shahzadi, T., Anwar, T., Bukhari, A., & Masaud, S. M. (2024). Synthesis and anti-inflammatory activity of benzimidazole derivatives; an in vitro, in vivo and in silico approach. *Heliyon*, *10*(9), e30102

[5] Ananta, M. F., Saha, P., Rahman, F. I., Spriha, S. E., Chowdhury, A. A., & Rahman, S. A. (2024). Design, synthesis and computational study of benzimidazole derivatives as potential anti-SARS-CoV-2 agents. *J Mol Struct, 1306*, 137940.

[6] Chedupaka, R., Audipudi, A. V., Sangolkar, A. A., Mamidala, S., Venkatesham, P., Penta, S., & Vedula, R. R. (2024). Design, synthesis, molecular docking, and dynamic studies of novel thiazole derivatives incorporating benzimidazole moiety and

assessment as antibacterial agents. *Mol Divers*, *28*(3), 565–1576.

[7]   Elwahy, A. H., Hammad, H. F., Ibrahim, N. S., Al-Shamiri, H. A., Darweesh, A. F., & Abdelhamid, I. A. (2024). Synthesis and antibacterial activities of novel hybrid molecules based on benzothiazole, benzimidazole, benzoxazole, and pyrimidine derivatives, each connected to N-arylacetamide and benzoate groups. *J Mol Struct*, *1307*, 137965.

[8]   Yadav, C. S., Azad, I., Khan, A. R., Nasibullah, M., Ahmad, N., Hansda, D., Ali, S. N., Shrivastav, K., Akil, M., & Lohani, M. B. (2024). Recent Advances in the Synthesis of Pyrazoline Derivatives from Chalcones as Potent Pharmacological Agents: A Comprehensive Review. *Results Chem*, *7*, 101326.

[9]   Serag, M. I., Tawfik, S. S., Badr, S. M., & Eisa, H. M. (2024). New oxadiazole and pyrazoline derivatives as anti-proliferative agents targeting EGFR-TK: Design, synthesis, biological evaluation and molecular docking study. *Sci Rep*, *14*(1), 5474.

[10] Batran, R. Z., Ahmed, E. Y., Awad, H. M., & Latif, N. A. A. (2024). Naturally based pyrazoline derivatives as aminopeptidase N, VEGFR2 and MMP9 inhibitors: Design, synthesis and molecular modeling. *RSC Advances*, *14*(31), 22434–22448.

[11] Benvie, A. M., Lee, D., Jiang, Y., & Berry, D. C. (2024). Platelet-derived growth factor receptor beta is required for embryonic specification and confinement of the adult white adipose lineage. *Iscience*, *27*(1).

[12] Yang, J., Pan, C., Pan, Y., Hu, A., Zhao, P., Chen, M., Song, H., Li, Y., & Hao, X. (2024). A Carbon 21 Steroidal Glycoside with Pregnane Skeleton from Cynanchum atratum Bunge Promotes Megakaryocytic and Erythroid Differentiation in Erythroleukemia HEL Cells through Regulating Platelet-Derived Growth Factor Receptor Beta and JAK2/STAT3Pathway. *Pharmaceuticals*, *17*(5), 628.

[13] Padhy, G. K., Panda, J., & Behera, A. K. (2016). Synthesis and characterization of novel benzimidazole chalcones as antibacterial agents. *Der Pharma Chem*, *8*, 235–241.

[14] Madej, T., Lanczycki, C. J., Zhang, D., Thiessen, P. A., Geer, R. C., Marchler-Bauer, A., & Bryant, S. H. (2014). MMDB and VAST+: tracking structural similarities between macromolecular complexes. *Nucleic Acids Res*, 1(42), D297–303.

[15] Huey, R., Morris, G. M., & Forli, S. (2012). Using AutoDock 4 and AutoDock vina with AutoDockTools: A tutorial. *The Scripps Research Institute Molecular Graphics Laboratory*, *10550*(92037), 1000.

[16] Pawar, S. S., & Rohane, S. H. (2021). Review on discovery studio: An important tool for molecular docking.

[17] Vichai, V., & Kirtikara, K. (2006). Sulforhodamine B Colorimetric assay for cytotoxicity screening. *Nature Protocol*, *14*(1).

3 Ionic gelation technique using biodegradable polymer for stabilized and sustained oral delivery of nano protein

Prasanna Parida[1,4,a], Amiya Kumar Prusty[2], Chinmaya Sahoo[3], Sujit Dash[4], Bikash Ranjan Jena[3], and Kirtimaya Mishra[3]

[1]Research Scholar (Pharmacy), Biju Patnaik University of Technology, Rourkela, Odisha, India
[2]Faculty of Pharmacy, C.V Raman Global University, Bhubaneswar, Odisha, India
[3] School of Pharmacy and Life Sciences, Centurion University of Technology and Management, Bhubaneswar, Odisha, India
[4]Institute of Pharmacy and Technology, Salipur, Cuttack, Odisha, India

Abstract: Ionotropic gelation forms nanoparticles, hydrogels, films and beads by cross-linking polyelectrolytes with counter ions (cation and anion) derived from biodegradable polymers. Because of nanoparticulate and hydrogel-stabilized drug delivery systems, the gelation approach has gained popularity in the innovative study area, resulting in an environment friendly pharmaceutical product development process. Hydrogels are having a three-dimensional structure and may absorb water in large volume or other synthetic fluids or natural fluids without losing shape. Such qualities can be utilized to encapsulate and maintain drug release action regarding protein and peptide medicines, resulting in stability in the stomach and increased bioavailability for non-invasive oral administration.

Keywords: Ionotropic Gelation, Nanoparticulate, Bioavailability, Hydrogel

1. Introduction

Nanoparticles made from natural or synthetic polymers, with a diameter of less than 1 mm, are known as polymer nanoparticles. They are widely utilized in drug delivery studies to transport medicines, genes, diagnostics and vaccinations into specific cells or tissues. The use of nanoparticles as a carrier system for drug and gene delivery is expanding for future preferences [1]. Many pharmaceutical methods of administration are designed to target certain cell types along with to achieve effectiveness [2]. Multiparticulate drug delivery systems are designed for oral, parenteral and topical formulations. These systems may be pellets, granules, beads, gel spheres, microcapsules, microspheres, lipospheres, microparticles and nanoparticles. In this strategy, the drug dosage is degraded into multiple subunits, which are typically made up of thousands of spherical particles having a predefined diameter range. To provide the recommended total dose, these subunits are

[a]litunar@gmail.com

DOI: 10.1201/9781003672869-3

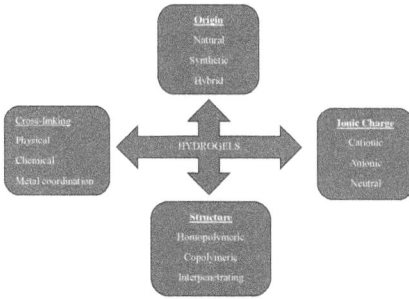

Figure 3.1. Classification of hydrogels.

Source: Cited from reference [4].

packaged in a sachet and encapsulated or compacted into a tablet [3].

Hydrogels are polymers that are connected together with hydrophilic groups, allowing them to absorb huge volumes of water without disintegrating. Ionically cross-linked hydrogels are important materials used in the pharmaceutical and medicinal fields. Controlling the rate of ionic crosslinking can be problematic since it depends on the inherent features of the ionotropic polymer and its ability to swiftly bind gel-forming ions. Introducing free cross-linking ions (CI) into the polymer solution makes it challenging to create consistent gels or delay the gelling process, limiting their use in applications requiring controlled on-demand or bulk gelation [4].

2. Ionic Gelation

Ionic gelation is an approach of gradually incorporating aqueous solutions that contain several ionic biopolymers, like low-methoxy pectin, sodium carboxymethyl cellulose, chitosan, gellan gum, sodium alginate etc., into an aqueous solution comprising adequate counter-ions [5].

Ionotropic gelation; IG is an approach containing polyelectrolytes; like polysaccharides mix with molecules; like cations, that are oppositely charged for achieving sol-gel transition. This leads to the generation of organized products like beads, films, nanoparticles and also hydrogels. There is substantial curiosity in employing polysaccharide-based hydrogels to encapsulate API. Hydrogels can be generated in many ways, like creating spherical or oval beads containing polymeric excipients & API [1, 3, 4]. A variety of factors influence droplet size and shape, including viscosity of the initial mixture, fluctuating relationships among droplets and the fluid around them (whether turbulent flow or laminar flow), surface tension and polymer concentration and molecular weight [6].

IG is a technique for producing nanoparticles as well as microparticles via electrostatic interactions involving multiple ionic species under specified conditions. At least one of the species should be a polymer. When implemented into the process, a medication or bioactive molecule can become entrapped within the polymeric chains, eventually encapsulating within the nanoparticle/ microparticle structure [7].

2.1. Mechanism of ionic gelation

Ionotropic gelation occurs when molecules with opposing charges react. For example, negatively charged divalent or multivalent ions can interact with positively charged polymer chains. This electrostatic interaction produces microstructured particles with interconnected nanofibrillar networks. These structures can be made utilising 3 methods, such as; internal, exterior and inverse gelation [8].

External approach, also referred as controlled diffusion method, is most widely used technique for ionotropic gelation. In this procedure, a polysaccharide solution is progressively added to a crosslinking solution. Crosslinking chemicals derived from the external continuous phase are integrated into the interior structure of polymer droplets, forming a bead matrix [9]. Hydrogel bead's outermost film initiates a rapid sol-gel transition, leading to instantaneous gel formation. As an outcome, counter-ions stream across the particle's several layers and into its core, resulting in a complicated gelation profile. The technique maximizes the link among polymer functional groups & ions at the surface while gradually diminishing towards the centre [10, 11]. Reverse gelation is an alternate method that includes gradually adding a medium containing gelling chemicals to a polymer solution. This approach is commonly used to produce microcapsules, those are polymer based soft shell microcapsules, containing oil form of emulsions. It utilizes modest quantities of biopolymer to generate a soft molecular shell. Characteristics of generated microcapsules differ depending on whether a W/O or O/W emulsion is used [7].

3. Material and Method

Chitosan, Sodium Alginate and Bovine serum albumin (BSA), TWEEN 60, PEG 4000, Sodium sulfate and Calcium chloride.

3.1. Hydrogel

Hydrogels are three-dimensional networks of crosslinked polymers. They are linked by covalent or noncovalent bonds like covalent bonds, hydrogen bonds, van der Waals forces and physical attractions. Their porous and entangled structure makes them ideal for carrying biomolecules and medicines, while cross-linking ensures their integrity. The covalent structure is generated through polymerisation. Nanosized gels can be made with functional groups for specific uses. The network in these gels allows them to hold more water [12].

3.2. Ca-alginate nanoparticle formulation

A 2M solution of 1.5 mL calcium chloride was poured in a drop by drop manner over 30 mL of sodium alginate solution; 0.2 wt%, having pH 5.0–5.5, while ultrasonication was conducted (35 kHz) in a continuous sequence. Following that, PEG 4000 (10 wt%) was introduced to the sodium alginate solution prior, being crosslinked with Ca2+ ions. The resulting combination was then kept at room temperature approximately for 24 hours.

3.3. Chitosan nanoparticle preparation

Chitosan is a biopolymer that has received a lot of focus and has been thoroughly studied for the manufacture of micro as well as nanoparticles. Its features, comprising low toxicity, biodegradability and strong biocompatibility, make it appropriate for application in biomedical as well as pharmaceutical products such as anti-inflammatory pharmaceuticals, antidiabetic medicines, enzyme and protein immobilisation and ophthalmology [13, 14].

0.2% w/v of Chitosan was submerged in 2% v/v of dilute acetic acid (3.0 pH). PEG 4000 (10 wt%) & TWEEN-60 (1.0 wt%) were added to the chitosan solution as stabilizers. 20 wt% Sodium sulphate was progressively introduced to a solution containing chitosan while rotating at a speed of 750 rpm (Figure 3.2).

Albumin from bovine serum (BSA) loaded nanoparticle preparation:

BSA-loaded particles were created by combining BSA comprising of a polysaccharide solution (0.2% alginate, 0.2% chitosan) at weight ratios ranging from 1:10–2:1. Chitosan nanoparticles and BSA-loaded alginate were manufactured utilising the same processes as that of unloaded nanoparticles (Figure 3.1). BSA contents in chitosan nanoparticles were measured employing Bradford approach [14, 15]. BSA had been enclosed in calcium alginate nanoparticles and its concentration was determined using UV-visible spectroscopy (wavelength = 595 nm), LAB INDIA Analytical UV-3200 UV/visible spectrophotometer. The effectiveness of encapsulation (EE) of BSA in nanoparticles was estimated using the formula [16]:

$$EE\ (\%) = \frac{BSA\ total - BSA\ supernatant}{BSA\ total} * 100$$

where,

BSA total = Total amount of BSA

BSA sup = BSA concentration in the supernatant

EE = Entrapment Efficiency

Nanoparticle Characterization:

Micelles, metal oxides, synthetic polymers and big biomolecules are all components of nanoparticles. Each of these materials has a distinct chemistry, which can be investigated via optical spectroscopy, Raman spectroscopy, X-ray fluorescence and absorbance and solid-state NMR. Nanoparticles› behaviour is greatly impacted by their nanometre size [17, 18]. To completely understand

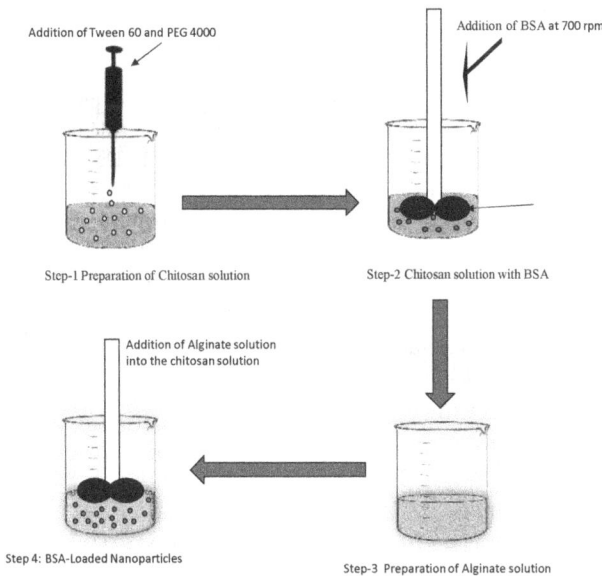

Figure 3.2. Schematic diagram of preparation of BSA-loaded nanoparticles.

Source: Author.

and predict nanoparticle behaviour, it is critical to characterize them by investigating their form, surface charge, size and porosity. Several characterization approaches have been developed to evaluate the distribution, shape, surface charge size and porosity of nanoparticles in various conditions [19].

Encapsulating BSA in Chitosan along with Alginate nanoparticles:

The protein, BSA was introduced into chitosan and alginate nanoparticles through first mixing them with a solution of polysaccharide. Table 3.1 shows that when the BSA content in the reaction mixture was 0.2 mg/mL, alginate nanoparticles encapsulated more than 99% of the sample. However, elevating the protein content from 0.2–0.8 mg/mL decreases the effectiveness of encapsulation by a rate of 15% [20].

4. Conclusion

A new approach has been devised for generating chitosan gel-like nanoparticles and stable aqueous colloids of alginate, without the need of an emulsion. This process involves combining protein with a polysaccharide solution before forming nanoparticles, which aids in the effective encapsulation of protein. The resultant polysaccharide nanoparticles demonstrate promise as carriers for physiologically active chemicals.

Table 3.1. BSA efficiency of entrapping within Chitosan nanoparticles

EE%	BSA conc. (Initially) (mg/ml)	Concentration of BSA in loaded nanoparticles (%)
25.0±1.0	0.2	0.02
45.0±2.0	0.5	0.08
53.0±3.0	0.8	0.19
24.0±2.0	1.1	0.23
13.0±1.0	1.4	0.23

Source: Author.

References

[1] Debnath, S., Kumar, R. S., & Babu, M. N. (2011). Ionotropic gelation–a novel method to prepare chitosan nanoparticles. *Research Journal of Pharmacy and Technology, 4*(4), 492–495.

[2] Grenha, A., Seijo, B., & Remuñán-López, C. (2005). Microencapsulated chitosan nanoparticles for lung protein delivery. *European Journal of Pharmaceutical Sciences, 25*(4–5), 427–437.

[3] Patil, P., Chavanke, D., & Wagh, M. (2012). A review on ionotropic gelation method: Novel approach for controlled gastroretentive gelispheres. *Int J Pharm Pharm Sci, 4*(4), 27–32.

[4] Nath, J., Sharma, K., Kumar, S., & Kumar, V. (2024). Ionotropic cross-linking methods for different types of biopolymeric hydrogels. In *Ionotropic Cross-Linking of Biopolymers*, (pp. 63–98). Elsevier.

[5] Bilati, U., Allémann, E., & Doelker, E. (2005). Development of a nanoprecipitation method intended for the entrapment of hydrophilic drugs into nanoparticles. *European Journal of Pharmaceutical Sciences, 24*(1), 67–75.

[6] Hasnain, M. S., Barik, H., Sahoo, R. N., Pattanayak, P., Panda, B. B., & Nayak, A. K. (2024). Ionotropic cross-linking of biopolymers: basics and mechanisms. In *Ionotropic Cross-Linking of Biopolymers* (pp. 3–31). Elsevier.

[7] Koukaras, E. N., Papadimitriou, S. A., Bikiaris, D. N., & Froudakis, G. E. (2012). Insight on the formation of chitosan nanoparticles through ionotropic gelation with tripolyphosphate.

Molecular Pharmaceutics, *9*(10), 2856–2862.

[8] Gadziński, P., Froelich, A., Jadach, B., Wojtyłko, M., Tatarek, A., Białek, A., Krysztofiak, J., Gackowski, M., Otto, F., & Osmałek, T. (2022). Ionotropic gelation and chemical crosslinking as methods for fabrication of modified-release gellan gum-based drug delivery systems. *Pharmaceutics*, *15*(1), 108.

[9] Racoviță, S., Vasiliu, S., Popa, M., & Luca, C. (2009). Polysaccharides based on micro-and nanoparticles obtained by ionic gelation and their applications as drug delivery systems. *Revue Roumaine de Chimie*, *54*(9), 709–718.

[10] Pavelková, M., Kubová, K., Vysloužil, J., Kejdušová, M., Vetchý, D., Celer, V., Molinková, D., Lobová, D., Pechová, A., Vysloužil, J., & Kulich, P. (2017). Biological effects of drug-free alginate beads cross-linked by copper ions prepared using external ionotropic gelation. *Aaps Pharmscitech*, *18*, 1343–1354.

[11] Zhang, H., Tumarkin, E., Peerani, R., Nie, Z., Sullan, R. M., Walker, G. C., & Kumacheva, E. (2006). Microfluidic production of biopolymer microcapsules with controlled morphology. *Journal of the American Chemical Society*, *128*(37), 12205–12210.

[12] Attia, M. S., El Nasharty, M. A., Rabee, M. M., Mohammed, N. N., Mohamed, M. M., Hosny, S. I., Abd El-Wahab, A. G., Mahmoud, A. G., Abd Elmaged, E. M., Afify, H. G., & Abdel-Mottaleb, M. S. (2024). Ionotropically cross-linked polymeric nanoparticles for drug delivery. In *Ionotropic Cross-Linking of Biopolymers* (pp. 301–353). Elsevier.

[13] Banerjee, T., Mitra, S., Singh, A. K., Sharma, R. K., & Maitra, A. (2002). Preparation, characterization and biodistribution of ultrafine chitosan nanoparticles. *International Journal of Pharmaceutics*, *243*(1–2), 93–105.

[14] Carreno-Gomez, B., & Duncan, R. (1997). Evaluation of the biological properties of soluble chitosan and chitosan microspheres. *International Journal of Pharmaceutics*, *148*(2), 231–240.

[15] Zohri, M., Nomani, A., Gazori, T., Haririan, I., Mirdamadi, S. S., Sadjadi, S. K., & Ehsani, M. R. (2011). Characterization of chitosan/alginate self-assembled nanoparticles as a protein carrier. *Journal of Dispersion Science and Technology*, *32*(4), 576–582.

[16] Masalova, O., Kulikouskaya, V., Shutava, T., & Agabekov, V. (2013). Alginate and chitosan gel nanoparticles for efficient protein entrapment. *Physics Procedia*, *40*, 69–75.

[17] Lama, S., Merlin-Zhang, O., & Yang, C. (2020). In vitro and in vivo models for evaluating the oral toxicity of nanomedicines. *Nanomaterials*, *10*(11), 2177.

[18] Jiang, C. C., Cao, Y. K., Xiao, G. Y., Zhu, R. F., & Lu, Y. P. (2017). A review on the application of inorganic nanoparticles in chemical surface coatings on metallic substrates. *RSC Advances*, *7*(13), 7531–7539.

[19] Modena, M. M., Rühle, B., Burg, T. P., & Wuttke, S. (2019). Nanoparticle characterization: What to measure?. *Advanced Materials*, *31*(32), 1901556.

[20] Xu, Y., & Du, Y. (2003). Effect of molecular structure of chitosan on protein delivery properties of chitosan nanoparticles. *International Journal of Pharmaceutics*, *250*(1), 215–226.

4 Overcoming hurdles in oral protein and peptide drug delivery: Strategies for future prospective

Prasanna Parida[1,2,a], Amiya Kumar Prusty[3], Rajat Kumar Prusty[4], Bikash Ranjan Jena[4], Debabrata Panda[4], and Sanjoy Das[5]

[1]Research Scholar (Pharmacy), Biju Patnaik University of Technology, Rourkela, Odisha, India
[2]Institute of Pharmacy and Technology, Salipur, Cuttack, Odisha, India
[3]Faculty of Pharmacy, C.V. Raman Global University, Bhubaneswar, Odisha, India
[4]School of Pharmacy and Life Sciences, Centurion University of Technology and Management, Bhubaneswar, Odisha, India
[5]Department of Pharmaceutical Sciences, Dibrugarh University, Dibrugarh, Assam, India

Abstract: Protein, as well as peptide-based medications, have a large pharmaceutical trade& prospective application since they are less risky than chemical pharmaceuticals in recent decades. Oral administration is generally the preferred method; however, because of their low absorption and breakdown in the gastrointestinal route, most proteins and peptides are currently delivered intravenously or subcutaneously. Numerous approaches to boost oral administration are being researched, such as transporter systems, stability enhancers, enzyme inhibitors and absorption enhancers. Research focused on creating novel strategies to circumvent the GI barriers of protein and peptide medicines. Some novel inventions have been patented, are in clinical testing and have already reached the market. This study examines formulation along with stability techniques regarding oral protein as well as peptide drug delivery systems using absorption enhancers, prodrugs, actively targeting Lymphatic transport, ionic liquid, mucus adherence, microneedle, pH regulation, enteric coating, nano-encapsulation, PEGylation, mucus penetration and peptide cyclization. Very few oral protein and peptide drugs are approved by the US FDA. Industries are facing more challenges when submitting applications to the Regulatory agencies concerned about bioavailability and stability.

Keywords: Absorption enhancer, stability, bioavailability, enzyme inhibitors, nano-encapsulation

1. Introduction

In recent years, USFDA has granted significantly increased number of new drugs. In 2018, 59 novel drugs were approved, with more than 50% of these approvals being for medicines for oral administration. Furthermore, more than one-third of the additional permissions granted were for biologic pharmaceuticals delivered by injection, such as intramuscular & intravenous routes [1].

Robert Bruce Merrifield synthesized first bioactive peptide in 1953. Since

[a]litunar@gmail.com

DOI: 10.1201/9781003672869-4

then, research into peptide and protein drugs has efficiently advanced, with a significant focus on drugs targeting multiple receptors. These medications possess a broad spectrum of beneficial clinical applications, as they can prevent the illnesses by modifying pathological mechanisms or physiological mechanisms. Medications containing Protein along with peptide are very important for treating cancer, autoimmune illnesses and cardiovascular problems. They have made significant contributions to the treatment of tumors and diabetes, resulting in substantial economic benefits [2]. The use of protein-based drugs is limited because of their physical as well as chemical instabilities, alike enzymatic breakdown within the gastrointestinal tract, low pH value, as well as rapid elimination from circulation. Injectable delivery techniques are ineffective, hence needle-free alternatives such as pulmonary, transdermal, oral, nasal and buccal delivery are being investigated. The primary issue with non-invasive protein delivery is to achieve exactly the same level of bioavailability similar to the injection technique. Efforts are being made to improve these delivery systems, which offer advantages such as avoiding the first-pass effect, being patient-friendly, reducing administration frequency and being painless [3].

The preferred method for drug delivery is the oral route. However, orally delivered proteins require various excipients such as permeation enhancers, enzyme inhibitors and polymeric carriers. Peptides have a greater potential for oral administration over proteins because of their lower size. Researchers are exploring methods to enhance the transport of hydrophilic peptides and proteins through closely connected cell layer junctions. Furthermore, researchers are investigating the use of mucoadhesion, nano-scale carriers and liposomes to enhance the delivery of large molecules. In the future, advanced oral protein formulations may utilize smart polymeric nanoparticles to safeguard and direct the proteins to specific locations, releasing them as required. The method of administration is a critical consideration in any treatment, as it impacts how the drug functions in the body.

Challenges such as poor absorption, metabolism and clearance make it difficult for protein and peptide-based treatments to be effective when taken orally or via non-oral mucosal routes. Therefore, such drugs are frequently delivered by injections. As these drugs have a short lifespan in the body and often need to be taken regularly, frequent injections can be inconvenient for patients. Researchers are developing new delivery systems to improve the administration of these drugs [4, 5].

Despite the difficulties of delivering medications through the digestive system, the oral route is still the most extensively researched for protein and peptide-based treatments due to its simplicity and widespread popularity among patients. Oral formulations can also save money for the pharmaceutical industry because they do not require specialized production facilities or direct engagement of healthcare experts [6].

Oral delivery has various advantages over other methods of administration. It is more widely tolerated by patients and enables less frequent and less painful administration, which can help to enhance the management of the disease. However, the efficacy of oral delivery has limitations due to its inadequate bioavailability. The relevance of this arises from the fact that proteases along with stomach acids within the digestive system can degrade pharmaceuticals, therefore these drugs frequently struggle to pass the intestinal wall. Irrespective of how the medication passes through the gastrointestinal tract, it will encounter stability and absorption problems. The fundamental problem associated with orally delivered peptide & protein medicines is to enhance the drug's absorption rate. Advances in drug formulation and technology have significantly improved the effectiveness of protein as well as peptide therapeutics. The therapeutic use of protein medications has been significantly impacted by nanotechnology in recent years [7]. Table 4.1 outlines the benefits of utilizing nanoparticles as delivery systems for peptide and protein medications. However, the application of nanoparticles in conjunction with peptide and protein drugs remains infrequent, as such medications are presumably in the early stages of studying and developing.

Table 4.1. Benefits of utilizing Nanoparticles to deliver peptide and protein medications

Sl No.	Advantages
1	Both the processes of preparing material are simple and clear.
2	Can be targeted by modified ligands to prolong retention time at specific absorption sites.
3	It reduces enzymatic degradation and agglomeration of peptide along with protein medications within the gastrointestinal tract, enhancing absorption across small intestine.
4	alters the spatial distribution of the medicine in the circulatory system.
5	It accomplishes the therapeutic impact of controlled release along with targeted therapy for illness.

Source: Cited from references [7, 8].

Non-parenteral modes of delivery for protein & peptide drugs include buccal, vaginal, rectal, nasal, transdermal, ophthalmic and oral. In the absence of an absorption enhancer, these methods are far less effective than parenteral administration. For instance, compared to parenteral treatment, oral delivery of LHRH required a dosage that was three thousand times more. The effective dosages of LHRH administered via the other routes were 400 times higher for rectal, 100 times greater than for nasal and 600 times higher for vaginal administration with respect to the parenteral route [8].

1.1. Types of barriers

1.1.1. Limitations to oral delivery system

Gastrointestinal tract is having a crucial role in breaking down carbohydrates, proteins and other nutrients into amino acids and simple sugars, while also serving as a barrier against harmful microorganisms. Consequently, intact peptides and proteins generally exhibit extremely low oral bioavailability, often less than 1% and occasionally even less than 0.1%. Oral medications encounter numerous biological hurdles before absorption, as described [9].

1.1.2. Biochemical barriers

There are two types of biochemical hurdles for proteins: pH along with enzymatic. Proteases, coupled with other enzymes, may quickly disintegrate proteins at specific splitting sites and are found across the gastrointestinal system. The breakdown of proteins can easily occur when the pH level is far from neutral, leading to the proteins unfolding and becoming inactive. The method of digestion originates from the mouth, whereby slightly acidic conditions (pH around 6.5) including the presence of lysozymes and salivary amylases initiate the decomposing of peptidoglycans and carbohydrates. However, the buccal cavity is not regarded as an important barrier to medication delivery because the residence length of a tablet or capsule is short, resulting in limited exposure to the compound [10].

The intestine and stomach have very active metabolic barriers that influence the digestion and absorption of orally consumed proteins. The gastric glands produce the protein-digesting pepsin, as well as hydrochloric acid along with mucus. The stomach comprises the most acidic surroundings of the body, with a pH of 1–2. Pepsin performs best in very acidic environments. Pepsin, a major endopeptidase found in the stomach, breaks peptide connections of aromatic residues like phenylalanine, tryptophan and tyrosine. Lipase enzymes aid in the digestion of oils, triglycerides, & fats, in the stomach. Digestion stays ongoing in the small intestinal tract, which contains digestive enzymes generated by the pancreas. Chymotrypsins, Carboxypeptidases, Trypsins and Elastases are the most commonly detected enzymes. Enterocytes in the small intestine produce carboxypeptidases, aminopeptidases, γ-glutamyl transpeptidases and endopeptidases [11–13].

The small intestine is made up of three main sections; jejunum, ileum and duodenum. Each of these components has distinct properties that influence how nutrients are absorbed. Food is exposed to varying levels of acidity as it digests and travels through the gastrointestinal tract. The stomach has a very acidic pH of 1.0–2.0 and the pH progressively increases as the meal passes via duodenum with pH 4–5.5, rectum with pH 7.0–7.5, jejunum with pH 5.5–7.0, ileum with pH 7.0–7.5 and colon. These changes regarding pH, along with variations in how quickly the stomach empties and how the intestines move, strongly impact how drugs taken by mouth are processed in the body. Protecting drugs from the effects of these enzymes and pH changes is crucial for delivering protein-based drugs effectively by mouth [14, 15].

2. Physiochemical Considerations

It is critical to evaluate the drug itself, as its ability to interact with GI tract is heavily influenced by its physicochemical features, such as particle size, solubility and chemical structure. Owing to the pH partition hypothesis, the gastrointestinal membranes operate as a lipoid barrier, allowing nonionized drugs to be absorbed more efficiently. Theophylline, for instance, is persistent and mildly ionized within the physiological pH range. It is entirely and quickly absorbed from uncoated tablets and solutions and it is barely metabolized during the first pass through the liver. However, it has been established that the rate of medication absorption is decreased when an immediate-release theophylline formulation is given concurrently [16].

2.1. Mucus barrier

The physical barrier that mucus creates keeps dangerous substances and pathogens from piercing the epithelial surface. Mucins are large glycoproteins that make up the majority of the mucus gel. Mucins are the main glycoprotein constituents of mucous, which coats the skin of certain amphibians as well as the surfaces of the cells lining the digestive, urogenital and respiratory systems. They defend epithelial cells against infection, water retention and physical or chemical damage while also making it easier for objects to flow through a tract. Various species create many structurally unique mucins; a single mucin can be found in more than one function. The size of groups within the mucin family might vary substantially. Most are tiny, having only a few hundred amino acid residues; however, others have a few thousand residues and are among the largest identified proteins. Irrespective of their size, all mucin polypeptide chains contain threonine and serine-rich categories, with hydroxyl groups linked to oligosaccharides via O-glycosidic connections. Furthermore, these groups consist of tandemly paired sequences that differ in number, length and amino acid sequence from one mucin to the next [17]. Carbohydrates in mucin can contribute for up to 90% of its total weight. Additionally, there are only two types of mucins: membrane-bound and secreted. Two human mucins are linked to membranes (MUC1 and MUC4), while four secrete mucus (MUC2, MUC5AC, MUC5B and MUC7) [18–23]. The remaining mucins (MUC6, MUC8 and MUC3) cannot be categorized [24–27]. Every single human mucin is same in other animals. Each human mucin has a primate counterpart. Thus, Porcine Submaxillary Mucin (PSM), one of the most fully researched mucins, demonstrates distribution among tissues and structure identical to MUC5B [28].

2.2. Enzyme suppressors

Protease inhibitors are chosen based on the therapeutic drug's structure. Understanding the selectivity of proteases is critical for guaranteeing drug

stability in the GI tract. The quantity of co-administered inhibitor(s) is critical for the intestinal permeability of a peptide or protein medication [29].

α-chymotrypsin, thiol metalloproteinase and Trypsin are the enzymes those cause degradation of insulin. Insulin stability is being studied in the context of medications that prevent the enzymes from degrading. α-chymotrypsin and Trypsin blockers comprise Camostat mesylate, soybean, aprotinin, pancreatic, FK-448. Bacitracin, p-chloromeribenzoate and 1,10-phenanthroline are insulin-degrading enzyme inhibitors. A study discovered that the combination of a stimulant (sodium cholate) with a protease blocker resulted in a 10 percent rise in rat intestinal glucose absorption [30].

In addition, cysteine's attachment to polycarbophil significantly increased the polymer's inhibitory impact on carboxypeptidase B, chymotrypsin, carboxypeptidase A. This compound demonstrated much more inhibitory efficacy than unaltered polycarbophil on obtained aminopeptidase N on undamaged mucosa in intestine [31–33].

Another way of enzyme suppression is to adjust the pH to disable local digestive enzymes. Trypsin, chymotrypsin and elastase can be inactivated by employing a pH-lowering buffer in sufficient amounts to drop the local intestinal pH values below 4.5.

2.3. The impact of physical and chemical characteristics of peptide/protein drug-loaded NPS

2.3.1. Nanoparticles
Nanoparticles, classified as polymer and solid lipid nanoparticles, have been widely employed for the oral administration of peptide and protein drugs. They

are composed of commercial and organic parts such as hyaluronic acid, cellulose, polycaprolactones, gelatin, chitosan, cyclodextrins, poly (lactic-co-glycolic) acid (PLGA), polyanhydrides [34, 35]. Natural polymer materials release and degrade drugs quickly, whereas manufactured polymer materials release drugs act slowly, lasting from few days to few weeks [36]. Polymer based nanoparticles can be implemented as a standard for assessing the transition of peptide along with protein therapeutics:

1. Such drugs can be encapsulated and protected from digestive enzymes.
2. Nanoparticles' shape, size and distribution should meet perticular specifications.
3. Entrapment efficiency and the loading rate of drug were greater.
4. The drug's release time should meet clinical medication standards.
5. The carrying material needs to be non-toxic and biodegradable

2.4. Diameter

The size of nanoparticles plays a critical role in how they are absorbed. The size of the particles must be appropriate for absorption. The size of the nanoparticles determines their uptake by M cells and enterocytes [37–39]. Enterocytes can take up nanoparticles smaller than 50 nm by endocytosis, whereas M cells are capable of carrying particles between 20 and 100 nm, as well as 100 and 500 nm. Drugs can be absorbed in the small intestine via two pathways; Transcellular and Paracellular [40–44].

2.5. The Surface charge

The electrostatic coupling among Chitosan; positively charged nanoparticles and Mucin glycosides; negatively charged can end up in significant mucosal adherence. This causes the nanoparticles to bind closer towards epithelial cells, increasing the uptake of intestine agglomerate of Peyer›s patches. Because the small intestine›s mucus layer has a negative charge and nanoparticles have positive charge. Chitosan nanoparticles, are mostly adhere on it and stay there for a greater duration of time [45]. However, the ion exchange between positively charged nanoparticles & the mucus layer inhibits the nanoparticles from penetrating deeper into the mucus layer and reaching at the surface of epithelial cells. This influences nanoparticle reuptake by epithelial cell.

Such surface charge of nanoparticles affects how easily they can be eliminated from the body. Positively charged nanoparticles are easier to remove than negative or neutrally charged nanoparticles, although neutral nanoparticles remain in the body for longer. Introducing PEG to the nanoparticle surface may render the particles more water-friendly while also shielding their surface charge, bringing them closer to electric neutrality. The size and amount of PEG employed have a significant impact on the particles' surface charge. Measuring the particle's action potential can assist understand how well PEG shields the nanoparticle surface and forecast how it passes through mucus [46, 47].

The carrier substance's surface charge influences nanoparticle adherence through the intestinal mucosa. The mucus is having a negatively charged sugar group. Such carrier polymers concerning a positive charge may interact with mucus via electrostatic forces, increasing nanoparticle retention duration on mucosal surface. Furthermore, bonding among mucus layers and cationic nanoparticles can limit nanoparticles' permeation by the mucous layer up to the surface of epithelial cells, influencing nanoparticle uptake through epithelial cells.

2.6. Approaches with respect to delivery of protein and peptide drugs orally (NPS)

When such nanoparticles are being administered orally, gastrointestinal tract (GIT) absorbs them. However, inadequate absorption continues to pose a significant hurdle in developing oral nanotechnology drug delivery systems. Further inquiry regarding the absorption mechanism has resulted in approaches for improving nanoparticle absorption in the gastrointestinal tract. These techniques cause intestinal epithelial cells to adhere and target M cells. Molecules can travel through cell membranes in four ways: transcellular, carrier-mediated, paracellular and receptor-mediated transport. Chemical modifications, absorption enhancers, mucosal adhesion systems and nanoparticles are all strategies for increasing protein bioavailability. This section summarizes regarding the enhanced bioavailability orally with respect to polypeptide as well as protein medicines.

2.7. Absorption boosters

Factors accelerating absorption can promote drug absorption by influencing the cell or bypassing routes. Table 4.2 lists the processes via which absorption enhancers work. The cell route may include absorption enhancers, which interfere with the cell's outermost

membrane structure or promote membrane protein loss, so boosting medication absorption. The intercellular route facilitates drug transfer by establishing exactly linkages among the cells. The gastrointestinal system's epithelial cells' membranes are pushed together, significantly limiting oral absorption [48, 49].

The uptake of peptides and protein medicines through the digestive tract faces two major challenges: degradation by gastrointestinal enzymes and the limited permeability of intestinal cells [50]. Absorption enhancers can be employed in nano-drug delivery systems to boost the absorption of peptides and proteins across the intestinal barrier, hence increasing their bioavailability [51]. Permeability enhancers allow peptides to enter cells by modulating tightly bound junctions [52]. Because of their large size and strong bonding, peptides and protein medicines may temporarily impact the intestinal barrier by changing its tightly bound junction, increasing intercellular gaps, or disturbing lipid bilayer stability. This increases the permeability of the biological membrane, resulting in better absorption of protein-peptide medicines from the gastrointestinal system into the bloodstream [53, 54].

Table 4.2. Mechanisms of absorption enhancers

Sl No.	Mechanism
1	Opening of tight junctions between epithelial cells
2	Increased membrane fluidity
3	Temporary disruption of the intestinal barrier integrity
4	Reduction of mucous layer viscosity

Source: Cited from references [48–51].

2.8. Mucoadhesive

The mucous layer may hinder medication absorption, yet it is advantageous for the creation of the bio-adhesive mucosal drug delivery system [55]. This technique primarily uses the mucous membrane to produce adhesion among mucus and polymer material, resulting in increased drug residence duration, mucosal adhesion and higher medication efficacy. Nanoparticles lengthen the retention duration in the gastrointestinal system by hydrophobic interaction, electrostatic contact, polymer chain interaction, Van der Waals force and penetration, therefore improving drug absorption [56, 57]. Mucous adhesives may directly influence on the flexibility of mucous epithelium, therefore enhancing the accessibility of peptide & protein medications [58]. Furthermore, mucosal adhesion system may avoid protein-peptide impairment, strengthen equilibrium, facilitate tightly bonded connections among epithelial cells, increase trans-membrane permeation, reduce administration frequency and manage the dissociation rate of protein polypeptides. Mucus adherence nanoparticles (MNP) have been discovered to be an exciting nanocarrier for protein and peptide therapeutics because of a better absorption, longer persistence period in the GI tract and easily sticking with the mucus layer for raising the drug conc. gradient [59–61]. After oral treatment, diabetic rats showed sustained hypoglycemia effects with a speedier onset. Orally administered conjugate-loaded MNPs demonstrated a pharmacological availability of 17.98 % ± 5.61% over an insulin solution infused subcutaneously, yielding in a two fold enhancement across MNPs treated with native insulin.

Certain studies show, chitosan is having a positive charge along with it can adhere with negatively charged mucous layer, resulting in mucous adsorption. This can increase the time a medicine stays in the small intestine [62]. Additionally, chitosan may rupture the tightly bound junction of Caco-2 a single-layer cells, lowering membrane permeability and increasing cell permeability (63–64). The relative bioavailability was 11.78% when administered orally to diabetic mice (50 IU/kg), with no significant systemic effects. Insulin can be delivered orally using pH-sensitive nanoparticles such as polymethacrylic acid, chitosan or polyethylene glycol. The nanoparticles' encapsulation effectiveness must achieve 99.9%, having an average particle size of 172 nano meter. Gut releases twice as much insulin than the stomach [65, 66]. When the pH was higher than 7, these nanoparticles gradually grew and disintegrated. Rebuilding the close bonding of intestinal epithelial cells could lengthen the time frame of hypoglycemia and also boost insulin absorption in the small intestine [67–70].

2.9. Additional approaches that encourage delivery of peptide and protein drug nanoparticles orally

2.9.1. Alteration of target molecule

Such target nanoparticles are intended to attach to particular molecules or clusters on the outermost surface of the desired molecule. Nanoparticles may attach with the relevant receptors upon the tiny intestinal epithelial cell's surface, directing those molecules or assemblies for recognition. This lowers mucus trapping and clearance effects on nanoparticles, improving medication absorption. Commonly employed target compounds include lectins, intrusive elements and vitamin B12 [71].

2.9.2. Protein Nano crystallization technology

Protein medications are frequently freeze-dried and solidified to extend their shelf life. However, this process can lead to protein aggregation and structural alterations, resulting in a loss of biological activity. Nanocrystallization is a strategy for keeping proteins stable and biologically active. It uses crystallisation technology to create protein nanoparticles ranging in size from 50 to 500 nm. These stable nanoparticles are created by using surfactants to stabilize them via charge or space. Protein nano crystallisation effectively protects proteins' biological activity, is biocompatible and is easily degradable and assembled.

2.9.3. Nanoparticles with Cell-Permeating Peptides

Cell-permeating peptides are short peptides comprised of amino acid structures, those are charged positively. Such peptides can cross cell membranes as well as transport large molecules or nanoparticles into cells. However, the specific technique by which cell-penetrating peptides increase their consumption of large molecules remains uncertain.

When certain penetrating peptides are physically combined with insulin, they enhance insulin absorption within the intestinal mucosa. To promote insulin absorption, additional penetrating peptides must covalently bond to insulin. The complex produced by insulins those are negatively charged & positively charged penetration by electrostatic adsorption is

then enveloped in pHPMA to produce nanoparticles. The pHPMA encapsulation triggers the surface of hydrophilic nanoparticles, which may fill up the positive charge of the permeation, resulting in increased efficiency in penetrating the mucus layer. During mucus layer penetration, pHPMA eventually breaks from such nanoparticles' surfaces, releasing the penetrating insulin complex. This method significantly improves insulin's ability to get into intestinal epithelial cells. Mucus-derived epithelial cells uptake nanoparticles approximately 20 times more than free insulin or unbound insuline. Pharmacological potency of such nanoparticles in rats with diabetes adhering with intragastric treatments is 6.61% in in comparison with insulin delivery subcutaneously [72–74].

3. Conclusion and Potential Futures for Peptide and Protein Drugs

Chemical modifications to protein polypeptides can enhance membrane permeability, medicine stability, immunogenicity, bioactivity and medicine stability. Furthermore, absorption enhancers can boost protein polypeptide absorption in the small intestine, although they are lacking protein specificity and gastrointestinal toxicity may resulted. Such mucosal adhesion system can lengthen retention duration of polypeptide & protein medications in GIT, raising bioavailability; however, it does not get improved oral drug permeability or prevented small intestinal mucosal clearance.

Lastly, information is fully integrated and implemented through an understanding of the gastrointestinal tract's absorption barrier and mechanism, the biological processes regarding absorption of protein polypeptide, the enhancement of oral delivery system, along with the nanotechnology of the protein polypeptide.

4. Acknowledgement

All authors express their heartfelt gratitude to School of Pharmacy & Life Sciences, Centurion University of Technology & Management, Bhubaneswar worth granting scientific environment to complete the review work.

References

[1] Rekha, M. R., & Sharma, C. P. (2013). Oral delivery of therapeutic protein/peptide for diabetes–future perspectives. *Int J Pharm, 440*, 48–62.

[2] Sachdeva, S., Lobo, S., & Goswami, T. (2016). What is the future of non invasive routes for protein- and peptide-based drugs? *Ther Deliv,7*, 355–357.

[3] US Food and Drug Administration. (2018). Novel drug approvals for 2018. *FDA* https://www.fda.gov/drugs/new-drugs-fda-cders-new-molecular-entities-and-newtherapeutic-biological-products/novel-drug-approvals-2018.

[4] Antosova, Z., Mackova, M., Kral, V., & Macek, T. (2009). Therapeutic application of peptides and proteins: parenteral forever?. *Trends in Biotechnology, 27*(11), 628–635.

[5] Muheem, A., Shakeel, F., Jahangir, M. A., et al. (2014). A review on the strategies for oral delivery of proteins and peptides and their clinical perspectives. *Saudi Pharm J.* doi:10.1016/j.jsps.2014.06.004.

[6] Maher, S., & Brayden, D. J. (2012). Overcoming poor permeability: translating permeation enhancers for oral peptide delivery. *Drug Discov Today Technol,9*, e71–e174.

[7] Liu, C., Kou, Y., Zhang, X., Cheng, H., Chen, X., & Mao, S. (2018). Strategies and industrial perspectives to improve oral absorption of biological macromolecules. *Expert Opin Drug Deliv, 15*, 223–233.

[8] Zhou, X. H., & Po, A. L. W. (1991). Peptide and protein drugs: II. Nonparenteral routes of delivery. *Int. J. Pharm, 75*,117–130.

[9] Maher, S., & Brayden, D. J. (2012). Overcoming poor permeability: Translating permeation enhancers for oral peptide delivery. *Drug Discov Today Technol, 9*, e71–e174.

[10] Fábián, T. K., Hermann, P., Beck, A., Fejérdy, P., & Fábián, G. (2012). Salivary defense proteins: Their network and role in innate and acquired oral immunity. *Int J MolSci, 13*, 4295–4320.

[11] Allen, A., & Carroll, N. J. (1985). Adherent and soluble mucus in the stomach and duodenum. *Dig Dis Sci, 30*, 55S–62S.

[12] Allen, A., Flemstrom, G., Garner, A., & Kivilaakso, E. (1993). Gastroduodenal mucosal protection. *Physiol Rev, 73*, 823–857.

[13] Whitcomb, D. C., & Lowe, M. E. (2007). Human pancreatic digestive enzymes. *Dig Dis Sci, 52*, 1–17.

[14] Masaoka, Y., Tanaka, Y., Kataoka, M., Sakuma, S., & Yamashita, S. (2006). Site of drug absorption after oral administration: Assessment of membrane permeability and luminal concentration of drugs in each segment of gastrointestinal tract. *Eur J Pharm Sci, 29*, 240–250.

[15] Kararli, T. T. (1995). Comparison of the gastrointestinal anatomy, physiology, and biochemistry of humans and commonly used laboratory animals. *Biopharm Drug Dispos, 16*, 351–380.

[16] Daugherty, A. L., & Mrsny, R. J. (1999). Transcellular uptake mechanisms of the intestinal epithelial barrier Part one. *Pharm Sci Technol, 2*, 144–151.

[17] Gendler, S. J., & Spicer, A. P. (1995). Epithelial mucin genes. *Annual Review of Physiology, 57*, 607–634.

[18] Gendler, S. J., Lancaster, C. A., Taylor-Papadimitriou, J., Duhig, T., Peat, N., Burchell, J., Pemberton, L., Lalani, E. N., & Wilson, D. (1990). Molecular cloning and expression of human tumor-associated polymorphic epithelial mucin. *Journal of Biological Chemistry, 265*(25), 15286–15293.

[19] Moniaux, N., Nollet, S., Porchet, N., Degand, P., Laine, A., & Aubert, J. P. (1999). Complete sequence of the human mucin MUC4: A putative cell membrane-associated mucin. *Biochemical Journal, 338*(2), 325–333.

[20] Gum, J. R., Jr, Hicks, J. W., Toribara, N. W., Siddiki, B., & Kim, Y. S. (1994). Molecular cloning of human intestinal mucin (MUC2) cDNA. Identification of the amino terminus and overall sequence similarity to prepro-von Willebrand factor. *Journal of Biological Chemistry, 269*(4), 2440–2446.

[21] Li, D., Gallup, M., Fan, N., Szymkowski, D. E., & Basbaum, C. (1998). B. Cloning of the amino-terminal and 5′-flanking region of the human MUC5AC mucin gene and transcriptional up-regulation by bacterial exoproducts. *Journal of Biological Chemistry, 273*(12), 6812–6820.

[22] Desseyn, J. L., Aubert, J. P., Van Seuningen, I., Porchet, N., & Laine, A. (1997). Genomic organization of the 3′ region of the human mucin Gene-MUC5B. *Journal of Biological Chemistry, 272*(27), 16873–16883.

[23] Bobek, L. A., Tsai, H., Biesbrock, A. R., & Levine, M. J. (1993). Molecular cloning, sequence, and specificity of expression of the gene encoding the low molecular weight human salivary mucin (MUC7). *Journal of Biological Chemistry, 268*(27), 20563–20569.

[24] Gum, J. R., Jr, Hicks, J. W., Toribara, N. W., Rothe, E. M., Lagace, R. E., & Kim, Y. S. (1992). The human MUC2 intestinal mucin has cysteine-rich subdomains located both upstream and downstream of its central repetitive region. *Journal of Biological Chemistry, 267*(30), 21375–21383.

[25] Toribara, N. W., Ho, S. B., Gum, E., Gum, J. R., Lau, P., & Kim, Y. S. (1997). The carboxyl-terminal sequence of the human secretory mucin, MUC6: Analysis of the primary amino acid sequence. *Journal of Biological Chemistry, 272*(26), 16398–16403.

[26] Shankar, V., Pichan, P., Eddy, R. L. Jr, Tonk, V., Nowak, N., Sait, S. N., Shows, T. B., Schultz, R. E., Gotway, G., Elkins, R. C., & Gilmore, M. S. (1997). Chromosomal localization of a human mucin gene (MUC8) and cloning of the cDNA corresponding to the carboxy terminus. *American Journal of Respiratory Cell and Molecular Biology, 16*(3), 232–241.

[27] Pigny, P., Guyonnet-Duperat, V., Hill, A. S., Pratt, W. S., Galiegue-Zouitina, S., d'Hooge, M. C., Laine, A., Van-Seuningen, I., Degand, P., Gum, J. R., & Kim, Y. S. (1996). Human mucin genes assigned to 11p15. 5: Identification and organization of a cluster of genes. *Genomics, 38*(3), 340–352.

[28] Eckhardt, A. E., Timpte, C. S., DeLuca, A. W., & Hill, R. L. (1997). The complete cDNA sequence and structural polymorphism of the polypeptide chain of porcine submaxillary mucin. *Journal of Biological Chemistry, 272*(52), 33204–33210.

[29] Bernkop-Schnürch, A. (1998). The use of inhibitory agents to overcome the enzymatic barrier to perorally administered therapeutic peptides and proteins. *Journal of Controlled Release, 52*(1–2), 1–6.

[30] Rick, S. (2005). Oral protein and peptide drug delivery. In Binghe, W., Teruna, S., Richard, S., eds. *Drug delivery: Principles and applications* (). New Jersey: Wiley Interscience.

[31] Hutton, D. A., Pearson, J. P., Allen, A., Foster, S. N. (1990). Mucolysis of the colonic mucus barrier by faecal proteinases: Inhibition by interacting polyacrylate. *Clin Sci, 78*, 271.

[32] Lueben, H. L., de Leeuw, B. J., Perard, D., Lehr, C. M., de Boer, A. G., Verhoef, J. C., et al. (1996). Mucoadhesive polymers in peroral peptide drug delivery: I, Influence of mucoadhesive excipients on the proteolytic activity of intestinal enzymes. *Eur J Pharm Sci, 4*, 117–128.

[33] Andreas, B. S., Alexander, H. K., Verena, M. L., & Thomas, P. (2004). Thiomers: Potential excipients for non-invasive peptide delivery systems. *Eur J Pharm Biopharm, 58*, 253–263.

[34] Mir, M., Ahmed, N., & Rehman, A. U. (2017). Recent applications of PLGA based nanostructures in drug delivery. *Colloid Surface B, 159*, 217–231.

[35] Kovalainen, M., Monkare, J., Riikonen, J., Pesonen, U., Vlasova, M., Salonen, J., et al. (2015). Novel delivery systems for improving the clinical use of peptides. *Pharmacol Rev,67*, 541–561.

[36] des Rieux, A., Fievez, V., Garinot, M., Schneider, Y. J., & Preat, V. (2006). Nanoparticles as potential oral delivery systems of proteins and vaccines: A mechanistic approach. *J Control Release, 116*, 1–27.

[37] Jao, D., Xue, Y., Medina, J., & Hu, X. (2017). Protein-based drug-delivery materials. *Materials,10*, 517.

[38] Lee, K. Y., & Yuk, S. H. (2007). Polymeric protein delivery systems. *Prog Polym Sci, 32*, 669–697.

[39] Zhao, Y. N., Xu, X. Y., Wen, N., Song, R., Meng, Q. B., Guan, Y., et al. (2017). A drug carrier for sustained zero-order release of peptide therapeutics. *Sci Rep-Uk, 7*, 5524.

[40] Frohlich, E., & Roblegg, E. (2016). Oral uptake of nanoparticles: Human relevance and the role of in vitro systems. *Arch Toxicol, 90*, 2297–2314.

[41] Griffin, B. T., Guo, J. F., Presas, E., Donovan, M. D., Alonso, M. J., & O'Driscoll, C. M. (2016). Pharmacokinetic, pharmacodynamic and biodistribution following oral administration of nanocarriers containing peptide and protein drugs. *Adv Drug Deliv Rev, 106*, 367–380.

[42] Desai, M. P., Labhasetwar, V., Amidon, G. L., Levy, R. J. (1996). Gastrointestinal uptake of biodegradable microparticles: Effect of particle size. *Pharmaceutical Research, 13*, 1838–1845.

[43] Florence, A. T., Hillery, A. M., Hussain, N., & Jani, P. U. (1995). Nanoparticles as carriers for oral peptide absorption: studies on particle uptake and fate. *Journal of Controlled Release, 36*(1–2), 39–46.

[44] des Rieux, A., Fievez, V., Garinot, M., Schneider, Y. J., & Preat, V. (2006). Nanoparticles as potential oral delivery systems of proteins and vaccines: A mechanistic approach. *J Control Release, 116*, 1–27.

[45] Park, J. H., Saravanakumar, G., Kim, K., & Kwon, I. C. (2010). Targeted delivery of low molecular drugs using chitosan and its derivatives. *Adv Drug Deliv Rev, 62*, 28–41.

[46] Chang, T. Y., Chen, C. C., Cheng, K. M., Chin, C. Y., Chen, Y. H., Chen, X. A., et al. (2017). Trimethyl chitosan-capped silver nanoparticles with positive surface charge: Their catalytic activity and antibacterial spectrum including multidrug-resistant strains of Acinetobacter baumannii. *Colloid Surface B, 155*, 61–70.

[47] Susa, M., Iyer, A. K., Ryu, K., Choy, E., Hornicek, F. J., Mankin, H., et al. (2010). Inhibition of ABCB1 (MDR1) expression by an siRNA

Nanoparticulate delivery system to overcome drug resistance in osteosarcoma. *PLoS One. 5*, e10764.

[48] Balda, M. S., & Matter, K. (2000). Transmembrane proteins of tight junctions. *Semin Cell Dev Biol, 11*, 281–289.

[49] Jao, D., Xue, Y., Medina, J., & Hu, X. (2017). Protein-based drug-delivery materials. *Materials, 10*, 517.

[50] Balda, M. S., & Matter, K. (2000). Transmembrane proteins of tight junctions. In *Seminars in Cell & Developmental Biology* (Vol. 11, No. 4, pp. 281–289). Academic Press.

[51] Richard, J. (2017). Challenges in oral peptide delivery: Lessons learnt from the clinic and future prospects. *Therapeutic Delivery, 8*(8), 663–684.

[52] Lundquist, P., & Artursson, P. (2016). Oral absorption of peptides and nanoparticles across the human intestine: Opportunities, limitations and studies in human tissues. *Advanced Drug Delivery Reviews, 106*, 256–276.

[53] Makhlof, A., Werle, M., Tozuka, Y., & Takeuchi, H. (2011). A mucoadhesive nanoparticulate system for the simultaneous delivery of macromolecules and permeation enhancers to the intestinal mucosa. *J Control Release, 149*, 81–88.

[54] Lanke, S. S., Gayakwad, S. G., Strom, J. G., & D'Souza, M. J. (2009). Oral delivery of low molecular weight heparin microspheres prepared using biodegradable polymer matrix system. *Journal of Microencapsulation, 26*(6), 493–500.

[55] Bagan, J., Paderni, C., Termine, N., Campisi, G., Lo Russo, L., Compilato, D., & Di Fede, O. (2012). Mucoadhesive polymers for oral transmucosal drug delivery: A review. *Current Pharmaceutical Design, 18*(34), 5497–5514.

[56] Shaji, J., & Patole, V. (2008). Protein and peptide drug delivery: Oral approaches. *Indian Journal of Pharmaceutical Sciences, 70*(3), 269.

[57] Duggan, S., Cummins, W., O'Donovan, O., Hughes, H., & Owens, E. (2017). Thiolated polymers as mucoadhesive drug delivery systems. *European Journal of Pharmaceutical Sciences, 100*, 64–78.

[58] Smart, J. D. (2005). The basics and underlying mechanisms of mucoadhesion. *Advanced Drug Delivery Reviews, 57*(11), 1556–1568.

[59] Sheng, J., He, H., Han, L., Qin, J., Chen, S., Ru, G., Li, R., Yang, P., Wang, J., & Yang, V. C. (2016). Enhancing insulin oral absorption by using mucoadhesive nanoparticles loaded with LMWP-linked insulin conjugates. *Journal of Controlled Release, 233*, 181–190.

[60] Khutoryanskiy, V. V. (2011). Advances in mucoadhesion and mucoadhesive polymers. *Macromolecular Bioscience, 11*(6), 748–764.

[61] Liu, Y., Zhang, J., Gao, Y., & Zhu, J. (2011). Preparation and evaluation of glyceryl monooleate-coated hollow-bioadhesive microspheres for gastroretentive drug delivery. *International Journal of Pharmaceutics, 413*(1–2), 103–109.

[62] Bravo-Osuna, I., Vauthier, C., Farabollini, A., Palmieri, G. F., & Ponchel, G. (2007). Mucoadhesion mechanism of chitosan and thiolated chitosan-poly (isobutyl cyanoacrylate) core-shell nanoparticles. *Biomaterials, 28*(13), 2233–2243.

[63] Yeh, T. H., Hsu, L. W., Tseng, M. T., Lee, P. L., Sonjae, K., Ho, Y. C., & Sung, H. W. (2011). Mechanism and consequence of chitosan-mediated reversible epithelial tight junction opening. *Biomaterials, 32*(26), 6164–6173.

[64] Mukhopadhyay, P., & Kundu, P. P. (2015). Chitosan-graft-PAMAM–alginate core–shell nanoparticles: A safe and promising oral insulin carrier in an animal model. *RSC Advances, 5*(114), 93995–94007.

[65] Rostamizadeh, K., Rezaei, S., Abdouss, M., Sadighian, S., & Arish, S. (2015). A hybrid modeling approach for optimization of PMAA–chitosan–PEG nanoparticles for oral insulin delivery. *RSC Advances*, *5*(85), 69152–69160.

[66] Su, F. Y., Lin, K. J., Sonaje, K., Wey, S. P., Yen, T. C., Ho, Y. C., et al. (2012). Protease inhibition and absorption enhancement by functional nanoparticles for effective oral insulin delivery. *Biomaterials*, *33*, 2801–2811.

[67] Liu, H., Tang, R., Pan, W. S., Zhang, Y., & Liu, H. (2003). Potential utility of various protease inhibitors for improving the intestinal absorption of insulin in rats. *J Pharm Pharmacol*, *55*, 1523–1529.

[68] Ibraheem, D., Elaissari, A., & Fessi, H. (2014). Administration strategies for proteins and peptides. *Int J Pharm*, *477*, 578–589.

[69] Bernkop-Schnürch, A. (1998). The use of inhibitory agents to overcome the enzymatic barrier to perorally administered therapeutic peptides and proteins. *Journal of Controlled Release*, *52*(1–2), 1–6.

[70] Bernkop-Schnurch, A., & Walker, G. (2001). Multifunctional matrices for oral peptide delivery. *Critical Reviews™ in Therapeutic Drug Carrier Systems*, 18(5).

[71] Yin, Y., Chen, D., Qiao, M., Lu, Z., & Hu, H. (2006). Preparation and evaluation of lectin-conjugated PLGA nanoparticles for oral delivery of thymopentin. *J Control Release*, *116*, 337–345.

[72] Kamei, N., Morishita, M., Ehara, J., & Takayama, K. (2008). Permeation characteristics of oligoarginine through intestinal epithelium and its usefulness for intestinal peptide drug delivery. *J Control Release*, *131*, 94–99.

[73] He, H. N., Sheng, J. Y., David, A. E., Kwon, Y. M., Zhang, J., Huang, Y. Z., et al. (2013). The use of low molecular weight protamine chemical chimera to enhance monomeric insulin intestinal absorption. *Biomaterials*, *34*, 7733–7743.

[74] Shan, W., Zhu, X., Liu, M., Li, L., Zhong, J., Sun, W., et al. (2015). Overcoming the diffusion barrier of mucus and absorption barrier of epithelium by self-assembled nanoparticles for oral delivery of insulin. *ACS Nano*, *9*, 2345–2356.

5   The impact of mirror therapy in improving hand motor function following stroke

DattaSai Pamidimarri[1,a], Nagarjuna Narayansetti[2], Preety Kumari[3], SriKavya Grandhi[1], Soumya Saswati Panigrahi[1], and Ravi Kumar Kalari[4]

[1]Assistant Professor, Department of Physiotherapy, School of Paramedics and Allied Health Science, Centurion University of Technology and Management, Bhubaneswar, Odisha, India

[2]Professor, Sai Sri College of Physiotherapy, Eluru, Andhra Pradesh, India

[3]Assistant Professor, Department of Physiotherapy, CT University, Ludhiana, Punjab, India

[4]Professor, Sree Rama Educational Trust College of Physiotherapy, MIMS Hospital, Vizianagaram, Andhra Pradesh, India

Abstract: This study investigates the impact of Mirror Therapy (MT) in improving hand motor function in individual recovering from stroke. Conducted at MIMS General Hospital in India, included 32 participants who were split into two groups: Group A received 30 minutes MT sessions. At the same time, Group B received 45 minutes sessions, with both the groups receiving therapy five days a week for four weeks. MT, which activates mirror neurons by observing patients' reflections, was paired with conventional rehabilitation treatments. Motor function improvement was evaluated by using Fugl Meyer Assessment Upper Extremity (FMA-UE) before and after the intervention. Results showed significant improvement in hand motor function after the MT intervention in both groups. However, the between-group analysis did not reveal substantial differences, although Group B demonstrated slight but significant improvement after longer, cumulative therapy sessions (900 minutes). The study concludes that MT is valuable to stroke rehabilitation programs due to its cost-effectiveness and simplicity, making it accessible for broader use in clinical settings. While no significant differences were found in between the groups, the results highlight the potential of MT to enhance motor recovery. Further research is needed to standardize MT protocols and explore its long-term benefits in stroke rehabilitation.

Keywords: Mirror therapy, hand rehabilitation, stroke, C.V.A., Fugl-Meyer Assessment

1. Introduction

Impairments in the upper limb such as challenges with fine, gross and coordinating movement may severely impact daily tasks, such as feeding and toileting and are commonly caused by strokes or cerebrovascular accidents. As such, rehabilitation efforts often focus on strengthening arm function to improve patient quality of life. Several treatment procedures exist, including training regimes, equipment and techniques [1]. One such technique is mirror therapy, which has emerged as a particularly

[a]ez.dattasai@gmail.com

DOI: 10.1201/9781003672869-5

effective technique for improving hand motor function. This therapy can help patients in alleviating pain from Complex regional and phantom limb [2] and hand injuries by engaging mirror neurons in the brain – the premotor cortex and the inferior parietal lobe. Altschuler et al. [3] were among the first to introduce mirror therapy in stroke patients in improving the motor function. Despite its proven effectiveness, a standardized protocol for its application remains absent and thus (ranging from 15 to 90 minutes per session) hinders its widespread adoption. Therefore, it is critical to evaluate the effectiveness of this technique using appropriate measures. This research seeks to evaluate the impact of mirror therapy on hand function utilizing the Fugl-Meyer Upper Extremity Assessment (FMA-UE) to measure motor recovery [4]. With a standardized protocol, mirror therapy can be a game-changer for patients with upper limb impairments, promoting functional restoration and improving their overall quality of life.

2. Materials and Methods

This study was conducted at MIMS General Hospital in Andhra Pradesh, India, following ethical approval from the Institution Research Board of Sree Rama Educational Ethical Committee in accordance with the ethical guidelines set forth in ICMR's Biomedical Research on Human Subjects, 2000, New Delhi. 32 participants were selected based on a set of criteria (Table 5.1) and through convenient sampling with random allocation into groups. The study spanned over a year with a primary objective to assess the impact of mirror therapy

Table 5.1. Selection criteria for the study

S.NO	Inclusion Criteria	Exclusion Criteria
1	Clinically and Radiologically diagnosed	Chronic stroke conditions
2	+/=24 score on the MMSE scale.	Poor cognition
3	Age between 40 to 65 years	Hearing and visual deficit and history of Hand Fractures

MMSE–Mini-Mental state examination

Source: Author.

(MT) and conventional treatment (CT) on individuals who had experienced a stroke. Participants who had been diagnosed with stroke (both clinically and radiologically) within a year received either 30 minutes (Group A) or 45 minutes (Group B) of MT sessions, five days a week for four weeks.

Mirror Therapy (MT) involves participants gazing into a mid-sagittal placed mirror, observing reflections of sound/unaffected limb movements as if impaired. This technique activates mirror neurons responsible for imitating actions and processing sensory information. Initially, participants spend 1–2 minutes visually engaging with the mirror reflection of the unaffected limb without movement, actively visualizing the mirrored image as the affected limb. To enhance the illusion, participants were instructed to imagine viewing their affected limb through a window instead of a mirror. Then, bilateral synchronous tactile stimulation through gentle stroking to stimulate the mirror neurons further. Subsequently, simple movements and exercises are initiated, such as motor exercises without objects; including unilateral and bilateral free movements, therapist-guided movements for the affected arm and simultaneous guidance of both arms, later with the objects. These structured exercises aim to leverage neuroplasticity to improve motor function during rehabilitation focusing on all the osteokinematic movements of elbow, wrist and Hand. Task-oriented movements include holding a water glass with or without water, holding a pen as if writing, gripping a cricket ball, holding an A4-size paper using the thumb and index finger, holding a key with the thumb and lateral part of the index finger and holding an empty water bottle.

Conventional treatment strategies were adopted to address each participant unique needs and clinical conditions, contributing to a comprehensive and patient-specific rehabilitation program. Conventional treatment was administered to both groups, encompassing various therapeutic strategies, including patient positioning, shoulder complex and hip complex exercises, mobilization and stretching, weight-bearing exercises and subluxation prevention. Patient positioning involved encouraging the side lying on the affected side to improve spatial and temporal awareness and prevent synergy development. Supine and side-lying on the sound side helps in reducing the muscle tone. Shoulder complex and hip complex exercises included implementing a side-lying position with scapular protraction to improve mobility and rhythm of the scapula thoracic area. Double arm elevation in a supine position was used to augment the range of motion, including pelvic bridging. Mobilization and stretching were performed in simple passive movements. During the flaccid stage, gentle passive exercises and stretching of various bi-articular muscles

were employed. In the spastic stage, gentle rhythmic passive movement and sustained gradual stretching were applied either manually or using splints. 15 to 20 minutes of prolonged icing on the bulk of spastic muscles were done and electrical stimulation using Faradic current was employed on weak muscles, especially antagonist muscles, which helps in reducing spasticity by reciprocal inhibition. Weight-bearing exercises such as bridging, elbow lying, sitting position with extra weight bearing on the affected side and supported standing within the patient's medical limitations were crucial in promoting muscle tone development and maintaining calcium absorption into the bones. Subluxation prevention was addressed through proper positioning and handling, providing shoulder slings or Bobath splints to prevent subluxation and implementing taping techniques effectively to prevent subluxation, provide room for free movement and sensory feedback by tactile stimulation and help in the development of muscle tone in shoulder musculature.

Outcome Measurement: Fugl Meyer Assessment scale is a well-established and validated tool was used to evaluate upper limb function [5, 6]. Pre-test evaluations were conducted on the first day of the study before any intervention or treatment, while post-test assessments were performed on the last day after treatment completion.

3. Results

Thirty-two participants – Group A (16) and Group B (16). The participants' FMA (Fugl-Meyer Assessment) Pre and Post-intervention data were collected and analyzed using paired statistics (Table 5.2, Figure 5.1). The results showed a

Table 5.2. Within group evaluation of Pre-test and Post-test data

Group	Test Data	Mean +/- S.D	t- value	P value[a]
A(16)	Pre Test	27.88 +/-4.78	-19.095	0
	Post Test	40.63 +/-3.89		
B(16)	Pre Test	26.75 +/- 3.47	-17.398	0
	Post Test	41.00 +/-2.68		

Values are presented as mean ± standard deviation (range).

Source: Author.

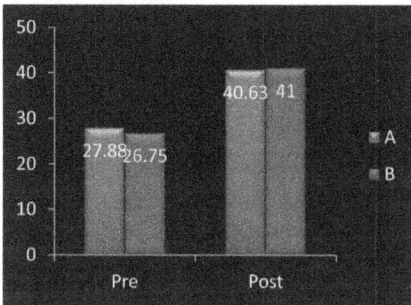

Figure 5.1. Pre and post intervention data—within group analysis

Source: Author.

statistically significant improvement in both groups, with t values of −19.095 and −17.398, respectively.

Blue Histogram–Group A; Red Histogram–Group B. Numerical Data inside the Histogram is indicating to the Mean value of the respective group.

Furthermore, unpaired statistical evaluation was conducted between the post-intervention data (Table 5.3, Figure 5.2). The analysis indicates no statistically significant improvement was observed, with a p-value of 0.536. Additionally, non-parametric evaluation

Table 5.3. In-between group evaluation of Post-test data

Group	Mean +/- S.D	t- value	P value[a]
A(16)	40.63+/-3.89	-0.317	0.536 (Not Sig)[a]
B(16)	41.00+/-2.68		

Values are presented as mean ± standard deviation (range).
[a]T-test (Unpaired Statistics)

Source: Author.

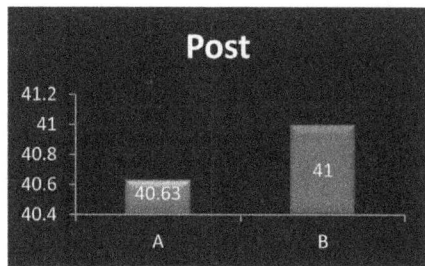

Figure 5.2. Post intervention data—in-between group analysis.

Source: Author.

of post-intervention data showed no significant improvement, with a p-value of 0.838.

Blue Histogram–Group A; Red Histogram–Group B. Numerical Data inside the Histogram is indicating to the Post Intervention mean value of the respective group.

Overall, the study suggests that, while pre- and post-intervention FMA scores improved significantly in both Groups, there was no statistical significant difference in postintervention data between the two groups.

4. Discussion

Mirror therapy is an innovative, highly effective, cost-effective technique for improving upper extremity function in post-stroke hemiparesis/cerebrovascular accident patients. This therapy involves Applying a mirror to establish visual feedback, which has emerged as a powerful catalyst for neuroplasticity within the lesioned brain. Mirror therapy is a practical way to improve arm function in relation to daily living, particularly beneficial in countries such as India; where most stroke survivors do not have access to proper rehabilitation services due to various reasons, including a lack of proper symptoms or the high cost of long rehabilitation sessions.

The main objective of this research was to assess how well mirror treatment works to restore hand motor function following a stroke with variation of treatment duration, involved 32 participants who were assessed for hand motor function using the FMA-UE on Day 1 before treatment and on last day after a 4-week treatment session of mirror therapy.

The study's findings show considerable improvement within the group (pre- and post-comparison of Group A and Group B), which is consistent with prior research reports by Kim et al. demonstrating that mirror therapy is a practical approach to improving arm function in relation to daily living. However, when the post-intervention values of Group A and B are compared, no significant difference is seen. Nevertheless, if we carefully examine the data, Group B, which underwent a treatment of 45 minutes per sessions five days a week for four weeks, that is in total equals to 900 minute, exhibited slight, though insignificant, differences from Group A, which underwent a treatment of 30 minutes per session five days a week for four weeks that is equal to 600 minutes.

Studies by Dorcas BC et al. [7], Arya et al. [8], Kim et al. [9] Samuel Kamlesh Kumar et al. [10] and Thieme et al. [11] highlight the beneficial impact of Mirror therapy in improving motor functions and activities of daily living, even months after the intervention. Improves the activity in both the primary motor cortex on the same and opposite sides of the body, leading to better coordination between the two sides. This therapy also engages the precuneus and posterior cingulate cortex during two-handed movements, which can help increase activity in the cerebellum of the affected brain. Furthermore, Mirror therapy induces a balance in hemispheric activation, favoring the lesioned brain and establishing crucial interconnections between the somatosensory and motor cortex. Mirror therapy also stimulates cognition-related regions, such as the dorsolateral prefrontal cortex, highlighting its potential for comprehensive neurorehabilitation strategies. Mirror therapy also shows improvement in stroke patients even outside the critical window recovery period which is the first three to six weeks of the post-impact [12]. All of these findings demonstrate the immense potential of mirror therapy in improving motor function in patients with post-stroke hemiparesis.

5. Conclusion

Mirror therapy is a rehabilitation technique that has shown considerable promise in improving the motor function of the hand in individuals who have suffered from a stroke. This cost-effective and straightforward intervention has been found to positively impact neuroplasticity, making it an invaluable addition to rehabilitation programs. This work contributes to the expanding body of research in this field by providing additional data of mirror treatment in stroke rehabilitation.

The study's findings emphasize the need for further research on mirror therapy, particularly its efficacy in treating different subtypes and severity levels of stroke. Given its potential to induce neuroplasticity changes that enhance its integration into clinical practice, exploring mirror therapy in various stroke cases can lead to a better understanding of its benefits and limitations.

References

[1] Anwer, S., Waris, A., Gilani, S. O., Iqbal, J., Shaikh, N., Pujari, A. N., & Niazi, (2022). Rehabilitation of upper limb motor impairment in stroke: A narrative review on the prevalence, risk factors, and economic statistics of stroke and state of the art therapies. *Healthcare, 10*, 190.

[2] Ramchandran, V. S., & Rogers-Ramchandran, D. (1996). Synaesthesia in phantom limbs induced with mirrors. *Proceedings of the Royal Society of London. Series B. Biological Sciences, 263*, 377–386.

[3] Altschuler, E. L., Wisdom, S. B., Stone, L., et al. (1999). Rehabilitation of hemiparesis after stroke with a mirror. *The Lancet, 353*, 2036.

[4] Fugl-Meyer, A. R., Jaasko, L., Leyman, I., Olsson, S., & Stegling, S. (1975). The post-stroke hemiplegic patient: A method for evaluation of physical performance. *Scandinavian Journal of Rehabilitation Medicine, 7*, 13–31.

[5] Gladstone, D. J., Danells, C. J., & Black, S. E. (2002). The Fugl-Meyer Assessment of motor recovery after stroke: A critical review of its measurement properties. *Neurorehabilitation and Neural Repair, 16*(3), 16.

[6] See, J., Dodakian, L., Chou, C., Chan, V., McKenzie, A., Reinkensmeyer, D. J., & Cramer, S. C. (2013). A standardized approach to the Fugl-Meyer assessment and its implications for clinical trials. *Neurorehabilitation and Neural Repair, 27*(8), 732–741.

[7] Gandhi, D. B., Sterba, A., Khatter, H., & Pandian, J. D. (2020). Mirror therapy in stroke rehabilitation: Current perspectives. *Therapeutics and Clinical Risk Management, 16*, 1.

[8] Arya, K. N. (2016). Underlying neural mechanisms of mirror therapy: Implications for motor rehabilitation in stroke. *Neurology India, 64*(1), 38–44.

[9] Kim, K., Lee, S., Kim, D., Lee, K., & Kim, Y. (2016). Effects of mirror therapy combined with motor tasks on upper extremity function and activities of daily living of stroke patients. *Journal of Physical Therapy Science, 28*, 483–487.

[10] Samuel Kamalesh, K., Reethajanet Sureka, S., Pauljebaraj, P., Benshamir, B., Padankatti, S. M., & David, (2014). Mirror therapy enhances motor performance in the paretic upper limb after stroke: A pilot randomized controlled trial. *Archives of Physical Medicine and Rehabilitation, 95*, 2000–2005.

[11] Thieme, H., Mehrholz, J., Pohl, M., Behrens, J., & Dohle, C. (2013). Mirror therapy for improving motor function after stroke. *Stroke, 44*, e1–e2.

[12] Jaafar, N., Che Daud, A. Z., Ahmad Roslan, N. F., & Mansor, W. (2021). Mirror therapy rehabilitation in stroke: a scoping review of upper limb recovery and brain activities. *Rehabilitation Research and Practice, 2021*(1), 9487319.

6 Modifying the triphasic liver CT scan protocol to achieve optimal vascular contrast

Aswathi P.[1,2,a], Abhijith S.[2,3,b], Anushka Valeska Fernandes[1,c], and Jayeeta Dandapat[4,d]

[1] Department of Radiology, School of Paramedics and Allied Health Science, Centurion University of Technology and Management, Bhubaneswar, Odisha, India

[2]Medical Radiology and Diagnostic Imaging, School of Allied Health Sciences, REVA University, Bangalore, Karnataka, India

[3]Medical Imaging Technology, Department of Radiology, K S Hegde Medical Academy, NITTE University, Mangalore, Karnataka, India

[4]Department of Optometry, School of Paramedics and Allied Health Science, Centurion University of Technology and Management, Bhubaneshwar, Odisha, India

Abstract: Introduction: Advancements in helical CT technology, offering broader anatomical coverage and faster scan times, have transformed liver imaging. The entire liver can now be assessed in a single breath hold, eliminating respiratory misregistration. However, it is now essential to optimize the timing of the scan.

Aim: This study aimed to Optimize scan timing for triple-phase liver CT by utilizing bolus tracking techniques.

Methodology: 20 patients who were prescribed for a Triple-phase abdomen CT scan were randomly selected for this study. They were divided into groups, A B, of 10 patients. They were administered 1.5ml/ kg of 370 mg/mL ultravist at injection rate of 3.0ml/s with a pressure injector. Using the bolus tracking technique, scans were performed at 5, 19, 44 s and 8, 22 and 47 s for the first, second and third phases, for groups A and B, respectively. The CT numbers (Hounsfield Units) were measured in the aorta, hepatic artery, portal vein and liver parenchyma.

Results: In the hepatic arterial phase, the hepatic artery showed better enhancement in group B (8s). There were no significant differences in any of the vessel enhancements in the portal venous phase. In the hepatic venous phase, liver parenchyma was better enhanced in group B patients (47 s).

Conclusion: In the post-threshold early arterial phase, a delay of 8 s was found to be optimal for enhancement and liver parenchyma showed maximum enhancement at a 47 s delay.

Keywords: Contrast sensitivity, helical CT, phase contrast, tomography, X-ray computed

1. Introduction

Advancements in helical CT technology have greatly improved liver imaging by increasing anatomic coverage and reducing scanning times. This lowers the chance of respiratory misregistration by allowing the liver to be scanned in

[a]aswathipuliyakkara@gmail.com, [b]abhijithshirlal1999@gmail.com, [c]anushka.fernandes@cutm.ac.in, [d]Jayeeta.dandapat@cutm.ac.in

DOI: 10.1201/9781003672869-6

a single breath-hold. The liver's blood supply comes from two primary sources: the arterial and the portal venous system. When a contrast agent is rapidly injected, it enhances the liver during two distinct phases: initially, the arterial phase (Hepatic) which is followed by the portal venous phase. Precise timing is essential to optimize the scan window for accurate imaging [10].

Techniques like the single-bolus method and automated systems have certain drawbacks, including higher costs, limited availability and longer review times, which can impact patient throughput. Consequently, fixed timing delays have been used in various imaging protocols, including those for liver imaging. The development of bolus tracking techniques has improved scan timing accuracy, enabling personalized adjustments to account for variations in individual patients' circulation [1].

Optimizing scan delays in the bolus tracking technique for triple-phase liver CT was the aim of this work. With advancements in helical CT technology, detailed liver imaging has become more refined. The liver's blood supply is mainly divided between the arterial and portovenous systems and the rapid injection of a contrast bolus enables the capture of two essential phases: the hepatic arterial phase and the Portovenous phase [6]. Multi-detector CT (MDCT) has demonstrated significant effectiveness in imaging the liver and pancreas, offering multiplanar and multi-phase views that enhance visualization of vascular structures and allow for clear differentiation of parenchymal lesions [10].

MDCT protocols typically include three- or four-phase liver scans. After non-contrast imaging, contrast is rapidly administered using a pressure injector.

Vascular cancers and lesions can be detected by imaging at the peak hepatic arterial phase. In contrast to the liver's lower-attenuation background, tumors such as hepatocellular carcinoma, which become more noticeable during the arterial phase, look brighter. Because the liver is increased in enhancement during the portovenous phase, hypo-vascular lesions appear as low-attenuation regions encircled by enhanced parenchyma [6]. The portal vein supplies around 67% of the liver's supply of blood, with the arterial phase coming first and followed by peak parenchymal enhancement. Additional delayed-phase imaging, taken 3–4 minutes after contrast administration, is essential for diagnosing conditions such as hemangiomas or cholangiocarcinoma [8].

2. Materials and Methodology

This study was done in the Department of Radiodiagnosis at K S Hegde Hospital, involving 20 patients who were referred for a triphasic liver study between June 2021 and August 2022. Approval was granted by the institutional ethics committee. We excluded patients with known portal vein thrombosis, simple liver cysts, liver lesion size, below 1 cm or above 10 cm, compromised renal function and known contrast media allergies. After obtaining informed consent and relevant clinical history, each patient underwent triphasic of the liver.

A GE Revolution 126-slice MDCT scanner was used for all scans. With a pitch of 12 mm, the gantry rotation time was set at 0.5 seconds, gap was 1.5 mm and slice thickness was 2.5 mm. Using a pressure injector, a 20G catheter was used to provide a contrast medium

(Ultravist 370 mgI/ml) at a flow rate of 3.0 mL/s and a dose of 1.5 mL/kg of body weight. After the contrast injection, the scan was started using bolus tracking. Scans for groups A and B were performed at 5, 19 and 44 seconds and 8, 22, 47 seconds, respectively, for the 1st, 2nd and 3rd phases which is mentioned in Table 6.1. A region of interest (ROI) was positioned in the aorta, slightly above the right hemidiaphragm, to initiate a pre-monitoring scan at 100 Hounsfield units (HU). HU measurements were obtained from the liver parenchyma, hepatic artery, hepatic vein, portal vein and abdominal aorta at each phase.

The Contrast Enhancement Index (CEI) was determined by subtracting the pre-contrast HU values by HU values of post-contrast in the same structure and comparisons were made between the two groups.

3. Statistical Analysis

An unpaired Student's t-test was employed to compare the two groups' varied CEI. P-values below 0.05 were regarded as statistically significant.

4. Results

According to this study, a scan delay of 8 seconds produced noticeably greater contrast enhancement in the hepatic artery during the hepatic arterial phase than a 5-second delay. For any of the structures, there were no significant variations in CEI across the different scan delays in the portovenous phase.

Table 6.1. Scan timings for group A and group B

Phase	Group A	Group B
Arterial	5s	8s
Porto-Venous	19s	22s
Venous	44s	47s

Source: Author.

Table 6.2. Contrast enhancement index in first (arterial) phase

Anatomic Structures	First phase (CEI)		P value
	Group A (5s)	Group B (8s)	
Abdominal aorta	226.43±36	234.22±50	0.6433
Hepatic artery	186.20±16.19	232.98±31.18	0.0107*
Portal vein	19.3±10.95	38.38±33.74	0.1480
Liver parenchyma	6.28±4.96	15.48±10.2	0.0218*

Source: Author.

Table 6.3. Contrast enhancement index in the second (portovenous) phase

Anatomic Structures	Second phase (CEI)		Significance
	Group A (19s)	Group B (22s)	
Abdominal aorta	129.75±32.6	148.21±37.2	0.1418
Hepatic artery	117.52±40	118.52±20.57	0.9314
Portal vein	139.56±32.75	127.69±24.92	0.1924
Liver parenchyma	55.06±15.8	66.8±22.1	0.3097

Source: Author.

Table 6.4. Contrast enhancement index in third (venous) phase

Anatomic Structures	Third phase (CEI)		Significance
	Group A (44s)	Group B (47s)	
Abdominal aorta	106.6±17.4	98.99±25.1	0.5133
Hepatic artery	91.54±26.99	96.20±24.66	0.7490
Portal vein	110.72±25.33	100.77±24.94	0.4451
Liver parenchyma	50.77±9.16	67.06±6.94	0.0235

Source: Author.

However, in the hepatic venous phase, a 47-second delay produced significantly higher enhancement in the liver parenchyma compared to a 44-second delay. Overall, the findings indicate that the 8-second delay in the arterial phase improves hepatic artery visualization, while the 47-second delay enhances liver parenchyma contrast in the venous phase, in portovenous phase with no significant differences observed.

5. Discussion

In order to accurately characterize the lesions, the triple-phase MDCT of the liver aims to improve the contrast between the lesions and the surrounding liver tissue. The tumor is more visible when there is a greater difference in CT attenuation between the tumor and healthy liver tissue. The significance of multi-phasic CT in detecting the hepatomas was initially described by Ohashi et al. [9]. Dynamic biphasic CT was performed, with early phase images captured 28 seconds after administration of contrast and delayed phase images obtained 100 sec post contrast. A additional study by Foley et al. [3] It was demonstrated that by initiating the scan at the moment the contrast bolus reaches the aorta, three distinct circulatory phases can be clearly identified using the triple-phasic hepatic CT technique

with the help of multi-row detector scanner. The initial flow is referred to as the hepatic arterial phase, while the second run is known as the portal venous inflow phase also called late arterial phase. The last or third, occurring 60 seconds after the start of the contrast administration, enhances the hepatic veins and is called the hepatic venous phase [3]. Since the majority of metastases of the liver are hypovascular, the portal venous phase is when they are easiest to find. The hepatic artery provides the majority of the blood supply for hypervascular primary cancers, such as hepatocellular carcinoma and certain metastases, including pancreatic islet cell carcinomas, carcinoids, melanomas, pheochromocytomas, choriocarcinomas and sarcomas. As a result, they become more visible in images taken during the hepatic arterial phase because they augment before the surrounding liver tissue [4, 7].

The study demonstrated that an 8-second post-threshold delay was optimal for achieving maximum enhancement of the arterial system in the liver. These findings are consistent with those reported by Sween et al. [10] and Goshima et al. [5], who also observed similar results which is mentioned in Table 6.5. As many clinical centers currently employ minimal delay techniques for the arterial phase, tailored delay settings may be necessary to optimize outcomes.

Table 6.5. Comparison between present study and other studies

Author	Iodine Conc.	Flow rate	Phase	Anatomical structures	CEI / +C HU
Present study	370 mgI	3 ml/s	Arterial (8 s)	Hepatic artery	232.98±31.18
			Venous (47 s)	Liver parenchyma	67.06±6.94
Sheoran Sween et al. [10]	300 mgI	3 ml/s	Arterial (8 s)	Hepatic artery	207±43.65
			Venous (48 s)	Liver parenchyma	40±8.49
Satoshi Goshima et al. [5]	300 mgI	4 ml/s	Arterial (10 s)	Abdominal aorta	336.2±61.8

Source: Author.

Additionally, in the venous phase, a 47-second delay yielded superior enhancement of the hepatic venous system compared to a 44-second delay. This finding aligns with the results of Sween et al. [10], suggesting that a delay exceeding 44 seconds should be considered for optimal venous enhancement in the liver.

The volume and flow rate of contrast media, in addition to the delay settings, can have a major impact on the vascular system's peak enhancement index timings. In particular, the volume of contrast medium exhibits a direct link with the time to maximal contrast enhancement from the beginning of injection, whereas the flow rate displays an inverse correlation. These factors must also be carefully considered when tailoring delay times to optimize imaging results [2].

6. Conclusion

In this study, using contrast media of 370 mg/ml, with dosage of 1.5 ml/kg, having a flow rate of 3 ml/s, an 8-second post-threshold delay was found to be optimal for arterial phase imaging, while a 47-second delay was optimal for the venous phase. Additional cofactors, such as flow rate and contrast media

volume, which may influence the peak contrast enhancement index time, should be considered when tailoring imaging protocols.

7. Acknowledgment

I would like to express my deepest gratitude to all contributors for the successful completion of this study. I sincerely thank K S Hegde Medical Academy for providing the resources and support necessary for this research. I also extend my appreciation to the patients and medical staff involved in the study, whose cooperation was essential.

References

[1] Bae, K. T., Heiken, J. P., & Brink, J. A. (1998). Aortic and hepatic contrast medium enhancement at CT. Part II. Effect of reduced cardiac output in a porcine model. *Radiology*, *207*(3), 657–662.

[2] Bae, K. T. (2010). Intravenous contrast medium administration and scan timing at CT: Considerations and approaches. *Radiology*, *256*(1), 32–61.

[3] Foley, W. D., Mallisee, T. A., Hohenwalter, M. D., Wilson, C. R., Quiroz, F. A., & Taylor, A. J. (2000). Multiphase hepatic CT with a multirow

detector CT scanner. *American Journal of Roentgenology, 175*(3), 679–685.

[4]   Forner, A., Vilana, R., Ayuso, C., Bianchi, L., Solé, M., Ayuso, J. R., ... & Bruix, J. (2008). Diagnosis of hepatic nodules 20 mm or smaller in cirrhosis: prospective validation of the noninvasive diagnostic criteria for hepatocellular carcinoma. *Hepatology, 47*(1), 97–104.

[5]   Goshima, S., Kanematsu, M., Kondo, H., Yokoyama, R., Miyoshi, T., Nishibori, H., ... & Moriyama, N. (2006). MDCT of the liver and hypervascular hepatocellular carcinomas: optimizing scan delays for bolus-tracking techniques of hepatic arterial and portal venous phases. *American Journal of Roentgenology, 187*(1), W25-W32.

[6]   Kurelli, S. C. G., Wang, J., Abbas, S., Kanduri, H. K., & Liu, W. (2020). Hepatobiliary and hepatic vascular anatomy evaluated with computed tomography and magnetic resonance imaging: the current status. *Radiology of Infectious Diseases, 7*(1), 1–6.

[7]   Kim, M. J., Choi, J. Y., Lim, J. S., Kim, J. Y., Kim, J. H., Oh, Y. T., ... & Kim, K. W. (2006). Optimal scan window for detection of hypervascular hepatocellular carcinomas during MDCT examination. *American Journal of Roentgenology, 187*(1), 198–206.

[8]   Kim, T., Murakami, T., Hori, M., Takamura, M., Takahashi, S., Okada, A., ... & Nakamura, H. (2002). Small hypervascular hepatocellular carcinoma revealed by double arterial phase CT performed with single breath-hold scanning and automatic bolus tracking. *American Journal of Roentgenology, 178*(4), 899–904.

[9]   Ohashi, I., Hanafusa, K., & Yoshida, T. (1993). Small hepatocellular carcinomas: Two-phase dynamic incremental CT in detection and evaluation. *Radiology, 189*(3), 851–855.

[10] Sween, S., Samar, C., & Binu, S. (2018). Triple-phase MDCT of liver: Scan protocol modification to obtain optimal vascular and lesional contrast. *Indian Journal of Radiology and Imaging, 28*(03), 315–319.

7 Network pharmacology and molecular docking analysis for elucidation of mechanism of action and molecular targets of Commiphora wightii in treatment of atherosclerosis

Sangeeta Chhotaray[1,a], Kanika Verma[2,b], Pralaya Kumar Sahoo[1], and Soumya Jal[1,c]

[1]School of Paramedics and Allied Health Sciences, Centurion University of Technology and Management, Odisha, India
[2]Natural Products for Neuroprotection and Anti-ageing Research Unit, Chulalongkorn University, Bangkok, Thailand

Abstract: Atherosclerosis is a persistent inflammatory cardiovascular disease marked by arterial constriction resulting from plaque formation. The production of atheromatous plaques results from arterial damage, which incites inflammation of endothelial cells, promotes the adhesion and infiltration of monocytes and leads to the creation of macrophage foam cells stimulated by oxidised low-density lipoprotein (oxLDL). The management of this illness often necessitates prolonged pharmacotherapy. Commiphora wightii, usually referred to as guggulu, is an oleo-gum resin exuded when its bark is damaged. It has been employed in Ayurvedic medicine to treat multiple conditions, including obesity, rheumatism, gout, inflammation and lipid metabolism disorders. It comprises a mixture of terpenoidal constituents, including monoterpenoids, sesquiterpenoids, diterpenoids and triterpenoids, as well as steroids and volatile oil. This study employed network pharmacology to discover the potentially beneficial components and targets of *C. wightii* for the treatment of atherosclerosis. We examined the mechanism of *C. wightii* in the treatment of atherosclerosis by developing a network that illustrates the relationships between its components and targets, together with a protein-protein interaction (PPI) network. We conducted functional analysis of the targets and executed molecular docking. Following the establishment of a PPI network utilising targets, the study examined 23 potentially effective components and 25 targets. Among the 25 targets, 8 hub genes (CRP, HMGCR, PCSK9, APOC3, LCAT, LPA, NOS3 and OLR1) have been identified for enhanced therapeutic applications. Twenty-five targets were identified as the principal objectives of the PPI network. Thereafter, the targets underwent investigation using Gene Ontology (GO) enrichment and Kyoto Encyclopaedia of Genes and Genomes (KEGG) pathway assessments. The hub genes are mostly associated with the calcium signalling network, cholesterol metabolism, MAPK signalling pathway, PI3K-Akt signalling pathway, lipid metabolism and atherosclerosis, cAMP signalling pathway and arginine biosynthesis. The molecular docking method is employed to investigate the interaction between targets and selected active compounds. The cited research illustrates that the method by which *C. wightii* addresses atherosclerosis is complex, primarily affecting cholesterol metabolism, anti-inflammatory processes and the reduction of nitric oxide generation, among other routes. Utilising network pharmacology and molecular docking methodologies to investigate the mechanism of *C. wightii* in the treatment of atherosclerosis would strengthen the scientific basis for the

[a]sangeetachhotaray797@gmail.com, [b]Kanika.honey.verma@gmail.com, [c]soumya.jal@cutm.ac.in

DOI: 10.1201/9781003672869-7

integration of traditional and Western medicine. This analysis will help improve the quality of *C. wightii* and ascertain its future potential.

Keywords: Atherosclerosis, *commiphora wightii*, molecular docking, network pharmacology

1. Introduction

Atherosclerosis is a persistent inflammatory condition characterized by narrowing of an artery's lumen as a result of plaque buildup [1]. It is the primary factor that leads to cardiovascular disease (CVD), which encompasses conditions such as myocardial infarction (MI), heart failure, stroke and claudication. Atherosclerosis mostly occurs in the intima of several medium-sized and large arteries, particularly at sites of vascular bifurcation [2]. Monocytes released into the bloodstream are drawn to the inner layer of blood vessels, where they undergo differentiation into macrophages. Following this, macrophages absorb both normal and altered low-density lipoproteins (LDLs) in excessive amounts and undergo a transformation into lipid-rich 'foam cells'. Foam cell development is a classic feature of atherosclerotic plaques [3] (p. 342). It is probable that this is regulated by the characteristics of the blood flow, as parts that are subjected to typical shear stress appear to be shielded. In these areas, endothelial cells produce genes that protect against the development of atherosclerosis [4]. Despite advances in interventional surgery and pharmaceutical treatment options, it persists as the leading cause of death and morbidity in developed nations [5]. According to data from the World Health Organisation (WHO), an estimated 17.7 million individuals die due to cardiovascular illnesses annually [6]. Hence, it is imperative to develop novel approaches to comprehend the fundamental processes of atherosclerosis in order to unveil innovative treatment interventions. Consequently, simple targeting of a single gene may not be sufficient to accomplish the objective of identifying and treating complex diseases. Nevertheless, it is imperative to design novel strategies to specifically address the complete biological networks that are responsible for the disease [7]. Hence, understanding the molecular processes that regulate illness prognosis is crucial in combating complex disorders [8].

Therapeutic adaptation of plants and herbs has a lengthy historical background [9]. Humans have been utilising plants for therapeutic purposes since the middle Palaeolithic era, some 60,000 years ago. Currently, the WHO approximates that more than 80% of the total population in underdeveloped countries dependent on traditional treatments, such as herbs, for sustenance and improving health conditions [10]. The use of therapeutic flora has evolved gradually over an extended duration and currently plays a crucial part in optimal health results. In the realm of traditional herbal medicine, blood stasis has significant importance as a pathogenic characteristic observed in a range of illnesses, such as hyperviscosity, hyperlipidaemia, inflammation, ischaemic brain damage and atherosclerosis [11]. Moreover, phytochemicals have a lengthy track record of being used for the prevention and management of CVDs [9].

At present, natural products constitute a significant ratio of advance pharmacological agents, particularly in the field of disease treatment [12]. The current research of herbal medicine mechanisms has embraced network pharmacology as a practical approach to map the established connections and analyse networks between natural medications and target genes. This approach aims to find possible targets by using multi-compound and multi-target theory [13]. And network pharmacology, on the other hand, has been developed to thoroughly evaluate the biological mechanisms of herbal pharmacological effects. It is a potential method that can speed up drug development and help us understand how numerous target component's function [14]. It uses experimental techniques in addition to computational tools like bioinformatics and network analysis and it combines data from many sources. Traditional herbal therapy has great promise in 'multi-compound, multi-target' treatments and recent work using network pharmacology methods to study the systems-level processes of plants and herbal formulations has demonstrated this [15]. Researchers can use molecular docking analysis to confirm the possible connections between active components and target genes defined by network pharmacology analysis [16]. Molecule docking is a method that includes superimposing small ligands onto the structures of macromolecular targets in order to reveal possible interactions between molecules that might occur at the binding site. Through such approach, a more comprehensive comprehension of precise connections between active components and the genes linked to different diseases may be achieved, therefore facilitating

the advancement of possible therapeutic approaches [17].

Our objective in this work is to examine the target and signalling pathways of bioactive compounds from *C. wightii* specifically with regard to the prevention of atherosclerosis using a network pharmacology method. Through the use of several databases, the intricate interconnections among the disease, medications and targets are methodically hypothesized and examined. The possible methods by which *C. wightii* targets critical processes in the treatment of atherosclerosis are investigated. Molecular docking analysis are employed to validate the putative targets of *C. wightii* against atherosclerosis.

2. Materials and Methods

2.1. Pharmacologically active compounds found in Commiphora wightii

Identification of bioactive compounds of *C. wightii* was conducted by academic search in literature sources such as PubMed (https://pubchem.ncbi.nlm.nih.gov/) [18]. *C. wightii*, well-known as guggul, is a medicinal plant valued for its potent bioactive compounds, particularly guggulsterone, which has been extensively studied for its therapeutic properties. This guggul tree is the popular name for this shrub, which grows in dry regions of Pakistan, Bangladesh and India. Its Indian home states include Madhya Pradesh, Karnataka, Assam, Gujarat and Rajasthan [19]. The investigation of bioactive compounds in *C. wightii* involves isolating and characterizing phytochemicals such as sterols, flavonoids and terpenoids, which exhibit anti-inflammatory, hypolipidemic and antioxidant activities. Inflammation, gout, rheumatism, obesity and lipid metabolic abnormalities are among the many conditions that guggulu has traditionally helped alleviate in India's traditional medical system [20]. Recent studies aim to isolate and characterize these compounds to explore their potential in treating chronic diseases such as cardiovascular disorders.

2.2. Target gene prediction in atherosclerosis

Predicting target genes related with atherosclerosis is an essential method for finding genes that might potentially contribute to the development and advancement of the disease. Three well recognized databases were used to identify target genes linked to atherosclerosis GeneCards (http://www.genecards.org), CoreMine (https://www.coremine.com/medical/) and Theraputic Target Database (TTD) (https://idrblab.net/ttd/) [21–24]. To establish a thorough and non-duplicative list, the investigation was carried out using the query 'atherosclerosis' as the main key terminology and any duplicate genes were eliminated.

2.3. Venn diagram construction

A Venn diagram is a type of visual representation employed to depict the interconnections among several collections of data. Analysis of putative target genes for atherosclerosis was conducted using the Venny tool. The result was presented as a Venn diagram showing the genes that may be targets across the various atherosclerosis datasets.

2.4. PPI network construction and hub gene identification

Identified genes from Venny plots were added to the STRING database at (https://string-db.org/) and Homo sapiens was chosen as the species [25]. A high confidence level (interaction score > 0.400) was used to construct PPI networks comprising 25 genes, while ensuring that unnecessary targets were eliminated. The outcome obtained from the STRING database was downloaded and imported into Cytoscape 3.7.2 U.S., available at (https://cytoscape.org/) [26]. Using Cytoscape, we identified the hub genes, sets of genes with strong interconnections [27].

2.5. Gene Ontology (GO) and pathway enrichment analysis

GO and Pathway Enrichment Analysis are two synergistic methodologies employed in bioinformatics to elucidate extensive datasets, such as those derived from gene expression or proteomics investigations. These strategies assist in identifying biological activities or pathways that are disproportionately represented in a specific group of genes or proteins. Here GO and KEGG pathway enrichment analysis was performed to elucidate the molecular processes contributing to atherosclerosis. The analysis utilized the Database for Annotation, Visualisation and Integrated Discovery (DAVID) [28]. The program offers a comprehensive library for gene functional analysis, including rich information on biological processes (BP), cellular components (CC), molecular function (MF) and associated signalling pathways [29, 30].

2.6. Molecular docking

To investigate how the bioactive components *of C. wightii* interacted with the hub genes, the molecular docking strategy was used. Because it concede the prediction of binding affinities and modes between receptors and ligands, this computational approach is widely used in drug development [31]. Following the hub gene and pathway analysis, one gene was chosen for the molecular docking procedure. The crystal structures of the s gene acquired from the RCSB Protein Data Bank [32]. The three-dimensional structures of the *C. wightii* phytocompounds were obtained from the PubChem database [33]. Both the compounds and the hub genes functioned as ligands and receptors, respectively. After the receptors were produced, any remaining ligand and water molecules were removed. The selected ligands for docking were a range of phytoconstituents known for their potential therapeutic properties, including bornyl acetate, cadinene, cineole, guggulsterol I-IV, mansumbinoic acid, myrrhanol A-C, quercetin and α-camphorene. The optimized ligands were then converted into the PDBQT format required for compatibility with the docking software. Molecular docking was conducted utilising AutoDock Vina, which was selected for its ability to efficiently predict binding interactions [34]. The active binding site of LCAT was carefully identified by examining the protein's known binding pocket. The grid box was set to encompass this active site, ensuring that the entire binding region was available for ligand interaction. Multiple docking simulations were performed, with parameters adjusted for

exhaustiveness and the number of dockings runs to ensure thorough exploration of the ligand-protein interactions [35]. After docking, the binding affinities of each ligand to LCAT were recorded in terms of binding energy (kcal/mol), with Lower energy numbers mean that the binding interactions are greater. The interactions between LCAT and the ligands were then analyzed using Biovia Discovery Studio and PyMOL for visualizing the docked complexes. This analysis focused on identifying key molecular interactions that contribute to the stability and specificity of ligand binding [36]. The potential for the ligand-receptor binding was evaluated using the binding energy, which is a result of molecular docking and usually less than or equal to 5 kcal/mol and the binding configuration between the two was forecasted using AutoDockTools [37, 38].

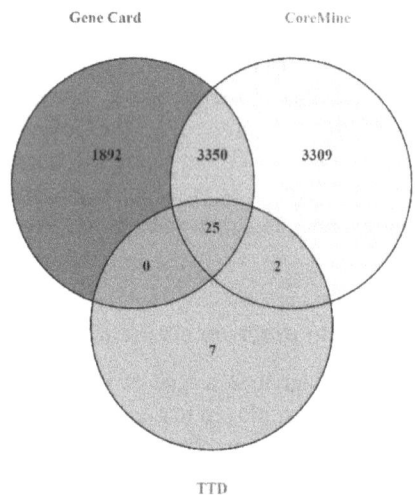

Figure 7.1. Venn diagram and intersection targets of atherosclerosis from different databases.

Source: Author.

3. Result

3.1. *Common targets of atherosclerosis*

Atherosclerosis is affected by several genetic variables, with several genes recognized as contributing to the disease's progression. These genes are often disseminated among many databases such as Gene Card, TTD and CoreMine. From those databases, potential atherosclerosis targets were retrieved. Upon querying the database for Atherosclerosis, 5,267 targets were identified in the Gene Card database, 6,686 targets in the Core Mine database and 34 targets in the TTD database. From the databases, 25 common genes were extracted for further investigation (Figure 7.1).

3.2. *PPI networking and hub gene identification*

To show how the atherosclerosis target chemicals interact with one another, a PPI network analysis was performed (Figure 7.2). The STRING database and Cytoscape were used to conduct this investigation. About 25 nodes and 48 edges made up the network. The degree values of nodes in a network were examined using the Network Analyser plugin in Cytoscape. Cytoscape was used to import data from the STRING database and create a visual PPI network for future research. The colour of the circle varied according to the degree value. A mean degree value of 3.84 was found by our research. Following PPI network analysis, the results were integrated into Cytoscape for the identification of hub genes. As a result, eight hub genes were found (Figure 7.3).

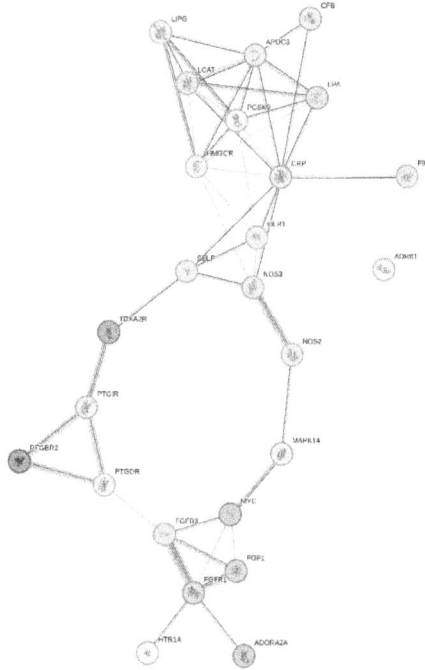

Figure 7.2. PPI network of shared genes of atherosclerosis.

Source: Author.

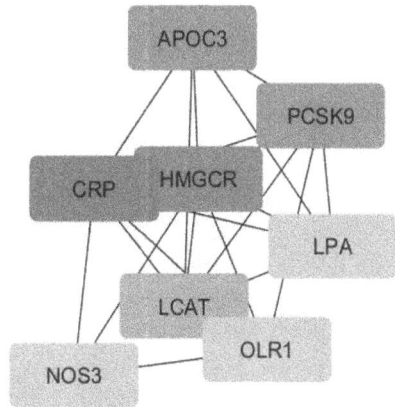

Figure 7.3. Hub genes associated with atherosclerosis.

Source: Author.

3.3. GO and KEGG pathway analysis

To obtain deeper insights into the BP, CC, MF and pathways related to the targeted genes associated with atherosclerosis, GO and KEGG pathway enrichment analyses were performed utilising the DAVID bioinformatics resources. The GO analysis results (Table 7.1) indicated that the primary biological processes linked to the targets pertained to lipopolysaccharide response and inflammatory response. The cellular component analysis indicated that the targets were predominantly associated with the plasma membrane and the extracellular region. The most significant enrichment keywords related to molecular function included nitric oxide synthase activity, low-density lipoprotein particle binding and apolipoprotein binding. The KEGG pathway analysis (Figure 7.4) revealed that numerous atherosclerosis targets were significantly enriched in various signalling pathways, including the hsa04020:Calcium signaling pathway, hsa04979:Cholesterol metabolism, hsa04080:Neuroactive ligand-receptor

Table 7.1. Top GO keywords for biological processes (BP), molecular functions (MF) and cellular components (CC) in the three areas of GO analysis

	Go Term	ID	P-value	Count
Biological Processes	Inflammatory response	GO:0006954	2.23E-08	9
	Response to lipopolysaccharide	GO:0032496	2.24E-05	5
	High-density lipoprotein particle remodeling	GO:0034375	1.91E-04	3
	Reverse cholesterol transport	GO:0043691	2.14E-04	3
	Lipoprotein metabolic process	GO:0042157	2.94E-04	3
Molecular Functions	Heparin binding	GO:0008201	7.95E-05	5
	NADP binding	GO:0050661	0.001297	3
	Serine-type endopeptidase activity	GO:0004252	0.001835	4
	Nitric-oxide synthase activity	GO:0004517	0.003725	2
	Receptor-receptor interaction	GO:0090722	0.004964	2
Cellular Components	Extracellular region	GO:0005576	8.85E-07	13
	Plasma membrane	GO:0005886	2.55E-04	16
	Extracellular space	GO:0005615	2.59E-04	10
	Early endosome	GO:0005769	0.005438	4
	Receptor complex	GO:0043235	0.027288	3

Source: Author.

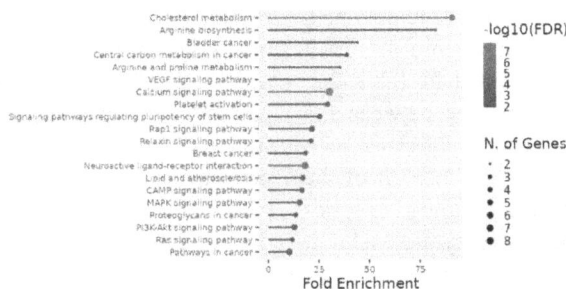

Figure 7.4. KEGG pathway enrichment analysis of the 25 targets associated with *C. wightii* and atherosclerosis.

Source: Author.

interaction, hsa04015:Rap1 signaling pathway, hsa04611:Platelet activation, hsa04550:Signaling pathways regulating pluripotency of stem cells, hsa04010:MAPK signaling pathway, hsa05200:Pathways in cancer, hsa04151:PI3K-Akt signaling pathway, hsa05230:Central carbon metabolism in cancer, hsa05417:Lipid and atherosclerosis, hsa04024:cAMP signaling pathway, hsa04926:Relaxin signaling pathway, hsa00220:Arginine biosynthesis and hsa05224:Breast cancer. Furthermore, KEGG indicated that the targets may exert a strong association with atherosclerotic effect directly via the lipid and atherosclerosis pathway and cholesterol metabolism.

3.4. Molecular docking analysis

Molecular docking studies were used to investigate the interactions between one potential target gene LCAT and twenty-three active compounds of *C. wightii*, including bornyl acetate, cadinene, cineole, guggulsterol I-IV, mansumbinoic acid, myrrhanol A-B, myrrhanone A, quercetin and α-camphorene. The molecular docking investigations were carried out using the AutoDock Tools program

Table 7.2. Top docking score of phytochemicals with LCAT

S.l No	Name	Score (kcal/mol)
1	Bornyl Acetate	−6.1
2	Cadinene	−7.0
3	Cineole	−5.8
4	Commipherol	−7.3
5	d-limonene	−5.8
6	d-α-phellandrene	−5.8
7	Eugenol	−5.6
8	Geraniol	−4.9
9	Guggulsterol-I	−8.9
10	Guggulsterol-II	−8.9
11	Guggulsterol-III	−8.4
12	Guggulsterol-IV	−8.1
13	Guggulsterol-V	−8.3
14	Linalol	−5.1
15	Mansumbinoic acid	−7.2
16	Mansumbinone	−7.2
17	Myrrhanol A	−5.9
18	Myrrhanol B	−13.1
19	Myrrhanone A	−12.7
20	Quercetin	−8.2
21	α-camphorene	−7.7
22	α-pipene	−5.9
23	α-terpineol	−5.7

Source: Author.

and 230 docking findings were obtained. The highest docking score of each docking process is listed in Table 7.2. A greater binding affinity between the chemical and the protein correlates with a lower docking score. Since all of the phytochemicals had binding energies less than -5 kcal/mol, they are useful in the treatment of atherosclerosis because they bind effectively to every target with an energy level below -5 kcal/mol. The majority of the active components bound strongly to the LCAT target gene, according to the results (Table 7.2).

4. Discussion

This study employed a network pharmacology and molecular docking methodology to thoroughly investigate the possible targeting of LCAT in atherosclerosis. Our database search revealed 25 shared targets among the atherosclerotic targets. PPI analysis identified 8 hub targets among the 25 shared targets. The data indicate that these eight hub targets are mostly linked to routes like hsa04020: Calcium signalling pathway, hsa04979: Cholesterol metabolism. hsa04080: Neuroactive ligand-receptor interaction, hsa04015: Rap1 signalling pathway, hsa04611: Platelet activation, hsa04550: Signalling pathways controlling pluripotency of stem cells, hsa04010: MAPK signalling pathway, hsa05200: Pathways in cancer. hsa04151: PI3K-Akt signalling pathway, hsa05230: Central carbon metabolism in cancer. hsa05417: Lipid Metabolism and Atherosclerosis hsa04024: cAMP signalling route, hsa04926: Relaxin signalling pathway, hsa00220: Arginine biosynthesis and hsa05224: Breast cancer. Our data indicate a relationship between the primary targets of LCAT and cholesterol metabolism, despite these targets not

being directly linked to lipid metabolism or arterial fat formation. The interconnection of these selected targets with lipid pathways and atherosclerosis may arise from the interaction of BP and signalling pathways within the context of atherosclerosis. Atherosclerosis is a multifaceted disease characterized by many dysregulations in the pulmonary vascular system, resulting in diverse cellular and molecular responses. Consequently, whereas the primary atherosclerotic targets identified in this work are directly associated with cholesterol metabolism, they also participate in signalling cascades or molecular interactions that indirectly influence this signalling pathway, so contributing to disease progression. Our research employed molecular docking analysis to investigate and assess the binding affinity between *C. wightii* and one of the eight possible hub targets for atherosclerosis. Our results suggested that *C. wightii* demonstrates strong binding affinity to these main targets. The findings underscore the need for further study to thoroughly clarify. The role of *C. wightii* in the prevention of atherosclerosis and the validation of its efficacy in both preclinical and clinical settings. The work relies on in silico analysis and bioinformatics data, necessitating experimental confirmation and further validation through in vitro and in vivo studies. Notwithstanding these constraints, the findings of this study establish a robust basis for further investigations in cardiovascular disorders and novel pharmacological advancements.

5. Conclusion

This study employed a network pharmacology technique to examine the probable mechanisms of *C. wightii's* bioactive components in atherosclerosis therapy. Our findings suggest that chemicals in *C. wightii*, including bornyl acetate, cadinene, cineole, guggulsterol I-IV, mansumbinoic acid, myrrhanol A-B, myrrhanone A, quercetin and α-camphorene, may significantly influence atherosclerosis by affecting essential biological targets. In conclusion, our work offers initial data about the molecular mechanism of *C. wightii* in atherosclerosis therapy. The results indicate that *C. wightii* may serve as a viable candidate for the formulation of novel therapeutics for atherosclerosis. Nonetheless, more in vitro and in vivo investigations are necessary to corroborate and refine these results and to comprehensively elucidate the function of *C. wightii* in atherosclerosis.

References

[1] Herrington, W., Lacey, B., Sherliker, P., Armitage, J., & Lewington, S. (2016). Epidemiology of atherosclerosis and the potential to reduce the global burden of atherothrombotic disease. *Circ Res, 118*(4), 535–546.

[2] Moore, K. J., Sheedy, F. J., & Fisher, E. A. (2013). Macrophages in atherosclerosis: A dynamic balance. *Nat Rev Immunol, 13*(10), 709–721.

[3] Moore, K. J., & Tabas, I. (2011). Macrophages in the pathogenesis of atherosclerosis. *Cell, 145*(3), 341–355. https://doi.org/10.1016/j.cell.2011.04.005

[4] Gimbrone Jr, M. A., Topper, J. N., Nagel, T., Anderson, K. R., & Garcia-Cardeña, G. (2000). Endothelial dysfunction, hemodynamic forces, and atherogenesis a. *Ann N Y Acad Sci, 902*(1), 230–240.

[5] Wang, T., Palucci, D., Law, K., Yanagawa, B., Yam, J., & Butany, J. (2012). Atherosclerosis: pathogenesis and pathology. *Diagn Histopathol, 18*(11), 461–467.

[6] Cardiovascular Diseases, WHO, 2021.

[7] Zuo, H. L., Lin, Y. C. D., Huang, H. Y., Wang, X., Tang, Y., Hu, Y. J., & Huang, H. D. (2021). The challenges and opportunities of traditional Chinese medicines against COVID-19: A way out from a network perspective. *Acta Pharmacol Sin, 42*(6), 845–847.

[8] Noor, F., Saleem, M. H., Aslam, M. F., Ahmad, A., & Aslam, S. (2021). Construction of miRNA-mRNA network for the identification of key biological markers and their associated pathways in IgA nephropathy by employing the integrated bioinformatics analysis. *Saudi J Biol Sci, 28*(9), 4938–4945.

[9] Petrovska, B. B. (2012). Historical review of medicinal plants' usage. *Pharmacognosy Reviews, 6*(11), 1–5.

[10] Hong, L., Guo, Z., Huang, K., Wei, S., Liu, B., Meng, S., & Long, C. (2015). Ethnobotanical study on medicinal plants used by Maonan people in China. *J Ethnobiol Ethnomed, 11*, 1–35.

[11] Prevention of cardiovascular disease: guidelines for assessment and management of total cardiovascular risk, WHO, 2007.

[12] Pal, S. K., & Shukla, Y. (2003). Herbal medicine: current status and the future. *Asian Pacific Journal of Cancer Prevention, 4*(4), 281–288.

[13] Zhang, R., Zhu, X., Bai, H., & Ning, K. (2019). Network pharmacology databases for traditional Chinese medicine: review and assessment. *Front Pharmacol, 10*, 123.

[14] Berg, E. L. (2014). Systems biology in drug discovery and development. *Drug Discovery Today, 19*(2), 113–125.

[15] Park, S. Y., Park, J. H., Kim, H. S., Lee, C. Y., Lee, H. J., Kang, K. S., & Kim, C. E. (2018). Systems-level mechanisms of action of Panax ginseng: A network pharmacological approach. *J Ginseng Res, 42*(1), 98–106.

[16] Liu, C., Liu, L., Li, J., Zhang, Y., & Meng, D. L. (2021). Virtual screening of active compounds from jasminum

lanceolarium and potential targets against primary dysmenorrhea based on network pharmacology. *Nat Prod Res*, *35*(24), 5853–5856.

[17] Nurhidayah, E. S., Ivansyah, A. L., Martoprawiro, M. A., & Zulfikar, M. A. (2018). A Molecular docking study to predict enantioseparation of some chiral carboxylic acid derivatives by methyl-β-cyclodextrin. In *Journal of Physics: Conference Series*, IOP Publishing. *1013*(1), 012203.

[18] White, J. (2020). PubMed 2.0. *Medical reference services quarterly*, 39(4), 382–387.

[19] Ravishankar, B., & Shukla, V. J. (2007). Indian systems of medicine: A brief profile. *African Journal of Traditional, Complementary and Alternative Medicines*, *4*(October), 319–337.

[20] Urizar, N. L., & Moore, D. D. (2003). GUGULIPID: A natural cholesterol-lowering agent. *Annual Review of Nutrition*, *23*(1), 303–313.

[21] Safran, M., Dalah, I., Alexander, J., Rosen, N., Iny Stein, T., Shmoish, M., & Lancet, D. (2010). GeneCards Version 3: The human gene integrator. *Database*, *2010*, baq020.

[22] Iyyappan, O. R., & Manoharan, S. (2022). Finding gene associations by text mining and annotating it with gene ontology. In *Biomedical Text Mining* (pp. 71–90). New York, NY: Springer US.

[23] Bader, G. D., & Hogue, C. W. (2003). An automated method for finding molecular complexes in large protein interaction networks. *BMC Bioinformatics*, *4*, 1–27.

[24] Wang, Y., Zhang, S., Li, F., Zhou, Y., Zhang, Y., Wang, Z., & Zhu, F. (2020). Therapeutic target database 2020: Enriched resource for facilitating research and early development of targeted therapeutics. *Nucleic Acids Res*, *48*(D1), D1031-D1041.

[25] Szklarczyk, D., Gable, A. L., Lyon, D., Junge, A., Wyder, S., Huerta-Cepas, J., Simonovic, M., Doncheva, N.T., Morris, J.H., Bork, P., & Jensen, L.J. (2019). STRING v11: Protein–protein association networks with increased coverage, supporting functional discovery in genome-wide experimental datasets. *Nucleic Acids Res*, *47*(D1), D607-D613.

[26] Otasek, D., Morris, J. H., Bouças, J., Pico, A. R., & Demchak, B. (2019). Cytoscape automation: empowering workflow-based network analysis. *Genome Biol*, *20*, 1–15.

[27] Chin, C. H., Chen, S. H., Wu, H. H., Ho, C. W., Ko, M. T., & Lin, C. Y. (2014). cytoHubba: identifying hub objects and sub-networks from complex interactome. *BMC Syst Biol*, *8*, 1–7.

[28] G Jr, D. E. N. N. I. S. (2003). DAVID: Database for annotation, visualization, and integrated discovery. *Genome Biology*, *4*(9), R60. https://doi.org/10.1186/gb-2003-4-9-r60

[29] Kanehisa, M., Furumichi, M., Sato, Y., Kawashima, M., & Ishiguro-Watanabe, M. (2023). KEGG for taxonomy-based analysis of pathways and genomes. *Nucleic Acids Res*, *51*(D1), D587-D592.

[30] Sherman, B. T., Hao, M., Qiu, J., Jiao, X., Baseler, M. W., Lane, H. C., Imamichi, T., & Chang, W. (2022). DAVID: A web server for functional enrichment analysis and functional annotation of gene lists (2021 update). *Nucleic Acids Res*, *50*(W1), W216-W221.

[31] Sahoo, R. N., Pattanaik, S., Pattnaik, G., Mallick, S., & Mohapatra, R. (2022). Review on the use of molecular docking as the first line tool in drug discovery and development. *Indian J Pharm Sci*, *84*(5).

[32] Berman, H. M., Westbrook, J., Feng, Z., Gilliland, G., Bhat, T. N., Weissig, H., Shindyalov, I. N., & Bourne, P. E. (2000). The protein data bank. *Nucleic Acids Res*, *28*(1), 235–242.

[33] Kim, S., Chen, J., Cheng, T., Gindulyte, A., He, J., He, S., Li Q, Shoemaker, B. A., Thiessen, P. A., Yu, B., & Zaslavsky, L. (2021). PubChem in

2021: New data content and improved web interfaces. *Nucleic Acids Res*, *49*(D1), D1388-D1395.

[34] Meng, X. Y., Zhang, H. X., Mezei, M., & Cui, M. (2011). Molecular docking: A powerful approach for structure-based drug discovery. *Curr Comput-Aided Drug Des*, *7*(2), 146–157.

[35] Pinzi, L., & Rastelli, G. (2019). Molecular docking: shifting paradigms in drug discovery. *International Journal of Molecular Sciences*, *20*(18), 4331.

[36] Jain, A. N. (2004). Virtual screening in lead discovery and optimization. *Current Opinion in Drug Discovery & Development*, *7*(4), 396–403.

[37] Morris, G. M., Huey, R., Lindstrom, W., Sanner, M. F., Belew, R. K., Goodsell, D. S., & Olson, A. J. (2009). AutoDock4 and AutoDockTools4: Automated docking with selective receptor flexibility. *J Comput Chem*, *30*(16), 2785–2791.

[38] Das, D. R., Kumar, D., Kumar, P., & Dash, B. P. (2020). Molecular docking and its application in search of antisickling agent from Carica papaya. *J Appl Biol Biotechnol*, *8*(01), 105–116.

8 Investigating the potential of medicinal plant essential oils as a powerful medication

B. Jyotirmayee, Ipsita Priyadarsini Samal, Sameer Jena, and Gyanranjan Mahalik[a]

Department of Botany, School of Applied Sciences, Centurion University of Technology and Management, Odisha, India

Abstract: Medicinal herbs utilized in ethnopharmacology contain plenty of compounds with different pharmacological effects. Research efforts worldwide have recently shifted their focus from developing new plant kinds with this much potential to discovering and developing new medicines and active compounds. Ethnopharmacological research has been instrumental in developing pharmaceuticals derived from natural sources, which modern treatment systems greatly appreciate. Much recent attention has been directed to studying these ancient therapeutic herbs' pharmacognostic, phytochemical and pharmacological characteristics. The ethnopharmacological perspective, which may have some basis in science, has the potential to aid in the development and identification of novel, safe and cost-effective pharmaceuticals. Essential oils are aromatic fragrances extracted by distillation from plants or other natural sources. One way to assess essential oil's biological activity is to consider its potential medical, nutritional, or cosmetic applications and then look for correlations with its aroma or storage techniques. It is of great interest in clinical research to study phytochemicals and their scientific information to develop safe, all-natural medications. A better immune system and fewer inflammation-related diseases and cancers may be within reach with the development of disease-specific medications that mimic or counteract cytokine activities. Essential oils possess remarkable antimicrobial, antioxidant and anti-inflammatory capabilities, which the integration of nanotechnological approaches might enhance. Research into essential oils is vital for developing novel bioactive delivery strategies due to the pharmaceutical industry's growing reliance on plants. More clinical trials are needed before essential oils can be seriously considered in the pharmacy or added to existing drugs.

Keywords: Medicinal herbs, ethnopharmacology, essential oil, biological activity, drug

1. Introduction

Herbalism uses medicinal plants because of their therapeutic capabilities, high nutrient content and bioactive nutrients. The active components derived from these plants are utilized to produce synthetic drugs, including antibiotics, laxatives, antimalarials, synergic medicines, therapy augmentation and preventative drugs [1]. India has around 45,000 medicinal plant species, most in the Western Ghats, the Eastern Himalayas and the Andaman and Nicobar Islands regions. People have been utilizing these plants for a very long time. As a result of extensive research into the plant's flavour, metabolic properties, purity, biological effects and potency, the area's Indigenous inhabitants created a Shastra known as Dravya Guna Shastra

[a]gyanranjan.mahalik@cutm.ac.in

DOI: 10.1201/9781003672869-8

[2]. Because of their diverse photochemical composition, these plants are great for bolstering the immune system and protecting it from many diseases. Understanding biological activity is crucial for future healthcare improvement and human welfare [3].

The 2017 Floral Statistics of India study was issued in Kolkata by the Botanical Survey of India, which estimates that 2,68,600 flowering plants were detected in the country [2]. Over three thousand plant species are known to have medicinal properties in India; two thousand five hundred plants are recognized as traditional medicines, with one hundred of them seeing regular use [1, 2]. Regarding plant richness, the Himalayas are a global biodiversity hotspot, with 18,440 species –25.3% of which are indigenous. Several Indigenous peoples rely on selling wild plants for subsistence, including the Bokshas, Tharus, Bhotias, Marchcha, Tolcha, Jaunasari, Kolta, Gangwal and Banw-rauat. These plants' abundance and quick decline provide some economic value to the indigenous communities [4]. Medicinal plants are full of potential because there are over 500,000 species of plants worldwide, the vast majority of which have not been researched for their potential medicinal uses. Medicinal herbs are an indirect source of many modern drugs, including aspirin [1, 4].

An example of a food crop with medicinal qualities is garlic. Researching therapeutic herbs helps keep people and animals safe from naturally occurring poisons. Herbal plants get their medicinal properties from secondary metabolites [3]. The World Health Organization (WHO) has also prepared plans, recommendations and criteria for herbal remedies, recognizing the value of traditional medicine [4]. Herbal remedies and medicinal plants can't be cultivated, processed, or manufactured without agro-industrial technology. Modern medicine uses therapeutic herbs indirectly [5]. 'Essential oils' are a volatile plant fraction containing fragrant compounds such as phenylpropanoids, mono- and sesquiterpenes, esters of fatty acids and aldehyde alcohols. According to estimates, oil content in plants ranges from 0.01 to 10%. About 300 of the 3000 essential oils found have a substantial economic impact on the medicinal, culinary, sanitary, perfume and agronomic industries [6]. Turmeric, tulsi, ginger, garlic, mint and onion are some of the most popular medicinal plants and their long history of usage in traditional medicine attests to their numerous beneficial properties. Nonetheless, additional investigation into these therapeutic plants can lead to the creation of drugs that better manage various diseases.

1.1. Turmeric–scientific name– Curcuma longa L.

The Zingiberaceae family, including this flowering plant, originates from southern Asia. This plant rhizome yields polyphenol curcumin, aiding cancer prevention and treatment. Due to its ability to inhibit the expression of numerous genes associated with cancer and angiogenesis, curcumin possesses preventative qualities. One way to stop cancer from progressing is to disrupt its signaling pathways [7]. Curcumin impacts several biochemical pathways that facilitate carcinogenesis, metastasis, mutagenesis, oncogene expression, cell cycle regulation and death [8].

1.2. Tulsi–scientific name– Ocimum sanctum Linn.

The Indian subcontinent is the original home of this Lamiaceae plant. Nowadays, it is grown almost everywhere in tropical Southeast Asia. Ocimum has substances that can combat chemically induced malignancies of the skin, liver, mouth and lungs. These compounds include eugenol, rosmarinic acid, apigenin, luteolin, sitosterol and carnosic acid. These compounds achieve this by changing gene expression, increasing antioxidant activity, triggering cell death and reducing the likelihood of new blood vessel formation and cancer metastasis [9].

1.3. Ginger–scientific name– Zingiber officinale roscoe.

Plants of the Zingiberaceae family, of which this blossom is a member, first appeared in the tropical rain forests of southern Asia and the Indian subcontinent. The anticancer properties of [6]-gingerol and [6]-shogaol apply to gastrointestinal cancer. Many types of cancer, including skin, ovarian, colon, breast and cervical cancers, are inhibited in their development and induced apoptosis by it [10].

1.4. Garlic–scientific name–Allium sativum L.

This plant, a member of the Amaryllidaceae family, is indigenous to the area of central Asia between the Mediterranean and China. The methanolic extract of *A. sativum* contains anticancer properties effective against bladder cell carcinoma, MCF 7, A549 and DU145. It has the potential to prevent cancers of the colon, rectal, stomach and prostate, in addition to bladder and prostate cancers [11].

1.5. Mint–scientific name–Mentha spicata L.

The Lamiaceae family, to which this belongs, can be found in subcosmopolitan distributions across North America, Europe, Asia and Africa. On the other hand, L methanol selectively kills colon cancer cells while sparing healthy ones. Conversely, Menthe inhibits cancer cell proliferation and metastasis by halting their formation [12].

1.6. Onion–scientific name–Allium cepa L.

The Amaryllidaceae family includes this and it is found all around the globe. Anticancer and antibacterial properties were significantly enhanced in lacto-fermented onion extracts from all species compared to naturally fermented onions. Research has shown that onions' antibacterial or cytotoxic effects strongly correlate with their flavonoid content. Onions treated with lacto-fermentative bacteria may have anticancer properties [13].

Their beneficial pharmacological effects have prompted much research into these six therapeutic plants. There are several limitations to the broad usage of synthetic immunomodulators, including the fact that they raise the risk of infection and generally affect the immune system. Immunomodulators derived from plants or plant byproducts are a relatively new field of study. Therefore, cytokine investigations are required to deduce the precise mechanism of the observed immunostimulation. Phytochemicals like flavonoids and saponins in essential oils may be responsible for their immune-system-boosting and anti-cancerous benefits. There is evidence that certain phytochemicals can modulate immune responses. A safer alternative to radiation, chemotherapy and surgery is the use of medicinal herbs to alleviate their adverse effects. When combined with chemotherapeutic medications, the active chemicals increase the drug's efficacy while reducing its adverse effects.

Natural chemicals are in high demand as potential new anticancer treatments due to the high toxicity, undesirable side effects and drug resistance of certain chemotherapeutic medications. The area of drug development can benefit greatly from this since newly produced medication can be employed singly or in combination to achieve greater effectiveness. Investigating compounds that modulate the immune system and cancer could aid in the treatment of both mild and severe illnesses and it could also help in the fight against the emergence of drug resistance. The effects seen in the essential oil can result from the individual actions of these chemical components or their combined effects. For future research, it is necessary to extract the pure chemicals responsible for the plant essential oil's notable effects on the immune system and malignant cells.

The present research examines the potential immunomodulatory and anticancer properties of six of the most popular medicinal plants. Plants such as ginger (*Zingiber officinale*), turmeric (*Curcuma longa*), mint (*Mentha spicata*), garlic (*Allium sativum*) and onion (*Allium cepa*) were used in this investigation. The results demonstrated that the oils inhibited the development of cancer cells in a dose-dependent manner. Various plant oils were tested in vitro to see how they affected the responsiveness of human CBMCs to the generation of IFN-γ and IL10 cytokines. The chemical compounds' potential immunomodulatory and anticancer effects were determined using silico research. The study of protein-protein interactions, the development of compound-targeted networks, the treatment of complicated disorders and medication targeting are all areas that greatly benefit from using bioinformatics tools. One potential use for in silico technologies is to assess the target mechanism connected to cancer and to build more efficient treatments. These strategies are cost-effective and produce more accurate results.

2. Material and Methods

2.1. Collection of sample

In the Indian state of Odisha, a farmer in the village of Uttangara provided the plant, as mentioned earlier, as a specimen. The farmer organically cultivates these plants, refraining from the use of synthetic fertilizers. Typical surveys documented all plant details, including organic fertilizer dosages, duration, preparation method, administration mode, necessary precautions, etc.

Jajpur is located at these coordinates: 20° 51′ 0″ N and 86° 20′ 0″ E. The area receives a subtropical climate characterized by high levels of relative humidity. During the Kharif season, the annual precipitation is concentrated within three to four months. On average, it rains about 1014.5 mm every year. Typically, this area is home to six different kinds of soil. Red sandy, lateritic, alluvial and deltaic alluvial soils are included. The majority of the district is covered with deltaic-alluvial soil and alluvial soil. An abundance of fertile land characterizes the district. Temperatures rarely drop below 55° F or rise over 107° F throughout the year, with a typical range of 60 to 101°F. The study's subject was a GIS map showing the study area in Odisha, India (Figure 8.1) [14, 15].

2.2. Plant material

The study utilized organic turmeric rhizomes and leaves, ginger rhizomes, garlic and onion bulbs, tulsi and mint leaves. For the experiments, fully grown plant parts were collected. Samples were carefully gathered and preserved in sterile bags. The collected materials were then washed and sliced into small parts. Extracting the essential oils from fresh leaves, bulbs, rhizomes and cloves was done right away.

2.3. Authentication of the plant

An extensive literature search was conducted before the plants were selected. The recorded medical usage of these six pants dates back centuries. Numerous studies have shown that these medicinal plants can potentially treat various illnesses. The use of these medications has long been central to Ayurvedic therapy.

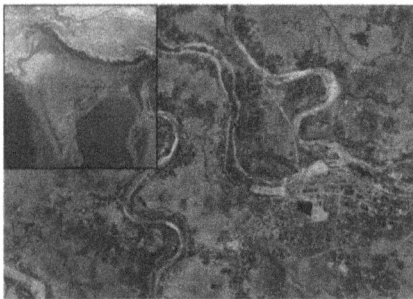

Figure 8.1. Representation of the area where the sample was taken using GIS.

Source: Author.

Research on the effectiveness of these species across multiple animal models, human subjects and laboratory settings was found through searches of electronic databases like Scopus and PubMed. The accuracy of the gathered plant sample was confirmed by Gyanranjan Mahalik, Associate Professor, Department of Botany of Centurion University of Technology and Management in Odisha, India. 'The Flora of Orissa, written by Saxena and Brahmam', 'Botany of Bihar and Orissa' by Bishen Singh Mahendra Pal Singh and 'Flora of Odisha with Ethnobotanical Applications' by Taranisen Panda were used as a reference to verify the documented morphological descriptions of plants. After recording detailed notes on each plant, the herbarium was examined for its taxonomic classification. The authenticity of the plants was confirmed by using these parameters.

2.4. Documentation

Various plant morphological parameters were examined, including the height, leaf count, rhizome weight, bulb weight, etc., of the turmeric, tulsi, ginger, garlic, mint and onion plants. All necessary information was recorded and the herbarium specimens were sent to the herbarium section of the Department of Botany, CUTM.

2.5. Essential oil extraction

A process called hydro distillation was employed to extract the essential oil. The device developed by Clevenger was used to hydro-distilled the essential oil. A round-bottomed flask was used to suspend 100g of chopped fresh leaves, rhizomes, cloves and bulbs in distilled water. For three hours, the apparatus was heated to 40–45°C to collect oil from leaves and for eight hours, it was heated to 50–60°C to collect oil from rhizomes, cloves and bulbs utilizing steam distillation. The essential oil was obtained from the aqueous layer by separating it using a sterile Eppendorf tube. Three independent batches were required to extract each essential oil [16, 17].

2.6. Oil yields

The overall oil content of all the samples was calculated. Rhizome oil yield per cent is calculated using amounts reported on a dry weight basis [16, 17].

$$\text{Rhizome oil } \% \text{ yield} \atop \text{(v/w) (dry weight)} = \frac{\text{Volume of essential oils (ml)} \times 100\,\%}{\text{Weight of raw materials}}$$

$$\text{Leaf oil } \% \text{ yield} \atop \text{(v/w) (Fresh weight)} = \frac{\text{Volume of essential oils (ml)} \times 100\,\%}{\text{Weight of raw materials}}$$

2.7. Immunomodulatory study

Isolation of CBMCs (Cord Blood Mononuclear Cells): Venipuncture collected umbilical cord blood from healthy full-term infants at 37 weeks of gestation. The blood was then deposited in sterile sodium heparin tubes right after birth. The whole heparinized blood sample was spun at 2000 RPM for 20 minutes at 15°C for the leukocyte ring. Ten millilitres of Phosphate-Buffered Saline (PBS) was added after its removal and a Falcon tube was utilized for the transfer. A medium-density gradient of Ficoll-PaqueTM PLUS (GE Healthcare) was then used to filter the homogenate. Centrifugation at 1000 RPM for 20 minutes at 15°C produced the leukocyte cloud. The cells were removed from interphase by washing them three times in PBS at 10 minutes, 1000 rpm and 15°C. After the last wash, the cells were kept in RPMI 1640 medium supplemented with 20 mm Hepes and 10% foetal bovine serum (FBS) using Himedia.

The number of CBMCs was determined using hemocytometers [18, 19].

Estimating cytokine: Using 5×106 CBMCs/ml, cultures were performed in triplicate. All of the oils—*C. longa, O. sanctum, Z. officinale, A. sativum, M. spicata* and *A. cepa*—were added to a 96-well tissue culture plate at a concentration of µg/ml. Adding ten microgrammes per millilitre of Con A raised the levels of IL-10 and IFN-γ. During cell development, no oil was applied to the control group. Plates were incubated at 37°C for 48 hours in a CO2 incubator with 5% carbon dioxide. A separate set of wells was utilized to make the cytokines. Cytokine quantification was performed by collecting, spinning and storing the culture supernatants at −70°C. The manufacturer's instructions were followed for the subsequent measurement of cytokines using sandwich ELISA kits from e-Bioscience for IFN-γ and IL-10. Typical interpolation curves were used to calculate picogrammes per millilitre [18, 19].

2.8. Anticancer activity

Cell culture: Human cervical cancer cells (HeLa, ATCC) and breast cancer cells (MCF 7, ATCC) were used to evaluate the anticancer effects of *C. longa, O. sanctum, Z. officinale, A. sativum, M. spicata* and *A. cepa* oil. The ATCC in the US is where the cell lines were acquired. All cell lines were cultured in DMEM media supplemented with 10% inactivated FBS and five mM L-glutamine in an incubator set at 37°C with 5% CO2 [20, 21].

MTT assay: Metabolically active cells are used in the MTT test to change the yellow tetrazolium salt-MTT into purple formazan crystals. Two million cells are added to each well of the 96-well plates using 100 µL of RPMI 1640. The plates are then placed in a CO_2 incubator set at 37°C with 5% CO_2 and allowed to grow for 24 hours. A new medium containing oil and extract concentrations between 0.1 and 100 g/mL is introduced after 48 hours. Throughout 24 to 48 hours, the cells were kept at 37°C with 5% CO_2. After the five-hour incubation period, twenty microlitres of MTT stock solution (5 mg/mL in PBS) is added to every well. The cells that made it through the process changed a yellow MTS molecule into a purple formazan. After the medium is emptied to dissolve the MTT metabolic output, 200 µL of DMSO is added to each well. After 5 minutes of shaking at 150 RPM, the optical density at 560 nm is determined [20, 21].

The proportion of growth inhibition was determined using the following calculation:

$$\% \text{ Cell inhibition} = 100 - \{(At\text{-}Ab)/(AcAb)\} \times 100$$

where At = Absorbance value of the test compound, Ab = Absorbance value of the blank and Ac = Absorbance value of the control.

2.9. Statistical analysis

The SPSS 25.0 statistical package was utilized for the analysis. We ran each experiment three times. Standard deviations (SDs) and means were employed to assess the data. The final report included the means and standard deviations from the three duplicated studies.

3. Results

3.1. Morphological analysis of the plants

Six distinct types of medicinal herbs were used in this research. Table 8.1 displays their morphological characteristics. In 5–7 months, turmeric plants reach a height of 51 ± 2.64 cm, with 5.3 ± 1.15 tillers and 7.2 ± 2 total leaves. The rhizome of turmeric weighs around 23.3 ± 3.05 grams. An essential oil content of approximately 0.5% for the leaves and about 1.2% for the rhizome is obtained from the fresh turmeric. In about five to seven months, a Tulsi plant will reach a height of 39.3 ± 7.02 cm with total leaves of 40 ± 8.18 cm. A yield of around 0.4% essential oil of leaves is achieved by subjecting fresh tulsi leaves to the Clevenger hydro-distillation process. In 5–7 months, the ginger plant grows to a height of (43 ± 2.88) cm, with 6.6 ± 0.6 tillers and 7 ± 0.57 complete leaves. A ginger rhizome weighs around 47.3 ± 6.11 grams.

An essential oil of about 0.7% is extracted from the ginger rhizome. The garlic plant grows to a height of 34 ± 4.35 cm in around 4 to 5 months, with 3.6 ± 0.57 tillers and 4 ± 1 total leaves. Garlic bulbs typically weigh around 27.35 ± 5.03 grams. An essential oil content of around 0.2% is extracted from the garlic bulb. The average height of a mint plant is 24 ± 2.64 cm after 3 to 4 months and its total leaf is 16.3 ± 1.52. When harvested young, Mint leaves contain around 2.62 per cent essential oil. The onion plant reaches a height of 28.6 ± 6.80 cm after 4 to 5 months, with a tiller number of 3 ± 1 and 4.3 ± 1.52 entire leaves. Onion bulbs typically weigh around 44.6 ± 6.11 grams. An essential oil of onion bulb concentration of around 1.7% is obtained by subjecting the onion bulb to the Clevenger hydro-distillation process [22].

3.2. Immunomodulatory activity through cytokine estimation

When human CBMCs were co-incubated with essential oils from these edible

Table 8.1. Assessment of medicinal plants based on their morphological traits

Sl. no.	Characters	Value (Mean ± SD) of plants					
		Turmeric	Tulsi	Ginger	Garlic	Mint	Onion
1.	Plant height (cm)	51 ± 2.64	39.3 ± 7.02	43 ± 2.88	34 ± 4.35	24 ± 2.64	28.6 ± 6.80
2.	Tiller no. per plant	5.3 ± 1.15	-	6.6 ± 0.6	3.6 ± 0.57	-	3 ± 1
3.	Total no. of leaves per plant	7 ± 2	40 ± 8.18	7 ± 0.57	4 ± 1	16.3 ±1.52	4.3 ±1.52
4.	Total rhizome weight (gm)	23.3 ± 3.05	-	47.3 ± 6.11	-	-	-
5.	Total bulb weight (gm)	-	-	-	27.3 ± 5.03	-	44.6 ± 6.11
6.	Total leaf oil (%)	0.5	0.4	-	-	2.62	-
7.	Total rhizome oil (%)	1.2	-	0.7	-	-	-
8.	Total bulb oil (%)	-	-	-	0.2	-	1.7

Source: Author.

medicinal plants, the release of IL-10 and IFN-γ cytokines was enhanced in various oils. Regarding this matter, after 48 hours of incubation, *C. longa, O. sanctum, A. sativum* and *A. cepa* oil demonstrated a dose-dependent rise in IFN-γ and a reduction in IL-10 compared to the control (0 μg/ml) (P <0.05). Comparing the control group to those who received oils extracted from *Z. officinale* rhizome and *M. spicata* leaves did not yield statistically significant results. There was a statistically significant difference between the control (16.57 μg/ml) and the *C. longa* rhizome oil (42.98 μg/ml) at 40 μg/ml (P <0.001). A higher concentration of IFN-γ (39.66 μg/ml) was seen in the experimental group when 40 μg/ml of *O. sanctum* leaf oil was used, in comparison to the control group's 15.7 μg/ml.

A. sativum bulb oil had a substantially greater concentration of IFN-γ at 33.54 μg/ml compared to the control group (16.20 μg/ml). Similarly, when comparing the control group (15.62 εg/ml) to the group treated with *A. cepa* bulb oil (40 μg/ml), the concentration of IFN-γ was 34.32 μg/ml more significant. When treated with 0–40 μg/ml doses of *Z. officinale* rhizome oil and *M. spicata* leaf oil, there was no noticeable alteration in the IFN-γ level. Table 8.2 shows that the oils of *C. longa, O. sanctum, A. sativum* and *A. cepa* all considerably decreased IL-10 release. At a concentration of 40 μg/ml, the concentration of IL-10 in the culture supernatant of *C. longa* was 10.85 μg/ml, *O. sanctum* was 10.56 μg/ml, *A. sativum* was 10.22 μg/ml and *A. cepa* was 10.45 μg/ml. When treated

Table 8.2. After 48 hours of incubation, the th1 and th2 cytokines IFN-γ and IL-10 expression pattern was examined using various essential oils at different doses

Cytokines	The concentration of oils in which cytokine was detected (μg/ml)	Mean cytokine level, μg/ml					
		C. longa	*O. sanctum*	*Z. officinale*	*A. sativum*	*M. spicata*	*A. cepa*
IFN-γ	0 (Control)*	16.57	15.7	15.16	16.20	16.12	15.62
	10	22.49	22.54	12.79	20.44	11.50	21.63
	20	25.33	27.14	11.98	24.50	11.71	26.44
	30	32.48	34.35	15.02	29.72	12.23	31.24
	40	42.98	39.66	16.88	33.54	12.92	34.32
IL-10	0 (Control)*	20.91	20.34	19.06	19.46	19.91	19.28
	10	17.25	18.55	17.43	18.81	18.28	16.32
	20	14.66	16.28	18.96	15.26	17.94	13.19
	30	11.24	13.65	14.98	14.46	16.56	11.56
	40	10.85	10.56	16.84	10.22	14.25	10.45

Source: Author.

with *Z. officinale* rhizome and *M. spicata* leaf oils, the IL 10 showed no noticeable change at dosages ranging from 0 to 40 μg/ml (Table 8.2) [23, 24].

3.3. Comparative anti-cancer efficacy of MCF-7 and HeLa breast and cervical cancer cell lines

Table 8.3 displays the results of the experiments conducted on Hela and MCF 7 cell lines, which demonstrated that *C. longa*, *O. sanctum*, *Z. officinale*, *A. sativum*, *M. spicata* and *A. cepa* bulb oil inhibited the proliferation of malignant cells. The preliminary screening for this activity involved evaluating the therapy effects of various oils using a one-dose assay. The low concentration was 10μg/ml, while the high concentration was 20μg/ml, tested against human cancer cell lines. A correlation between the oil concentration in different plants and their ability to prevent the proliferation of cancer cells was observed. Compared to the leaf oil, the essential rhizome of *C. longa* exhibited a higher percentage of inhibition against the HeLa and MCF 7 cell lines. At a concentration of 20μg/ml, the HeLa cell line was observed to be inhibited by (95 ± 1.5) and (98 ± 1.30) of turmeric leaf and rhizome oil, respectively. At 20μg/ml, the rhizome oil inhibited the MCF 7 cell line to a greater degree than the leaf oil (93 ± 1.4), with a value of 96 ± 1.5.

In comparison to the MCF 7 cell line, HeLa cells were significantly inhibited by *C. longa* leaf oil. The rhizome oil from this plant exhibited a more significant percentage of inhibition (98 ± 1.30) for the HeLa cell line and (96 ± 1.5) for the MCF 7 cell line, even at a dose of 20μg/ml. At 20μg/ml, the HeLa cell line was observed to have an inhibition

Table 8.3. The impact of several essential oils on the percentage of cell inhibition in two distinct cancer cell lines

Sl. No.	Plant	HeLa	MCF 7
		20 μg/ml	20 μg/ml
	Turmeric (Leaf)	95 ± 1.5	93 ± 1.4
	Turmeric (Rhizome)	98 ± 1.30	96 ± 1.5
	Tulsi (Leaf)	92 ± 1.7	87 ± 1.2
	Ginger (Rhizome)	89 ± 1.5	88 ± 1.20
	Garlic (Bulb)	97 ± 1.2	92 ± 1.3
	Mint (Leaf)	45 ± 1.6	41 ± 1.7
	Onion (Bulb)	67 ± 1.2	62 ± 1.6

Source: Author.

percentage of (92 ± 1.7) caused by *O. sanctum* leaf oil. For the MCF 7 cell line, however, leaf oil at 20μg/ml produced a value of (87 ± 1.2). In comparison to the MCF 7 cell line, HeLa cells were significantly inhibited by *O. sanctum* leaf oil. At a concentration of 20μg/ml, the percentage of inhibition against the HeLa cell line by *Z. officinale* rhizome oil was determined to be (89 ± 1.5). Rhizome oil produced an output of (88 ± 1.20) at a 20μg/ml concentration when tested on the MCF 7 cell line. Contrasted with the MCF 7 cell line, HeLa cells were significantly inhibited by *Z. officinale* rhizome oil.

Another time, the percentage of inhibition against the HeLa cell line by *A. sativum* bulb oil was (97 ± 1.2) at a dose of 20μg/ml. When tested in the MCF 7 cell line, bulb oil at a concentration of 20μg/ml yielded a result of (92 ± 1.3). *A. sativum* bulb oil had a much more potent inhibitory effect on HeLa cells than on the MCF 7 cell line. The HeLa cell line was found to have an inhibition percentage of 45 ± 1.6 when administered *M. spicata* leaf oil at a concentration of 20μg/ml. When tested on the MCF 7 cell

line, leaf oil at 20μg/ml produced a value of (41 ± 1.7). The HeLa cell line was inhibited by 67 ± 1.2 per cent by *A. cepa* bulb oil at a 20μg/ml concentration. At a concentration of 20μg/ml, bulb oil produced a result of (62 ± 1.6) in the MCF 7 cell line. When tested against HeLa cells, *A. cepa* bulb oil exhibited a significantly higher inhibition percentage than the MCF 7 cell line (Table 8.3) [25, 26].

4. Discussion

Current research focuses on the immunomodulatory and anticancer effects of these six vital medicinal herbs. Results demonstrated that during 48 hours of incubation, the oils of *C. longa, O. sanctum, A. sativum* and *A. cepa* exhibited immunomodulatory effects by measuring cytokines. Compared to the control group, which had a concentration of 0 μg/ml, there was a dose-dependent increase in IFN-γ and a drop in IL-10. Statistical analysis of the power compared to oils extracted from *M. spicata* leaves and *Z. officinale* rhizome revealed no significant results. Treatment with 0–40 μg/ml doses of *Z. officinale* rhizome oil and

M. spicata leaf oil did not result in any noticeable change in the IFN-γ level. *A. sativum, C. longa, O. sanctum* and *A. cepa* oils all worked together to suppress IL-10 secretion. No appreciable change was observed in the IL 10 levels when the *Z. officinale* rhizome and *M. spicata* leaf oils were treated, with doses ranging from 0 to 40 µg/ml (Figure 8.2). The percentage of cell inhibition on the HeLa and MCF 7 cell lines was higher for essential oil from the rhizome of *C. longa* compared to oil from the leaves. When tested against HeLa cells, *O. sanctum* leaf oil, *Z. officinale* rhizome oil, *A. sativum* bulb oil, *M. spicata* leaf and *A. cepa* bulb oil demonstrated a significantly higher inhibition percentage than the MCF 7 cell line (Figure 8.3).

Some malignancies, COX-2, NOS, Cyclin D1 and other genes with downstream products can be suppressed by turmeric. This impacts the development, metastasis and angiogenesis of tumours, as well as growth factor receptors and cell adhesion molecules [8, 27, 28]. Researchers have discovered that curcumin blocks cancer-causing signalling pathways called STAT3 and NF-B. Reducing Sp-1 activity and the expression of its housekeeping genes may also help stop cancer from spreading and invading new areas. It has been shown that curcumin can block the activation of Sp-1 and the genes it regulates in colorectal cancer cell lines [29]. These genes include ADEM10, calmodulin, EPHB2, HDAC4 and SEPP1. According to a recent study, Curcumin lowers Sp-1 DNA binding activity in non-small cell lung carcinoma cells and suppresses Sp-1 activity in bladder cancer cells. In a series of epithelial ovarian tumours, curcumin analogues induce cell death by autophagy and endoplasmic reticulum

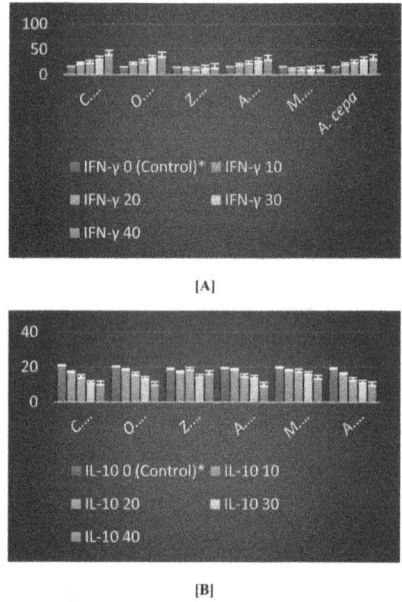

[A]

[B]

Figure 8.2. Essential oil effects at different doses are shown graphically. A refers to IFN-γ, while B refers to IL-10 expression.

Source: Author.

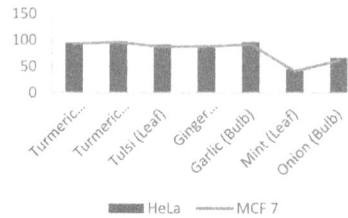

Figure 8.3. Two separate cancer cell lines used to illustrate the effect of various essential oils on the proportion of cell restriction.

Source: Author.

stress, among other potential pathways. Curcumin analogue-induced apoptosis could be worsened by inhibiting autophagy, which could cause severe endoplasmic reticulum stress [30].

It inhibits the production of immune response factors, including interleukin-2, NO, NF-Kappa B and PHA-induced T-cell proliferation. According to research, it makes NK cells more lethal. Cell proliferation and cytokine production are reduced when these immunological traits are not stimulated as much [31]. Many inflammatory disorders showed the anti-inflammatory benefits of CUR and *C. longa*. Serum levels of inflammatory mediators such as total protein and phospholipase A2 were positively affected, while white blood cell, neutrophil and eosinophil counts were reduced. In addition to lowering immunological abnormalities, CUR and *C. longa* treatment reduced transforming growth factor-beta, interferon-gamma, immunoglobulin E (Ig)E and the pro-inflammatory cytokine interleukin 4 (IL)-4 levels in patients [32]. When TH1, Th17 and Tregs, three crucial types of T helper (Th) cells, are out of whack, they produce inflammatory substances and lead to RA. A large number of autoantibodies are produced by B lymphocytes, which play an essential role in the progression of RA. Researchers have shown that curcumin helps alleviate RA symptoms in immune and synovial fibroblast cells by blocking pro-inflammatory mediators such as AP-1, NF-B and mitogen-activated protein kinases (MAPKs). A therapeutic promise for rheumatoid arthritis treatments is curcumin therapy, which reduces clinical symptoms without causing adverse effects [33].

Free radicals are neutralized by the powerful antioxidants found in the Tulsi plant's leaves. In the MCF 7 breast cancer cell line, it caused cell death [34]. Elixir of Tulsi has a multiplicative effect on reducing tumour cell proliferation and increasing longevity. This protects DNA from radiation, which might damage it [35]. To halt chemically-induced cancers of the mouth, skin, liver and lungs, Tulsi and its phytochemicals, including eugenol, rosmarinic acid, carnosic acid, luteolin, apigenin, myretenal and β-sitosterol increased antioxidant activity, altered gene expression, led to cell death and halted the development of new blood vessels and cancer's spread.

Additionally, apigenin, carnosic acid, rosmarinic acid and eugenol are phytochemicals that can shield DNA from radiation damage [36]. An IFN-alpha type Th1 immune response was induced by Tulsi extract. There has been a significant uptick in the proportion of NK (CD 16+, CD 56+) and T helper cells (CD 3+, CD 4+), which lend credence to the cytokine levels.

The flavonoids in Tulsi leaf extracts are responsible for their immunomodulatory actions [37]. The Tulsi plant's leaf extract may include various secondary components, including carbs, glucose, saponins, terpenoids, tannins, flavonoids, glycosides and fatty acids. Consequently, there are a lot of phenols in Tulsi leaves. Benzene, -farnesene, 1,2,4-triethyl cyclohexane and eugenol are the primary components of the methanolic extract. Furthermore, these phytochemicals possess antibacterial and antistress capabilities, analgesic effects, immunomodulatory and anti-inflammatory properties and antimicrobial properties. Herbal remedies, such as Tulsi, have several benefits over synthetic pharmaceuticals [38].

Ginger includes about 50 compounds, such as gingerols and diarylheptanoids. Ginger prevents cyclooxygenase, lipoxygenase and arachidonic acid metabolism. When applied to human breast cancer cell lines, [6]-gingerol inhibits their adhesion, invasion, motility and activity [39]. Ginger components extracted with ethanol destroyed panc-1 and other human pancreatic cancer cell lines. Two of its primary components, [6]-shogaol and [6]-gingerol, induce cell death in tumors and inhibit cell proliferation, respectively [40]. Studies have shown that F6 and the chemicals reduced the production of 'activation markers' CD25 and CD69, which hindered the activation of CD8+ cytotoxic T-lymphocytes [41]. The concentration and duration of time dictated how much ginger was toxic to PSC and cyst wall. Remarkably, ginger had a more significant impact than [6]-gingerol. In addition, neither of the extracts promoted NO production that the infection or IFN-γ caused. The effect is amplified when IFN-γ is present. This herbal remedy may further protect against host cell death from elevated NO levels. The [6]-gingerol found in ginger could be responsible for its effects [42].

Garlic polyphenols suppress human breast cancer cells in vitro and reduce animal disease cases. Dialyl disulfide (DADS) and other garlic oil-soluble molecules are superior to other soluble substances in lowering breast cancer [43]. Evidence suggests garlic's diallyl disulfide (DADs) may prevent colon cancer. Allicin and ajoene, two of garlic's active ingredients, are new and powerful cancer fighters. By preventing cell damage, flavonoids like allicin inhibit cancer. Some substances, including selenium and allyl sulfide, are anti-mutagens and may protect against cancer by damaging cells and DNA or encouraging the body to fix damaged DNA [44]. Several activities seem to enhance the immune system's efficacy, such as macrophage activation, regulation of cytokine release, stimulation of lymphocytes, dendritic cells, natural killer cells and immunoglobulin synthesis. Garlic aids in treating and preventing several diseases by regulating the immune system; these include cancer, obesity, metabolic syndrome, heart disease, stomach ulcers and metabolic syndrome. Many of the pharmacological effects of the herb are due to changes in the secretion of cytokines [45]. After being given to mice, the chemical stimulated NK activity, macrophage phagocytosis, spleen cell proliferation and cytokine secretion (including IFN-gamma and TNF-alpha). A study found that lymphocytes treated with AGE (Aged Garlic Extract) were more receptive to mitogens and other stimuli. In terms of nitric oxide generation, though, it lagged below PSK [46].

The subG0/G1 phase, linked to cell death, was significantly more prevalent in the cell cycle analysis. Adrenocortical tumour models, specifically SW13 cells, can have diminished viability, vigour and survival by a crude methanolic extract of wild mountain mint [47]. Phytochemicals derived from mints cannot inhibit the growth of numerous human cancers, including cervical, lung, breast and prostate. Additionally, they trigger cell death, hinder cancer cell migration and invasion, increase Bax and p53 gene expression, regulate TNF and IL-6, IFN-γ and IL-8 and bring about a senescent appearance [48]. Mentha species are rich in phenolic compounds, the most notable being flavonoids. Flavonoids, some of which are lipophilic, are present in mints. Among the many pharmacological actions of mint's constituents are those of an antioxidant, cytoprotective, hepatoprotective, chemopreventive and antidiabetic. Though mint has many beneficial uses, it also has certain species that harm humans [49].

In addition to being rich in vitamin C, onions are powerful antioxidants that ward off free radicals that might cause cancer. A lower risk of prostate cancer was linked with frequent allium consumption, but a lower risk of stomach and oesophageal cancer was related to the highest allium consumption [50]. Cardiovascular disease, cancer and other degenerative illnesses are some of the many that onions fight off with their abundance of flavonoids, sulfoxides of alkenyl cysteine and other non-nutrient antioxidant compounds [51]. Onions include several compounds with cancer properties, including quer. Combined with quercetin, ONA protects against epithelial ovarian, breast, brain, colon and lung cancers [52]. Thiosulfinates, thiosulfonates, capaenes, s-oxides and sulfoxides are some of the acid chemicals found in it. These compounds inhibit cell growth by damaging DNA and killing cancer cells [53]. Improved levels of cell CD4, IFN-γ and the ratio of IFN- γ to IL4 (Th1/Th1-Balance), together with the immunomodulatory effects of quercetin and the onion plant, were shown by decreased levels of Th2 cytokines, IL-5 and I L13, as well as IL-6, IL-8, IL-10, TNF-α and IGE [54].

Oleu europea, Salvia officinalis, Curcuma longa, Glycyrrhiza glabra and *Silybum marianum* were the plants with the highest number of P-C-T-P interactions. In vitro, cytotoxicity and anti-inflammatory studies showed that the top-scoring plants were comparable to piroxicam [28]. The medications showed promising results against breast cancer proteins in the in-silico investigation, with high glide scores and energies. There is a difference in the binding affinity of curcumin and resveratrol to proteins with different interaction sites

for amino acids [55]. Carvacrol docked more effectively with the estrogen receptor alpha. Because of its superior molecular docking performance, carvacrol is among the most promising drugs for use as estrogen receptor alpha-selective inhibitors [56]. The sulfur and phenolic compounds from *A. sativum* L. exhibited suitable binding energies and were non-toxic, making them ideal for α-glucosidase inhibition. Among these substances, the α-glucosidase enzyme was most inhibited by methionol and caffeic acid, which also had the lowest binding energy [57]. Based on the binding affinity analysis results, non-organosulfur compounds outperformed the organosulfur compounds. The chemicals found in garlic show great promise as potential inhibitors of the cancer-related receptors CCR5 and CXCR4 [58].

5. Conclusion

The rising costs of health care and the frequency of chronic diseases have piqued the attention of both researchers and consumers in essential oils as potential natural immunomodulators and anti-cancerous agents for various medical conditions. Traditional Indian medicine relies on multiple plants for different ailments, but little is known about these plants' immunomodulatory and anticancer effects. Research on cytokines is necessary to assess the phytochemicals' immunomodulatory effects. Nowadays, essential oils derived from medicinal herbs do not harm regular cells, owing to modern medicine. One of their peculiarities is the ability it has to control the immune system. Despite their helpful biological properties, these oils have the potential to induce a variety of adverse reactions, including allergies, cytotoxicity, hypersensitivity and immunological suppression. Therefore, scientific research on these oils and developing alternative drugs to treat infectious diseases are paramount. Many medicinal plants have anticancer properties but have not yet been studied. Additional research is necessary into the in-silico omics methods and potential modes of action of various medicinal plants, both known and unknown.

6. Acknowledgments

The authors would like to express their deepest gratitude to the Department of Botany at Centurion University of Technology and Management and the Institute of Life Sciences in Odisha, India, for funding and facilitating the execution of these studies. Furthermore, we would like to express our gratitude to the co-authors who played an essential part in producing the work. At last, we'd like to express our appreciation to Dr Shasank Sekhar Swain, Director and Research Head, Salixiras Research Private Limited, Bhubaneswar, Odisha, India, who has been an invaluable resource during this examination.

References

[1] Rasool Hassan, B. A. (2012). Medicinal plants (importance and uses). *Pharmaceut Anal Acta*, 3(10), 2153–2435.

[2] Patro, L. (2016). Medicinal Plants of India: With special reference to Odisha. *International Journal of Advance Research and Innovative Ideas in Education*, 2(5), 121–135.

[3] Poddar, S., Sarkar, T., Choudhury, S., Chatterjee, S., & Ghosh, P. (2020). Indian traditional medicinal plants: A concise review. *International Journal of Botany Studies*, 5(5), 174–190.

[4] Dwivedi, T., Kanta, C., Singh, L. R., & Prakash, I. (2019). A list of some important medicinal plants with their medicinal uses from Himalayan State Uttarakhand, India. *Journal of Medicinal Plants Studies*, 7(2), 106–116.

[5] Mahalik, G., Sahoo, S., & Satapathy, K. B. (2017). Evaluation of phytochemical constituents and antimicrobial properties of *Mangifera indica* L. Leaves against urinary tract infection-causing pathogens. *Asian Journal of Pharmaceutical and Clinical Research*, 10(9), 169–173.

[6] Panda, S., Sahoo, S., Tripathy, K., Singh, Y. D., Sarma, M. K., Babu, P. J., & Singh, M. C. (2020). Essential oils and their pharmacotherapeutics applications in human diseases. *Advances in Traditional Medicine*, 1–15.

[7] Gunnars, K. (2021, May). *Proven Health Benefits of Turmeric and Curcumin.*

[8] Wilken, R., Veena, M. S., Wang, M. B., & Srivatsan, E. S. (2011). Curcumin: A review of anti-cancer properties and therapeutic activity in head and neck squamous cell carcinoma. *Molecular Cancer*, 10, 1–19.

[9] Sridevi, M., Bright, J. O. H. N., & Yamini, K. (2016). Anti-cancer effect of Ocimum sanctum ethanolic extract in non-small cell lung carcinoma cell line. *Int J Pharm Pharm Sci*, 8, 8–20.

[10] Prasad, S., & Tyagi, A. K. (2015). Ginger and its constituents: role in prevention and treatment of gastrointestinal cancer. *Gastroenterology Research and Practice*, 2015(1), 142979.

[11] Umadevi, M., Kumar, K. S., Bhowmik, D., & Duraivel, S. (2013). Traditionally used anticancer herbs in India. *Journal of Medicinal Plants Studies*, 1(3), 56–74.

[12] Faridi, U., Dhawan, S. S, Pal, S., Gupta, S., Shukla, A. K., Darokar, M.

P., & Sharma, A. Mint can help fight cancer: OMICS. *A Journal of Integrative Biology.*

[13] Ravanbakhshian, R., & Behbahani, M. (2018). Evaluation of anticancer activity of lacto-and natural fermented onion cultivars. *Iranian Journal of Science and Technology, Transactions A: Science, 42,* 1735–1742.

[14] Satapathy, K. B., Mishra, P. K., & Jena, G. J. (2020). Medico-botany of plants used in rituals in Jajpur district of Odisha. *International Journal of Botany Studies, 5*(4), 01–08.

[15] Nayak, R. K., Manchala, M., Jena, B., Das, J., Mohanty, S., & Shukla, A. K. (2022). Crop Production Constraints Related to Secondary and Micro Nutrients in the Soils of Jajpur District, Odisha.

[16] Haro-González, J. N., Castillo-Herrera, G. A., Martínez-Velázquez, M., & Espinosa-Andrews, H. (2021). Clove essential oil (*Syzygium aromaticum* L. Myrtaceae): Extraction, chemical composition, food applications, and essential bioactivity for human health. *Molecules, 26*(21), 6387

[17] Dung, P. N. T., Dao, T. P., Le, T. T., Tran, H. T., Dinh, T. T. T., Pham, Q. L., Tran, Q. T., & Pham, M. Q. (2021). Extraction and analysis of chemical composition of *Ocimum gratissimum* L essential oil in the North of Vietnam. *In IOP Conference Series: Materials Science and Engineering,* 1092(1), 012092. IOP Publishing.

[18] Popa, M., Măruțescu, L., Oprea, E., Bleotu, C., Kamerzan, C., Chifiriuc, M. C., & Grădişteanu Pircalabioru, G. (2020). In vitro evaluation of the antimicrobial and immunomodulatory activity of culinary herb essential oils as potential perioceutics. *Antibiotics, 9*(7), 428.

[19] Ji, H. Y., Liu, C., Dai, K. Y., Yu, J., Liu, A. J., & Chen, Y. F. (2021). The extraction, structure, and immunomodulation activities in vivo of polysaccharides from Salvia miltiorrhiza. *Industrial Crops and Products, 173,* 114085.

[20] Najar, B., Shortrede, J. E., Pistelli, L., & Buhagiar, J. (2020). Chemical composition and in vitro cytotoxic screening of sixteen commercial essential oils on five cancer cell lines. *Chemistry & Biodiversity, 17*(1), e1900478.

[21] Perna, S., Alawadhi, H., Riva, A., Allegrini, P., Petrangolini, G., Gasparri, C., Alalwan, T. A., & Rondanelli, M. (2022). In vitro and in vivo anticancer activity of basil (Ocimum spp.): Current insights and future prospects. *Cancers, 14*(10), 2375.

[22] Juškevičienė, D., Radzevičius, A., Viškelis, P., Maročkienė, N., & Karklelienė, R. (2022). Estimation of morphological features and essential oil content of basils (*Ocimum basilicum* L.) grown under different conditions. *Plants, 11*(14), 1896.

[23] Thimmulappa, R. K., Mudnakudu-Nagaraju, K. K., Shivamallu, C., Subramaniam, K. J. T., Radhakrishnan, A., Bhojraj, S., & Kuppusamy, G. (2021). Antiviral and immunomodulatory activity of curcumin: A case for prophylactic therapy for COVID-19. *Heliyon, 7*(2), e06350.

[24] Afolayan, F. I. D., Adegbolagun, O., Mwikwabe, N. N., Orwa, J., & Anumudu, C. (2020). Cytokine modulation during malaria infections by some medicinal plants. *Scientific African, 8,* e00428.

[25] Paradkar, P. H., Juvekar, A. S., Barkume, M. S., Amonkar, A. J., Joshi, J. V., Soman, G., & Vaidya, A. D. B. (2021). In vitro and in vivo evaluation of a standardized *Curcuma longa* Linn formulation in cervical cancer. *Journal of Ayurveda and Integrative Medicine, 12*(4), 616–622.

[26] Osman, A. M. E., Taj Eldin, I. M., Elhag, A. M., Elhassan, M. M. A., & Ahmed, E. M. M. (2020). In-vitro anticancer and cytotoxic activity of ginger

extract on human breast cell lines. *Khartoum Journal of Pharmaceutical Sciences, 1*(1), 26–29.

[27] Mahajan, A., Badhe, P., & Mali, P. (2024). Integrating Network Pharmacology and Experimental Evaluation of *Ocimum tenuiflorum* (Tulsi) compounds targeting Breast Cancer Markers.

[28] Khairy, A., Ghareeb, D. A., Celik, I., Hammoda, H. M., Zaatout, H. H., & Ibrahim, R. S. (2023). Forecasting of potential anti-inflammatory targets of some immunomodulatory plants and their constituents using in vitro, molecular docking and network pharmacology-based analysis. *Scientific Reports, 13*(1), 9539.

[29] Widyananda, M. H., Puspitarini, S., Rohim, A., Khairunnisa, F. A., Jatmiko, Y. D., Masruri, M., & Widodo, N. (2022). Anticancer potential of turmeric (*Curcuma longa*) ethanol extract and prediction of its mechanism through the Akt1 pathway. *F1000Research, 11*, 1000.

[30] Vallianou, N. G., Evangelopoulos, A., Schizas, N., & Kazazis, C. (2015). Potential anticancer properties and mechanisms of action of curcumin. *Anticancer Research, 35*(2), 645–651.

[31] Yadav, V. S., Mishra, K. P., Singh, D. P., Mehrotra, S., & Singh, V. K. (2005). Immunomodulatory effects of curcumin. *Immunopharmacology and Immunotoxicology, 27*(3), 485–497.

[32] Memarzia, A., Khazdair, M. R., Behrouz, S., Gholamnezhad, Z., Jafarnezhad, M., Saadat, S., & Boskabady, M. H. (2021). Experimental and clinical reports on anti-inflammatory, antioxidant, and immunomodulatory effects of Curcuma longa and curcumin, an updated and comprehensive review. *BioFactors, 47*(3), 311–350.

[33] Mohammadian Haftcheshmeh, S., Khosrojerdi, A., Aliabadi, A., Lotfi, S., Mohammadi, A., & Momtazi-Borojeni, A. A. (2021). Immunomodulatory effects of curcumin in rheumatoid arthritis: evidence from molecular mechanisms to clinical outcomes. *Reviews of Physiology, Biochemistry and Pharmacology*, 1–29.

[34] Noorjahan, C. M., & Saranya, T. (2018). Antioxidant, anticancer and molecular docking activity of tulsi plant. *Agricultural Science Digest-A Research Journal, 38*(3), 209–212.

[35] Singh, D., & Chaudhuri, P. K. (2018). A review on phytochemical and pharmacological properties of Holy basil (*Ocimum sanctum* L.). *Industrial Crops and Products, 118*, 367–382.

[36] Baliga, M. S., Jimmy, R., Thilakchand, K. R., Sunitha, V., Bhat, N. R., Saldanha, E., Rao, S., Rao, P., Arora, R., & Palatty, P. L. (2013). *Ocimum sanctum* L (Holy Basil or Tulsi) and its phytochemicals in the prevention and treatment of cancer. *Nutrition and Cancer, 65*(sup1), 26–35.

[37] Mondal, S., Varma, S., Bamola, V. D., Naik, S. N., Mirdha, B. R., Padhi, M. M., Mehta, N., & Mahapatra, S. C. (2011). Double-blinded randomized controlled trial for immunomodulatory effects of Tulsi (*Ocimum sanctum* Linn.) leaf extract on healthy volunteers. *Journal of Ethnopharmacology, 136*(3), 452–456.

[38] Borah, R., & Biswas, S. P. (2018). Tulsi (*Ocimum sanctum*), excellent source of phytochemicals. *International Journal of Environment, Agriculture and Biotechnology, 3*(5), 265258.

[39] Ramakrishnan, R. (2013). Anticancer properties of *Zingiber officinale*–Ginger: A review. *Int J Med Pharm Sci, 3*(5), 11–20.

[40] Akimoto, M., Iizuka, M., Kanematsu, R., Yoshida, M., & Takenaga, K. (2015). Anticancer effect of ginger extract against pancreatic cancer cells mainly through reactive oxygen species-mediated autotic cell death. *PloS One, 10*(5), e0126605.

[41] Nordin, N. I., Gibbons, S., Perrett, D., & Mageed, R. A. (2013). Immunomodulatory effects of *Zingiber officinale* Roscoe var. Rubrum (Halia Bara) on inflammatory responses relevant to psoriasis. *In The Open Conference Proceedings Journal*, 4(1).

[42] Amri, M., & Touil-Boukoffa, C. (2016). In vitro anti-hydatic and immunomodulatory effects of ginger and [6]-gingerol. *Asian Pacific Journal of Tropical Medicine*, 9(8), 749–756.

[43] Nicastro, H. L., Ross, S. A., & Milner, J. A. (2015). Garlic and onions: Their cancer prevention properties. *Cancer Prevention Research*, 8(3), 181–189.

[44] Petrovic, V., Nepal, A., Olaisen, C., Bachke, S., Hira, J., Søgaard, C. K., … & Otterlei, M. (2018). Anti-cancer potential of homemade fresh garlic extract is related to increased endoplasmic reticulum stress. *Nutrients*, 10(4), 450.

[45] Arreola, R., Quintero-Fabián, S., López-Roa, R. I., Flores-Gutiérrez, E. O., Reyes-Grajeda, J. P., Carrera-Quintanar, L., & Ortuño-Sahagún, D. (2015). Immunomodulation and anti-inflammatory effects of garlic compounds. *Journal of Immunology Research*, 2015(1), 401630.

[46] Kyo, E., Uda, N., Suzuki, A., Kakimoto, M., Ushijima, M., Kasuga, S., & Itakura, Y. (1998). Immunomodulation and antitumor activities of aged garlic extract. *Phytomedicine*, 5(4), 259–267.

[47] Patti, F., Palmioli, A., Vitalini, S., Bertazza, L., Redaelli, M., Zorzan, M., Rubin, B., et al. (2020). Anticancer effects of wild mountain *Mentha longifolia* extract in adrenocortical tumor cell models. *Frontiers in Pharmacology*, 10, 1647.

[48] Tafrihi, M., Imran, M., Tufail, T., Gondal, T. A., Caruso, G., Sharma, S., Sharma, R., et al. (2021). The wonderful activities of the genus Mentha: Not only antioxidant properties. *Molecules*, 26(4), 1118.

[49] Mimica-Dukic, N., & Bozin, B. (2008). Mentha L. species (Lamiaceae) as promising sources of bioactive secondary metabolites. *Current Pharmaceutical Design*, 14(29), 3141–3150.

[50] Teshika, J. D., Zakariyyah, A. M., Zaynab, T., Zengin, G., Rengasamy, K. R. R., Pandian, S. K., & Fawzi, M. M. (2019). Traditional and modern uses of onion bulb (*Allium cepa* L.): A systematic review. *Critical Reviews in Food Science and Nutrition*, 59(sup1), S39-S70.

[51] Elberry, A. A., Mufti, S., Al-Maghrabi, J., Sattar, E. A., Ghareib, S. A., Mosli, H. A., & Gabr, S. A. (2014). Immunomodulatory effect of red onion (*Allium cepa* Linn) scale extract on experimentally induced atypical prostatic hyperplasia in Wistar rats. *Mediators of Inflammation*, 2014(1), 640746.

[52] WWW.Foodfacts.mercola.com/onion, 2016

[53] Upadhyay, R. K. (2016). Nutraceutical, pharmaceutical and therapeutic uses of *Allium cepa*: A review. *International Journal of Green Pharmacy (IJGP)*, 10(1).

[54] Marefati, N., Ghorani, V., Shakeri, F., Boskabady, M., Kianian, F., Rezaee, R., & Boskabady, M. H. (2021). A review of anti-inflammatory, antioxidant, and immunomodulatory effects of *Allium cepa* and its main constituents. *Pharmaceutical Biology*, 59(1), 285–300.

[55] Pushpalatha, R., Selvamuthukumar, S., & Kilimozhi, D. (2017). Comparative insilico docking analysis of curcumin and resveratrol on breast cancer proteins and their synergistic effect on MCF-7 cell line. *Journal of Young Pharmacists*, 9(4), 480.

[56] Farhad, M., Bhuiyan, I. H., Uddin, S. M., Chowdhury, H. M., Huda, N., Khan, M. F., Alam, S., Paul, A., Kabir, M. S. H., & Rahman, M. (2016). Anticancer potential of isolated phytochemicals from *Ocimum sanctum* against breast cancer: in silico molecular docking

approach. *World Journal of Pharmaceutical Research*, *5*(12), 1232–1239.

[57] Sadeghi, M., Moradi, M., Madanchi, H., & Johari, B. (2021). In silico study of garlic (*Allium sativum* L.)-derived compounds molecular interactions with α-glucosidase. *In Silico Pharmacology*, *9*(1), 11.

[58] Balqis, B., Lukiati, B., Amin, M., Arifah, S. N., Atho'illah, M. F., & Widodo, N. (2022). Computational study of garlic compounds as potential anti-cancer agents for the inhibition of CCR5 and CXCR4. *CMU J Nat Sci*, *21*(1), e2022012.

9 Profiling of bioactive compounds in methanolic leaf extract of Artemisia nilagirica (C.B.Cl) Pamp. through GC-MS and In-silico study of some selected compound

Sagorika Panda[1], Kalpita Bhatta[2,a], Himansu Bhusan Samal[3], and Pratikshya Mohanty[3]

[1]Department of Botany, Dhenkanal Mahila Mahavidyalaya, Dhenkanal, Odisha, India

[2]Department of Botany, School of Applied Sciences, Centurion University of Technology and Management, Bhubaneswar, Odisha, India

[3]Department of Pharmaceutics, School of Pharmacy and Life Sciences, Centurion University of Technology and Management, Bhubaneswar, Odisha, India

Abstract: Artemisinin is a compound of current interest in the treatment of vector borne disease such as malaria and dengue. It is a highly effective sesquiterpene lactone endoperoxide against vectors like Plasmodium, Ades and Anopheles. Determination of exact quantity and purity of Artemisinin in *A. nilagirica* is a challenging task, as the compound is present in a very low application. Artemisinin is thermolabile, highly unstable and lacks chromophoric groups. The aim of the investigation was to develop a simple protocol for quantification of Artemisinin in methanolic leaf extract using GC-MS spectroscopy analysis and identification of structurally important bio-active compounds of *A. nilagirica*. The result has reported presence of total 25 compounds in methanol leaf extract of *A. nilagirica*. This is the first report of chromatographic screening of *A. nilagirica* for Artemisinin derivatives.

Keywords: GC-MS analysis, *Artemisia nilagirica*, methanol leaf extract, Bio-active compound, sesquiterpene lactone

1. Introduction

The plant products have been used by human as a traditional medicine due to the efficacy of therapeutic compounds against diverse diseases. *A. nilagirica* (family Asteraceae), is an annual herb found in Westernghats and Nilgiri region of Tamilnadu, having significant pharmacological activities. Essential oil of *A. nilagirica* posses anti-fungal, anti-microbial, insecticidal and larvicidal activities [1]. They are used to reduce angiogenesis, antitumor, anticancer activities and also for the treatment of joint pain, liver problems and loss of appetite [2]. The plant contains significant amount of artemisinin and its derivatives. Artemisinin (MW-282 g/mmol) is an endoperoxide sesqueterpene lactone (SQL) effective against several strains of *Plasmodium*, malaria parasite and also round worm [3]. Artemisinin causes death of

[a]Kalpita.bhatta@cutm.ac.in

DOI: 10.1201/9781003672869-9

parasite by its mode of action through cleavage and reacting with heame and iron oxide, which forms free radicals cause damage susptible proteins [4]. Due to its limited content in plant Artemisia sp, its derivatives were search and analyzed. Several derivatives of Artemisinin are dihydroartemisinin, artemether, artesunate and arteether (Figure 9.1) were discovered and synthesized [5]. These derivatives showed better efficacy, tolerability and oral availability than Artemisinin, as well as minimal side effects. In 2015, YouyouTu was awarded with Nobel Prize in Medicine and Physiology for Novel therapy against Malaria [6]. The active principle of Artemisinin is a lactol endoperoxide, which produces artemether and artesunate which become active become metabolized in the blood into dihydro-artemisinin [7]. However, Artemisinin and its derivatives having limitations like poor availability, short half-life, poor water solubility and long term use may cause toxicity [8]. Recent studies have confirmed the potential use of artemisinin as a COVID-19 treatment. These compounds were also strong effective against SARS-CoV-2, the mode of action is it hinds the initial stages of infection by clogging both peplomer protein and TGF-β-dependent mechanisms. In next step, the drug impinges with various intracellular steps that are important for SARS-CoV-2 multiplication [9]. Research has shown that Artemisia extract and ACTs can inhibit SARS-CoV-2. Additionally, artemisinin and its derivatives have exhibited anti-inflammatory properties, including the suppression of interleukin-6 (IL-6), which is crucial in the

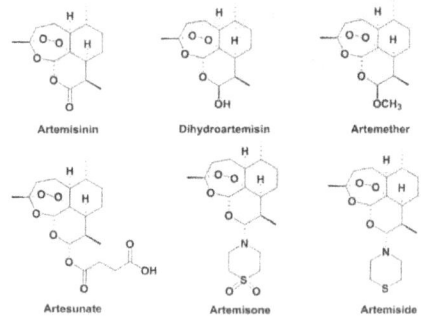

Figure 9.1. Molecular Structure of artemisinin and its derivatives.

Source: Author.

progression of severe COVID-19 [10, 11].

Artemisinin having mild adverse effects, they are more likely to take with another drug mefloquine in place of its sole therapy [12]. The threat of artemisinin hostility also emerging ACT (Artemisinin Combination Therapy) and APT (Artemisinin–enforced Praziquantel Treatment) should be crucial [13].

Various analytical techniques, including X-ray crystallography, spectrophotometry, mass spectroscopy, polyairthmatic analysis and gas chromatography, are employed to determine the accurate chemical composition of compounds. A combined analytical technique called gas chromatography-mass spectroscopy (GC-MS) is used to detect and quantify bioactive substances in plant samples, which are frequently found in trace amounts. For the first time, volatile organic components present in the plant's methanolic leaf extract have been studied in detailed. The results of this investigation will aid in the discovery of compounds that may find immense use in medicine.

2. Method

2.1. Collection and identification

Fresh leaves of *A. nilagirica* were accumulated from Nilgiri region of Tamilnadu. The plant material was recognized and authenticated by Dr. P. K. Swain from the Department of Botany, Utkal University, Bhubaneswar, Odisha, India. The voucher number is BOTU-10570. Fresh plant parts was washed under running tap water, air dries and powdered in electric blender, used for *in vitro* studies.

2.2. Preparation of plant materials

The dry leaf powder (250g) of *A. nilagirica* was eluted against methanol (500 ml) by Soxhlet apparatus for 48 hours. After collection of effluent, the extract was converted to greenish- black semi-solid by rotary evaporator. And stored at 4°C in airtight container for further use.

2.3. Gas chromatography-mass spectroscopy (GC-MS) analysis

The experiment utilized a combined gas chromatograph and mass spectrometer system (Agilent-19091-433HP1, USA), featuring a 7890 A gas chromatograph and a 5675C Inert MSD with Triple-Axis detector.

2.4. Identification of compounds

Bioactive components identification was activated on their retention indices and interpretation of mass spectrum was conducted using data base of Willey and NSIT libraries. This library database consists of 62,000 patterns of known compounds. The unknown bioactive compounds of *A. nilagirica* fraction obtained were composed with the std. mass spectra of known compounds stared in NIST Library (NIST II) and the results obtained have been tabulated.

3. Results

The various secondary metabolites present in the plant is vividly described in Table 9.1. Which has highly therapeutic values.

Screening of methanolic extracts for phytochemical composition of *A. nilagirica*. There are alkaloids, flavonoids, saponins, tannins, phenols, steroids and terpenoids with sesqueterpene lactones (SQL).

3.1. Gas chromatography-mass spectroscopy profiling of methanol leaf extract of A. nilagirica

The GC-MS analysis of the methanolic leaf extract from *A. nilagirica* revealed 13 peaks and several groups of phytochemical components. The highest Retention time (RT) was observed for 3-Eicosyne and Citronellylisobutyrate (41.139). The examination identified a total of 25 compounds in the methanol fraction of *A. nilagirica,* each demonstrating various biological activities. Figure 9.2 displays the chromatogram, while Table 9.1 presents the chemical components, including their retention time (RT), molecular formula, molecular weight and concentration (%) in the MFHAL. The GC-MS analysis of the methanol fraction derived from *A. nilagirica* leaves indicated the presence of multiple bioactive compounds. Those

Table 9.1. Phytocomponents confirmed in the methanol extract of A. nilagirica by GC-MS analysis

Bioactive compound	Mol. formula	Mol. Weight	RT	Peak Area %	Biological activity
Myo-inositol	$C_6H_{12}O_6$	180.16	31.943	39.66	Anti-cancer, anti-convulsant and anti-HIV activity
1,4-dioxan-2-yl-alpha-methylfuranoside	$C_9H_{18}O_6$	222.237	31.943	39.66	Carbohydrate derivatives, gap junction forming compounds
N,N-Dimethylvaleramide	$C_7H_{15}NO$	129.203	32.029	13.40	Analgesic activity
Thiophene, tetrahydro-2-methyl	$C_5H_{10}S$	102.198	32.029	13.40	Anti-inflammatory activity, Insecticidal activity
L-Aspargine	$C_4H_8N_2O_3$	132.119	32.029	13.40	Osmoprotectant
Aminopyrazine	$C_3H_5N_3$	95.105	15.539	10.44	Antimicrobial and anti-tubercular agents
Isoborneol	$C_{10}H_{18}O$	154.253	15.539	10.44	Antibacterial activity
Cyclopentene,1,2,3 trimethyl	C_8H_{14}	110.2	15.539	10.44	Anti-fungal activity
2H-1-Benzopyran-2-one	$C_9H_6O_2$	146.145	24.503	9.06	Component of anticoagulants, antimicrobial, anti-inflammatory and antioxidant activity
N- Acetylnorephedrine	$C_{11}H_{15}NO_2$	193.242	30.911	4.85	Insecticidal and mosquitocidal activity
1,3-propanediamine, N-(2- amino ethyl)	$C_5H_{15}N_3$	117.196	30.911	4.85	Insecticidal and antimalarial activity
L- Alanine, N-(1-oxopentyl)-methyl-1-ester	$C_7H_{15}NO_5$	193.197	30.911	4.85	Anti-cancer activity
Bicyclo (2.2.1)heptan-2-one,1,7,7-trimethyl (artemisinin derivative)	$C_{10}H_{16}O$	152.233	14.731	4.33	Insect repellent and insecticidal activity, Antimalarial activity
Caryophyllene	$C_{15}H_{24}$	204.357	23.84	4.20	Insecticidal activity
Bicyclo(3.1.1) heptane,2,4,6 -trimethyl	$C_{10}H_{18}$	138.249	35.04	3.19	Anti-cancer and anti-oxidant compounds Antimalarial activity
3,7,11,15-Tetramethyl-1-2-hexadecen-1-ol	$C_{20}H_{40}O$	296.531	35.04	3.19	Antioxidant activity, Insecticidal activity
3-Eicosyne	$C_{20}H_{38}$	278.524	41.139	2.43	Palmitic acid
Citronellylisobutyrate	$C_{14}H_{26}O_2$	226.355	41.139	2.43	Insect repellent activity

Bioactive compound	Mol. formula	Mol. Weight	RT	Peak Area %	Biological activity
1,6-Cyclodecadiene,1-methyl-5-methylene-8-(1-methylethyl)	$C_{15}H_{24}$	204.351	25.672	2.41	Mosquito larvicidal and Antimalarial activity
Napthalene,1,2.4a,5,6,8a-hexahydro-4,7-dimethyl-1-(1-methylethyl)	$C_{14}H_{24}$	201.357	25.672	2.41	Mosquito repellent
n-Hexadecanoic acid	$C_{16}H_{32}O_2$	985	38.292	2.30	Lipid derivative
Alpha-caryophyllene	$C_{15}H_{24}O$	220.356	24.855	2.08	Insecticidal and mosquitocidal activity
Tetraacetyl –d-xylonic nitrile	$C_{14}H_{17}NO_9$	343.288	29.526	1.65	Anti-asthematic and anti-inflammatory activity
3,3-Iminosprolamine	$C_{10}H_{25}N_3$	187.33	29.526	1.65	Anti-oxidant activity
3-methyl-3,5-(cyanoethyl) tetrahydro-4-thiopyranone	$C_{12}H_{16}N_2OS$	236.333	29.526	1.65	Insect repellent

Source: Author.

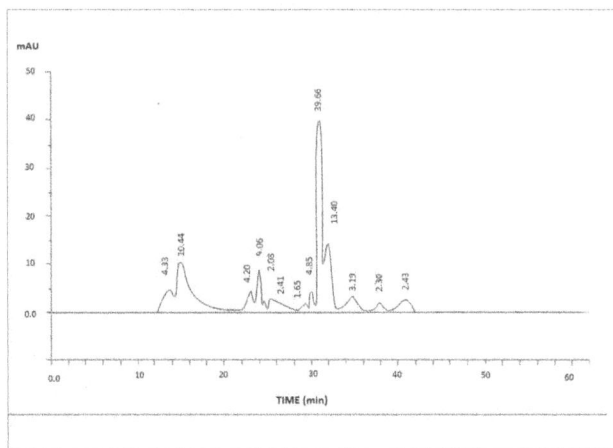

Figure 9.2. Chromatogram of methanolic leaf extract of *A. nilagirica* by GC-MS analysis.

Source: Author.

were Myo-inositol, 1,4-dioxin-2yl-alpha-methylfuranoside, N,N-Dimethylvaleramide, Thiophene tetrahydro-2 methyl, L-Aspargine, Aminopyrazine, Isoborneol, Cyclopentene 1,2,3 trimethyl, 2H-1-Benzopyran-2-one, N-Acetylnorephedrine, 1,3–propanediamine N-(2-aminoethyl), L-Alanine N-(1-oxopentyl) methyl-1-ester, Bicyclo(2.2.1) heptan-2-one 1,7,7–trimethyl, Caryophyllene, Bicyclo(3.1.1) heptanes 2,4,6-trimethyl, 3,7,11,15-Tetramethyl-1-2-hexadecen-1-ol, 3-Eicosyne, Citronellylisobutyrate, 1,6-Cyclodecadiene 1-methyl-5-methylene-8-(1-methylethyl), n- Hexadecanoic acid, Alpha-Caryophyllene,

Tetraacetyl-d-xylonic nitrile, 3,3-Imino-sprolamine, 3-Methyl-3,5(cyanoethyl) tetrahydro-4-thiopyranone.

4. Discussion

The qualitative analysis of compounds can be obtained by GC-MS [14]. The analysis of *A. nilagirica* leaves confirmed the presence of 25 compounds presented in 13 peaks. The identified compounds posses many medicinal and insecticidal properties of Plant. For instance, Bicyclo(2.2.1) heptan-2-one 1,7,7–trimethyl (RT-14.7), Bicyclo(3.1.1) heptanes 2,4,6-trimethyl(35.04) and 1,6-Cyclodecadiene 1-methyl-5-methylene-8-(1-methylethyl) (RT-25.67) having insecticidal and insect repellent activities. Also Citronellylisobutyrate (RT-41.13), Alpha-Caryophyllene (RT-24.85), 3-Methyl-3,5(cyanoethyl)tetrahydro-4-thiopyranone(RT-29.52) having insect repellent activities. Aminopyrazine (RT-15.53), 2H-1-Benzopyran-2-one (RT-24.50), L-Alanine N-(1-oxopentyl) methyl-1-ester (RT-30.91) having anti-microbial and anti-cancer activity. Phytol Diterpene (RT-19.67) is an anti-microbial, anti cancer and anti-leukotriene and water pill [15].

The GC-MS analysis of *Cassia italic* leaf methanol extract revealed the presence of N-Hexadecanoic acid, 3,7,11,15-tetramethyl-2-hexadecane-1-ol and methyl esters, which aligns with our current research findings. Asdadi et al. [16], investigated the chemical composition and properties of *A. herba alba* essential oils in relation to Covid-19, attributing the activity to Terpene Lactone, which is also a primary component in our study. Additionally, compounds such as methyl ester pentanoic acid and n-hexadecanoic acid were identified in the methanol fraction of Hebiscus asper leaves [11],

mirroring the compounds observed in our research. These compounds show alpha-reductase inhibitor, anti-fibrinolytic, hemolytic, anti- microbial and pesticide activity. The quantification of Artemisinin and Dihydro- Artemisinin indried leaves of *A. annua* was carried out by GC-MS analysis [18]. Bioavailability of Artemisinin via Cytochrome P450 inhibition and downstream efficacy was reported by [19]. In the present study, Bicyclo(2.2.1) heptan-2-one1,7,7 –trimethyl is also a derivative of Dihydroartemisinin. *A. nilagirica* also posses anti-tumor effect against MDA-MB-231 breast cancer cells. The anti cancer activity might be due to the presence of bio-active compounds like Myo-inositol, N-(1-oxopentyl)–methyl-1-ester and Bicyclo(3.1.1) heptane 2,4,6 trimethyl [20] explained the Artemisinin and its derivatives in *A. annua* extracts by capillary electrophoresis and Mass spectroscopy. In our study, a artemisinin derivative is found in peak-13. In another species like *A. absinthium* essential oil having wound healing property due to oxygenated monoterpenes and sesqueterpenes [21]. Sesqueterpene Lactones were also recorded in this study. Artemisininin very low concentration by ^{1}H-NMR spectroscopy reported in *A. annua* plant extract [22]. Very similar analytical protocol like HPTLC-ELSD also established for quantification of Artemisinin in *A. annua* extracts. Several bioactivity and efficacy of essential oils extracted from *A. annua* against *Tribolium casteneum* [23], which are quite similar to our insecticidal and mosquitocidal property of *A. nilagirica*. Different chemical composition of scented oils of 8 *Artemisia* species from Iran was studied by Ranjbar et al. [24].

5. Conclusion

The current research demonstrates that the methanolic leaf extract of Artemisia nilagirica contains various secondary metabolites with numerous medicinal properties, particularly insecticidal and repellent activities. Analysis using GC-MS identified 25 bioactive compounds, which contribute to a range of functions including antimicrobial, antioxidant, chemotherapeutic, inhibiting inflammation, anti-HIV, reducing malria, antiplasmodium, insecticidal and repellent effects. The presence of these plant-based compounds is responsible for the therapeutic and pharmaceutical benefits of A. nilagirica. To create novel medications using some of the bioactive substances present in this plant species, more study is required.

6. Acknowledgment

Kalpita Bhatta thanks PG Department of Botany, Utkal University and Department of Pharmaceutics, School of Pharmacy and Life Sciences, Centurion University of Technology and Management, for providing the facilities to carry out the work.

References

[1] Suresh, J., N. M. Mahesh, J. Ahuja, and K. S. Santilna. "Review on Artemisia nilagirica (Clarke) pamp." *Journal of Biologically Active Products from Nature* 1, no. 2 (2011): 97–104.

[2] Loo, C. S. N., Lam, N. S. K., Yu, D., Su, X.-Z., & Lu, F. (2017). Artemisinin and its derivatives in treating protozoan infections beyond malaria. *Pharmacological Research*, *117*, 192–217.

[3] Campbell, J. M., Keith Ellis, R., Nason, P., & Re, E. (2015). Top-pair production and decay at NLO matched with parton showers. *Journal of High Energy Physics*, *2015*(4), 1–33.

[4] Winzeler, Elizabeth A., and Micah J. Manary. "Drug resistance genomics of the antimalarial drug artemisinin." *Genome biology* 15 (2014): 1–12.

[5] Waknine-Grinberg, Judith H., Nicholas Hunt, Annael Bentura-Marciano, James A. McQuillan, Ho-Wai Chan, Wing-Chi Chan, Yechezkel Barenholz, Richard K. Haynes, and Jacob Golenser. "Artemisone effective against murine cerebral malaria." *Malaria journal* 9 (2010): 1–15.

[6] Andersson, Jan, Hans Forssberg, and Juleen R. Zierath. "Avermectin and artemisinin—Revolutionary therapies against parasitic diseases." *The Nobel Assembly at Karolinska Institutet* (2015).

[7] Liu, R., Dong, H.-F., Guo, Y., Zhao, Q.-P., & Jiang, M.-S. (2011). Efficacy of praziquantel and artemisinin derivatives for the treatment and prevention of human schistosomiasis: a systematic review and meta-analysis. *Parasites & Vectors*, *4*, 1–17.

[8] Aderibigbe, B. A. (2017). Design of drug delivery systems containing artemisinin and its derivatives. *Molecules*, *22*(2), 323.

[9] Uckun, Fatih M., Saran Saund, Hitesh Windlass, and Vuong Trieu. "Repurposing anti-malaria phytomedicine artemisinin as a COVID-19 drug." *Frontiers in Pharmacology* 12 (2021): 649532.

[10] Agrawal, P. K., Agrawal, C., & Blunden, G. (2022). RETRACTED: Artemisia Extracts and Artemisinin-Based Antimalarials for COVID-19 Management: Could These Be Effective Antivirals for COVID-19 Treatment?. *Molecules*, *27*(12), 3828.

[11] Fuzimoto, Andrea D. "An overview of the anti-SARS-CoV-2 properties of Artemisia annua, its antiviral action, protein-associated mechanisms, and

repurposing for COVID-19 treatment." *Journal of integrative medicine* 19, no. 5 (2021): 375-388.

[12] Konstat-Korzenny, E., Ascencio-Aragón, J. A., Niezen-Lugo, S., & Vázquez-López, R. (2018). Artemisinin and its synthetic derivatives as a possible therapy for cancer. *Medical Sciences*, 6(1), 19.

[13] Bergquist, R., & Elmorshedy, H. (2018). Artemether and praziquantel: Origin, mode of action, impact, and suggested application for effective control of human schistosomiasis. *Tropical Medicine and Infectious Disease*, 3(4), 125.

[14] Cong, Y., Banta, G. T., Selck, H., Berhanu, D., Valsami-Jones, E., & Forbes, V. E. (2011). Toxic effects and bioaccumulation of nano-, micron-and ionic-Ag in the polychaete, Nereis diversicolor. *Aquatic Toxicology*, 105(3–4), 403–411.

[15] Parveen, S., & Shahzad, A. (2010). TDZ-induced high frequency shoot regeneration in Cassia sophera Linn. via cotyledonary node explants. *Physiology and Molecular Biology of Plants*, 16, 201–206.

[16] Asdadi, A., Hamdouch, A., Gharby, S., & Idrissi Hassani, L. M. (2020). Chemical characterization of essential oil of Artemisia herba-alba asso and his possible potential against covid-19. *Journal of Analytical Sciences and Applied Biotechnology*, 2(2), Anal-Sci.

[17] Olivia, N. U., Goodness, U. C., & Obinna, O. M. (2021). Phytochemical profiling and GC-MS analysis of aqueous methanol fraction of Hibiscus asper leaves. *Future Journal of Pharmaceutical Sciences*, 7, 1–5.

[18] Elfawal, M. A., Towler, M. J., Reich, N. G., Golenbock, D., Weathers, P. J., & Rich, S. M. (2012). Dried whole plant Artemisia annua as an antimalarial therapy. *Plos One*, 7(12), e52746.

[19] Desrosiers, N. A., & Huestis, M. A. (2019). Oral fluid drug testing: Analytical approaches, issues and interpretation of results. *Journal of Analytical Toxicology*, 43(6), 415–443.

[20] Nagy, Á., Munkácsy, G., & Győrffy, B. (2021). Pancancer survival analysis of cancer hallmark genes. *Scientific Reports*, 11(1), 6047.

[21] Benkhaled, A., Boudjelal, A., Napoli, E., Baali, F., & Ruberto, G. (2020). Phytochemical profile, antioxidant activity and wound healing properties of Artemisia absinthium essential oil. *Asian Pacific Journal of Tropical Biomedicine*, 10(11), 496–504.

[22] Castilho, A., Pabst, M., Leonard, R., Veit, C., Altmann, F., Mach, F., Glössl, J., Strasser, R., & Steinkellner, H. (2008). Construction of a functional CMP-sialic acid biosynthesis pathway in Arabidopsis. *Plant Physiology*, 147(1), 331–339.

[23] Deb, A. A., Okechukwu, C. E., & Emara, S. (2019). Live kidney donor: Surgical aspect and outcome. *Urol Nephrol Open Access J*, 7(6), 133–141.

[24] Ranjbar, N., & Zhang, M. (2020). Fiber-reinforced geopolymer composites: A review. *Cement and Concrete Composites*, 107, 103498.

10 Design and development of Psidium Guajava Leaf extract loaded topical emulgel

Sk Hasibur Ali[1], Sonalika Mahapatra[2], Tamosa Mukherjee[3], Mamalisa Sahoo[2], Monali Priyadarshini Mishra[2], Himansu Bhusan Samal[1], and Yashwant Giri[1,a]

[1]School of Pharmacy and Life Sciences, Centurion University of Technology and Management, Odisha, India
[2]School of Paramedic and Allied Health Sciences, Centurion University of Technology and Management, Odisha, India
[3]School of Forensic Sciences, Centurion University of Technology and Management, Odisha, India

Abstract: The objective of the study is to identify the Psidium guajava leaf extract's *in-vitro* antimicrobial property and create a herbal antifungal emulgel for the treatment of skin infections. It has been found that the portion of Psidium guajava exhibits a broad spectrum of antimicrobial action. The leaves, stems, bark and roots of P. guajava have been found to contain more than 20 different chemicals. The TSA utilized the leaves as an antibiotic to heal wounds. Emulgel, a novel method for the topical administration of medication, is a blend of emulsion and gel. Its dual control release characteristic combines enhanced adhesion to the wound and good permeability. Methanol, ethanol and N-hexane were used as menstruum extraction of guajava and an *in-vitro* antibacterial study was performed to optimize the best extract in terms of antimicrobial activity. The methanolic extract was found to have significant antimicrobial activity against tested strains and was selected as an optimized extract. Emulgels were created using methanolic extract. For this purpose, the fusion method followed by high-speed homogenization was used. All the developed emulgels showed good *in-vitro* antimicrobial properties and good physicochemical behavior. The viscosity was found in the 5412 to 6346cPs range and the pH was 5.7 to 6.4, which is compatible with the skin pH and will not irritate the skin while applying. The developed emulgels also showed spreadability in the 31.12 to 38.34 g.cm/s, suggesting that the ability to spread the formulations was good and would spread quickly during application onto the skin. It can be concluded that the developed formulations were compatible and effective against bacterial strains responsible for skin infections and will be effective in treating skin infections.

Keywords: Psidium guajava, Methanolic extract, emulgels, *in-vitro* antibacterial study

1. Introduction

In many regions of the world, traditional medical practice is studied by indigenous cultures as an essential component of their culture and understanding of health. Indian Ayurveda with conventional Chinese medicine is two of the most ancient traditional medicinal practices still in practice today. These methods employ therapies utilizing native medicines derived from natural sources

[a]yashwant.giri@cutm.ac.in

DOI: 10.1201/9781003672869-10

to enhance wellness and standard of life. Different approaches are presently being employed to discover novel bioactive chemicals, as plants have historically served as traditional medicinal sources. Psidium guajava L., commonly referred to as guava, is a tiny tree that is a member of the Myrtle family. Despite guava trees are native to tropical regions ranging from southern Mexico to northern South America, they have been cultivated in many other nations with tropical and subtropical temperatures, enabling worldwide production. Preparations of the leaves have historically been utilized as anti-diarrhea remedies in traditional medicines throughout a number of nations. Moreover, other many uses have been recorded elsewhere on all continents, with the exception of Europe [1]. The application of the medicine is either topical or oral, depending on the condition. Tropical regions are home to the evergreen shrub Psidium guajava. The leaves and fruits of Psidium guajava, which has been neutralized in the Southeast Asia region, were taken from America. There have been several reports of the guava's broad range of therapeutic effects against human illnesses. More than 20 chemicals have been shown to be present in P. guajava's leaves, stems, bark and roots. Guava leaves were used to cure stomach problems and diarrhoea. In TSA, the leaves were applied topically or as a decoction to treat wounds, ulcers and toothaches as an antibiotic. Iron, calcium, phosphorus and vitamin C are also included in guava fruits. The Emulgel formulation is an experimental method of topical medication delivery that combines emulsion and gel. Similar to emulsion and gel, it features a twofold control release [2]. Gel is a novel class of formulation that distributes the prescription drug more quickly than

lotion, cream and ointment. Treatment of skin problems can be achieved by using a medication in an emulgel formulation [3]. Over alternative routes of administration, topical administration of medicinal medicines has several advantages. A traditional emulsion can be changed into an Emulgel through the addition of a gelling agent to the aqueous phase. For example, the usage of transparent gels has grown in both pharmacological and cosmetic preparations under the main category of semisolid preparations [4]. Emulgels' thixotropicity, greasiness, spreadablility, easily removal, soothing, less staining, lengthy shelf life, biocompatibility, opacity and pleasing qualities are some of their advantages for dermatological application [5–17]. This study aims to develop an optimized herbal emulgel containing psidium guajava leaf extract for the topical management of skin infections.

2. Materials and Method

2.1. Collection and identification of plant

Psidium guajava trees in Centurion University's Jatni garden produced the samples of leaves.

2.2. Drying

After the plant leaves were left in the shade for a full 12-day period, they were ground into a smaller size using a home mechanical grinder. The powder was then prepared for extraction using the maceration method.

2.3. Making a crude extract from the leaves

Powder of leaves was mixed with methanol and was taken in glass tight container, Maceration was the extraction process used to extract the crude extract from the leaf.

2.4. Qualitative chemical tests for phyto-constituents

2.4.1. Test for steroids

a. **Salkowski test:** 10 milligrams of extract diluted in one millilitre of chloroform were mixed with one millilitre of concentrated H_2SO_4. The existence of steroids is signified by the reddish-brown tone of the chloroform level and the fluorescent green colour of the acid surface.

b. **Liebermann test:** 2 milligrams of residue had been added and a small amount of acetic anhydride was heated. After the test tube's contents had cooled, 2 millilitres of concentrated sulphuric acid were added via the test tube's side. The emergence of blues colour provided evidence for the existence of sterols.

c. **Liebermann-Burchard test:** 2 ml of conc. H_2SO_4 was poured from the test tube's sides, 10 milligrams of the extract was solubilized in 1 ml of chloroform and 1 ml of acetic anhydride was supplemented. The development of a reddish violet hue at the intersection denotes the presence of steroids.

2.4.2. Test for saponins

Foam test: In a graduated cylinder, a 1 ml extract solution has been diluted to 20 ml with distilled water and swirled for 15 minutes to check for foam development. The presence of stable foam indicates the existence of saponins.

2.4.3. Test for alkaloids

Mayer's test: The extract (two millilitres) was placed in a test tube. Mayer's reagent (0.1 ml) and 0.2 ml diluted HCL were added. Alkaloids are present when a yellowish buff precipitate forms.

2.4.4. Test for glycoside

Million's test: Two millilitres of Million's reagent were applied to the extract. The existence of proteins and amino acids can be detected by the formation of a white precipitate that becomes red upon heating.

2.4.5. Test for tannins

a. **Ferric Chloride test:** To five millliliters of extract solution, one milliliter of a 5% ferric chloride solution was added. The presence of tannins is signified by a greenish-black colour.

b. **Lead Acetate test:** One millilitre of a 10% aqueous lead acetate solution was added to five millilitres of extract. Tannins are present when a yellow-colored precipitate starts to form.

c. **Potassium Dichromate Test:** One millilitre of a 10% aqueous potassium dichromate solution was added to five millilitres of extract. Tannins may be present if a yellowish-brown precipitate forms.

2.4.6. Test for proteins

a. **Biuret test:** A single milliliter of a 10% sodium hydroxide solution was incorporated into the extract and subsequently heated. A drop of 0.7% copper sulphate solution was incorporated to the aforementioned mixture. The emergence of a violet-purple tint indicates the presence of proteins.

b. **Xanthoproteic test:** Five millilitres of strong nitric acid were added to a little amount of test residue in two millilitres of water. Protein presence is shown by the formation of a yellow colour.

2.4.7. Test for amino acids

Ninhydrin test: The extract was boiled after being exposed to the Ninhydrin reagent at a pH of 4–8. Amino acid content is shown by the development of purple colour.

2.4.8. Test for carbohydrates

Molish test: In a test tube, two millilitres of extract solution were treated with a few drops of a 15% ethanolic alpha-naphthol solution. Next, two millilitres of concentrated sulphuric acid were carefully poured along the test tube's walls. Carbohydrates are present when a reddish-violet ring forms at the intersection of two layers.

2.4.9. Test for reducing sugars

Fehling's test: Five millilitres of extract solution and five millilitres of an equal combination of Fehling's solutions A and B were combined and heated. Development of brick red A precipitate suggests that reducing carbohydrates are present.

2.5. Preliminary solubility screening of the extract

It was done with water, coconut oil, tween-20 and span-20, propylene glycol; using a vortex mixer and ultrasonicator. It was found that the extract was soluble in propylene glycol.

2.6. Formulation of emulgels

The emulsion was created in two phases: an oil phase using Span 20 and clove oil and an aqueous phase made of Tween 20 and glycerin combined with distilled water. A separate carbopol mixture was prepared and 2 drops of triethanolamine was added into it and kept it aside overnight. Prepared emulsion was added into the gel with continuous stirring with

an overhead stirrer. Extract was added to propylene glycol and mixed using vortex mixture for about 5 minutes. Prepared extract mixture was added into previously developed emulgel with continuous stirring. Developed herbal emulgel was subjected to further evaluation (Table 10.1).

2.7. Evaluation of the developed herbal emulgels

2.7.1. Physical evaluation

The created emulgels' physical characteristics, including their colour, look and smoothness, were examined both visually and tactilely.

2.7.2. Measurement of pH

The emulgel's pH was determined using a digital pH meter that was calibrated. After dipping the electrode into the formulation, the pH was measured ten minutes later.

2.7.3. Viscosity

Using a rotational viscometer (LABMAN VISCOMETER) with spindle No. L2 at 10 RPM and room temperature, the viscosity of the generated emulgels was determined.

Table 10.1. Formulation table of emulgels

Ingridients	F1	F2	F3
Extract	1.0g	1.0g	1.0g
Clove oil	1.0ml	1.0ml	1.0ml
PG	6.5ml	5.5ml	5.5ml
Tween-20	1.0ml	1.2ml	1.5ml
Triethanolamine	0.1	0.1	0.1
Span-20	1.0ml	1.2ml	1.5ml
Carbapol 940	100mg	150mg	200mg
Distilled water	40ml	40ml	40ml

Source: Author.

2.7.4. Spreadability

The device, which consists of a wooden block supplied by a pulley at one end, was used to measure spreadibility. This approach measured spreadibility based on the emulgel's slide and drag properties. On the stationary slide, an excess of the gel under investigation (about 2g) was added. After that, the gel was encased in a second glass side with a hook and the same stationary slide dimensions. For five minutes, a 1 kg weighted was positioned atop each of the two slides to force out air and create a consistent layer of gel between them. The excess gel was removed by scraping off the edges. Next, a 50 gramme weight was used to tie the top plate. And allowed it to travel over the stationary slide. The time required to get detached from the stationary slide was recorded.

$$S = M*L/T$$

where, spreadability (S) equals
   M-Weight (bound in the upper slide) in the pan
   L-Length, Moved by the glass side
   T = Time (in seconds) required to split

2.7.5. Anti-bacterial activity of emulgels

Using the disc diffusion method, the emulgel extracts of leaves were tested for their ability to inhibit the growth of Gram-negative bacteria, specifically E. coli. When comparing the outcomes, amicacin was utilized as the positive reference standard and water was considered the negative. The bacterial inoculum was prepared using nutrient broth and the screening procedure was carried out using nutrient agar media.

3. Results and Discussion

3.1. Qualitative chemical tests for phyto-constituents

Different Chemical Tests for Phyto-Constituents were done and results are tabulated in Table 10.2.

3.2. Preliminary solubility screening of the extract

It was done with water, coconut oil, tween-20 and span-20, peopylene glycol; using a vortex mixer and ultrasonicator. It was found that the extract was soluble in propylene glycol (Figure 10.1).

Table 10.2. Results for phytochemical screening of extracts

Sl.no	Tests	Results
1	Alkaloid	Present
2	Flavonoid	Present
3	Terpenoid	Present
4	Saponins	Present

Source: Author.

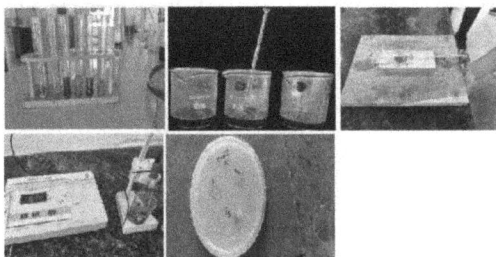

Figure 10.1. Phytochemical screening of extracts, Developed herbal emulgels, Spreadability test of developed emulgels, pH of emulgels, *in-vitro* anti-bacterial activity.

Source: Author.

3.3. Spreadability testing of the emulgels

It was done so as to measure that how easily a product can spread over a surface. So, from these above formulations, the F3 shows best spreadability results. 31.12, 34.22 and 38.34g.cm/s was the obtained results of spreadability for F1, F2 and F3, respectively.

3.4. pH of emulgels

pH of emulgels were found in a compatible range with the skin, which suggested that it will not irritate upon application. The results were6.4, 5.9 and 5.7 for F1, F2 and F3, respectively (Figure 10.1).

3.5. Viscosity

A rotational viscometer (LABMAN VISCOMETER) with spindle No. L2 was used to evaluate the viscosity of the created emulgels at 10 RPM, room temperature and the developed formulations showed satisfactory viscosity range of 5412 to 6346cPs, which is important for the adherence of the formulation after application on to affected area.

3.6. In-vitro anti-bacterial activity

All the three emulgels of Psidium guajava leaves eaxtract were assessed by disc diffusion method. It was found that F3 showed better antifungal activity among all (Figure 10.1).

4. Conclusion

Emulgel of Psidiumguajava Psidium guajava leaves demonstrated strong antibacterial activity against human pathogenic infection; hence, they may be utilized as a possible antimicrobial medication. Lupeol and Taraxerone are measure constituents of Hydro alcoholic extract of Psidium guajava leaves. *In-vitro* studies performed for activities like anti-microbial and it showed significant activity against tested microbs.

References

[1] Kokate C. K. (1994). Practical phramacognosy. *Vallabh Prakashan*, Fourth edition. New Delhi, *123*.

[2] Kaila, A. N. Textbook of industrial Pharmacognosy. CBS Publishers and Distributors. New Delhi, *268*.

[3] Mithal, B. M., & Saha, R. N. (2003). Handbook of cosmetics, first edition. *Vallabh Prakashan*, 1–10, 110–121.

[4] Jain, N. K., Roy, R., Pathan, H. K., Sharma, A., Ghosh, S., & Kumar, S. (2020). Formulation and evaluation of polyherbal aqueous gel from Psidium guajava, Piper betel and Glycerrhizaglabra extract for mouth ulcer treatment. *Research Journal of Pharmacognosy and Phytochemistry*, *12*(3), 145–148.

[5] Rajad, S., Karodi, R., Dhanake, K., Kohakde, S., & Bendre, S. (2023). Formulation and evaluation of polyherbal mouth ulcer gel containing bombax ceiba thorn extract and psidium guajava leaf extract. *Journal of Coastal Life Medicine*, *11*, 845–857.

[6] Ilomuanya, M. O., Ajayi, T., Cardoso-Daodu, I., Akhimien, T., Adeyinka, O., & Aghaizu, C. (2018). Formulation and evaluation of polyherbal antioxidant face cream containing ethanol extracts of Psidium Guajava and Ocimum Gratissimum. *Nigerian Journal of Pharmaceutical Research*, *14*(1), 61–68.

[7] Khan, A., Sohaib, M., Ullah, R., Hussain, I., Niaz, S., Malak, N., de la Fuente, J., Khan, A., Aguilar-Marcelino, L.,

Alanazi, A. D., & Ben Said, M. (2022). Structure-based in silico design and in vitro acaricidal activity assessment of Acacia nilotica and Psidiumguajava extracts against Sarcoptesscabiei var. cuniculi. *Parasitology Research*, *121*(10), 2901–2915.

[8] Patwardhan, B., & Hooper, M. (1992). Ayurveda and future drug development. *Int. J. Alternative Complement Med*, *10*, 9–11.

[9] Hook, D. J., Pack, E. J., Yacobucci, J. J., & Guss, J. (1997). Approaches to automating the dereplication of bioactive natural products. The key step in highthroughput screening of bio active materials from natural sources. *J Biomol Screening*, *2*, 145–152.

[10] Borris, J. (1996). Natural product research; perspectives from a major pharmaceutical company Merck Research laboratories. *J Ethnopharmacol*, *51*, 29.

[11] Chaudhari, S. R., Chavan, M. J., & Gaud, R. S. (2004). Anti-inflammatory and analgesic activity of Capparis zeylanica root extracts. *Indian J Nat Prod*, *20*(1), 36–39.

[12] Weber. (1902). Frederic Albert Constantin. *Bulletin du Muséum d'Histoire Naturelle*, *8*(3), 220–223, f. 1–2.

[13] Cowan, M. M. (1999). Plant products as anti-microbialagents. *Clin Microbiol Rev*, *12*, 564–82.

[14] Rol Prakash, R. P., & Rao, R. (2010). Lpharmaceutical and clinical research. *3*, 126–129.

[15] Saxena, M., Saxena, J., & Khare S. (2012). A brief review on: Therapeutical values of Lantana camara plant. *Int J Pharmacy & Life Sci (IJPLS)*, *3*(3), 1551–1554.

[16] Oyedeji, O. A., Ekundayo, O., & Konig, W. A. (2012). Volatile Lantana camara Linn: A RevieInter. *J of Phytotherapy*, *2*(2), 66–73.

11 Design and development of topical herbal antimicrobial gel loaded with cadamba leaf extract

Kuldip Singh[1], Tamosa Mukherjee[2], Sonali Bhujabal[1], Yashwant Giri[1,a], Gurudutta Pattnaik[1], Amulyaratna Behera[1], and Biswaranjan Mohanty[3]

[1]School of Pharmacy and Life Sciences, Centurion University of Technology and Management, Odisha, India
[2]Institute of Pharmacy and Technology, Salipur, Odisha, India
[3]School of Forensic Sciences, Centurion University of Technology and Management, Odisha, India

Abstract: The objective of the current research work was to detect the *in-vitro* antimicrobial activity of kadamba leaf extract and formulate and evaluate gels loaded with Neolamarckia cadamba leaf extract. Neolamarckia cadamba is believed to contain tonic, bitter, sweet, astringent, febrifugal, anti-inflammatory, antibacterial, digestive, carminative, diuretic, expectorant and antiemetic properties. Additionally, it is used to alleviate inflammation and fever. The flowers are consumed like a vegetable. Although the leaves have a minor disagreeable aroma, their decoction helps in the treatment of amenorrhea, bruises and ulceration. Various research studies indicate that cadamba leaf extract possesses potent antibacterial properties. Initial screening of these extracts revealed the presence of secondary metabolites from plants such as terpenoids, tannins and phenols. The formulated gels showed acceptable physicochemical properties. The viscosity of formulations was found in the range of 6492 to 7241 cPs and the pH was in the range of 6.3 to 7.1, which is considerably compatible with the skin pH and will not irritate the skin upon application. The developed gels also showed spreadability in the range of 39.16 to 46.24 g.cm/s, which indicates the excellent spreadability of emulgels. In-vitro antibacterial activity revealed that the created gel containing the methanolic extract of Neolamarckia cadamba leaves had much higher ZOI against bacterial and fungal species than other manufactured gels. Furthermore, the produced herbal gels were subjected to a short-term stability test and no significant deviation in chemical and physical evaluated parameters was discovered during the investigation, indicating that the products are both physical and chemical stable. Hence, it can be concluded that the formulated herbal gels of *Neolamarckia cadamba* leaves extract were safe and effective for topical application, with potent antimicrobial activity.

Keywords: Neolamarckia cadamba, antimicrobial, methanolic extract, herbal gels

1. Introduction

In many nations, a wide range of plants are employed as medicines. The history of Ayurvedic medicine can be traced back to the ancient times in India and its neighboring countries. As the highest producer of medicinal plants in the globe, India is recognized as the 'Botanical Paradise'. Plants having medicinal activities can treat a variety of illnesses and

[a]yashwant.giri@cutm.ac.in

DOI: 10.1201/9781003672869-11

diseases, including diabetes, cardiovascular disease and liver damage. In the past few years, there has been a rise in interest for alternative medicines and medicinal applications of natural materials, particularly those produced from plants. The interest in medicinal products of plant origin is due to many reasons, most notably, traditional treatments can be inefficient; abusive and/or inappropriate use of synthetic drugs leading to side effects and other challenges. In this context, antimicrobial potential of Neolamarckia cadamba against a wide range of microorganisms was studied. It is a perennial tropical tree found in nations such as Australia, Philippines, Papua New Guinea, Cambodia, Indonesia, Malaysia, Bangladesh, Nepal, Myanmar, Srilanka and India [1–3]. The species is mistakenly known as Anthocephalus chinensis because of its fragrance orange blooms, which grow in thick globe-shaped clusters and are used to make perfumes. It is used to manufacture timber and paper, as well as a decorative plant. It plays an important role in Indian mythology and religion [2]. Due to the great importance of the Cadamba tree to humanity, many religions in India believe that God lives within it. The Sanskrit shloka 'Ayi Jagadamba Mad Amba Kadamba Vana Priyavaasini Haasa Rate' indicates that Goddess Durga favours residing in a Cadamba tree forests. The Cadamba tree reaches approximately 45 meters in height and features a broad umbrella-shaped canopy with a straight cylindrical trunk. It expands swiftly in length but requires 6–8 years to augment in width. The trunk measures 100–160 cm in diameter, while the leaves range from 13 to 32 cm in length. Cadamba is reported for treating a variety of illnesses; the extract made from the bark and leaves is very beneficial [3–5]. Several researchers throughout the world have concentrated their efforts on identifying a variety of phytochemicals and secondary metabolites (saponins, indole and quinoline alkaloids, secoiridoids and triterpenes) of therapeutic significance in the Cadamba. The kadam tree has excellent antibacterial properties, making it ideal for treating wounds and abrasions. People use it extensively to treat conditions like skin infections, inflammation and blood sugar issues [5]. The kadam tree has the following health advantages. The goal of the current experiment was to create a topical antimicrobial gel formulation that had extract from Neolamarckia cadamba leaves for both effective antimicrobial and wound-healing properties [6, 7].

2. Materials and Method

2.1. Plant collection and identification

The leaves were collected from the Neolamarckia cadamba tree situated at the garden of Centurion University, Jatni.

2.2. Drying of leaves

The collected leaves were sun-dried for duration of twelve days, following which they underwent size reduction via a household mechanical grinder. The powder was then prepared for extraction through the use of soxhlet equipment [8–10].

2.3. Preparation of crude extract of the leaves

The dried leaf powder (500 g) was placed in a Soxhlet tube and 99.8% methanol extrapure AR was added to ensure it had a sufficient volume. Processing was done on the extraction procedure. It ran for roughly three to four cycles. Thick semi-solid was produced by concentrating the solvent that was acquired through filtration by evaporating the methanol. The concentrated crude extract was placed into a china plate and dried in desiccators for a week [11, 12].

2.4. Qualitative chemical tests for phyto-constituents

2.4.1. Test for steroidal Triterpenes
Salkowski Test: A small number of drops of conc. H_2SO_4 was incorporated in the chloroform extract and mixed by thoroughly shaking it. The lowest layer of the mixture turns crimson [13–16].

2.4.2. Tests for saponins
Foam Test: When a tiny amount of extract and water are mixed together, foam forms, which lasts for 10 minutes. It confirms the existence of saponins [13–16].

2.4.3. Tests for Alkaloids
Mayer test (Potassium-mercuric-iodidesolution): Alkaloids give cream colour precipitate with this reagent [13–16].

2.4.4. Test for carbohydrates
Molish's Test: Molish's reagent and concentrated sulphuric acid is added slowly around the edges of the test tube containing the extract. Formation of a reddish violet ring confirms the presence of carbohydrate [13–16].

2.4.5. Test for tannins
Lead Acetate Test: The test filtrate was combined with distilled water containing a 10% w/v solution of basic lead acetate. If a precipitate develops; the tannins are in the solution [13–16].

2.4.6. Test for glycosides
Keller-Kiliani Test: 1 ml concentrated H_2SO_4, 4.0 ml of glacial acetic acid and 10 ml aqueous extract were mixed with 1 drop of 2.0% $FeCl_3$. The presence of cardiac steroidal glycosides was evident as a brown ring formed within the various layers [13–16].

2.5. Preliminary solubility screening of the extract

It was done with water, coconut oil, tween-20 and span-20, propylene glycol; using a vortex mixer and ultra sonicator. It was found that the extract was soluble in propylene glycol [14, 15, 17].

2.6. Screening anti-bacterial activity

By employing the cup plate method, the antibacterial activity of the leaf was assessed, using methanol and aqueous extracts. The assessment was done against Gram-negative and positive bacteria (Staphylococcus aureus, Klebsiella pneumonia). To compare the outcomes, amicacin was utilized as the reference standard. To prepare the bacterial inoculum, nutrient broth was employed. Nutrient agar was also utilized for the screening process [18–22].

2.7. Preparation of herbal gels

150 mg of carbopol 940 was added in 12 ml of water and kept aside overnight. Two drops of triethanolamine was added to adjust the pH, Extract was added with ethanol and propylene glycol and vortexed for about 5 minutes. Prepared extract mixture was added into previously developed gel with continuous mixture [18–20]. Developed herbal gel was subjected to evaluation (Table 11.1).

Table 11.1. Formulation table of herbal gels

Ingridients	F1	F2	F3	F4
Extract (gm)	1	1	1	1
Carbopol-940 (mg)	150	150	150	150
Propylene glycol (ml)	4.5	4.0	3.5	0.3
Ethanol (ml)	2.5	3.0	3.5	4.0
Triethanolamine (ml)	0.04	0.04	0.04	0.04
Distilled water (ml)	12	12	12	12

Source: Author.

2.8. Evaluation of the developed herbal gel

2.8.1. Physical evaluation
Developed formulations were tested for colour, odour, greetiness and homogeneity. Washability of the formulation was also checked under running tap water [13–15].

2.8.2. Spreadability test
The purpose of the measurement was to gauge the ease with which a product might cover a surface. Spreadability is defined as the duration of time, measured in seconds that two slides take to break away from the gel when placed in between them and subjected to a given stress. To compress the uniformly thick glass slides, an excess of sample was sandwiched between the two and a specific amount of weight was applied to them [20]. After adding a weight of 70 g, the time that was required for detaching both slides was documented.

The spreabability was computed employing the following equation:

$$S = M.L/T$$

where, M = weight knotted to mobile slide, L = size of the slide, T = time required to get detached.

2.8.3. pH
pH was tested using a electronic pH meter to assess the levels of acidity and basicity in the samples. The instrument calibration was done using pH 4, 7 and 9 buffer solutions; prior of each usage. Ten minutes before the room temperature reading was taken, the electrode was placed into the sample [13].

2.8.4. Viscosity
The viscosity of the compositions was assessed with a Brookfield Viscometer (DV-I PRIME, USA). At 0.3, 0.6, 1.5 rotations per minute, the gels were rotated. The viscosity was calculated by multiplying the given factor in the viscometer manual with the respective dial reading [13–15].

2.8.5. Stability
All the formulations were subjected to stability studies. The formulations were stored for three months at two different temperatures, $4 \pm 2°C$ and $30 \pm 2°C$, 65 RH. Following three months, the formulations' pH and viscosity were measured and compared to the original values [13–15].

3. Results and Discussion

3.1. Antimicrobial activity

The cup plate method was utilized, for evaluating the antibacterial activity of Neolamarckia cadamba leaf extracts and optimized formulations. The assessment was done against Gram-positive and negative bacteria (Staphylococcus aureus and Klebsiella pneumonia). Significant zone of inhibition was seen in all samples against tested strains of Klebsiella pneumonia and Staphylococcus aureus.

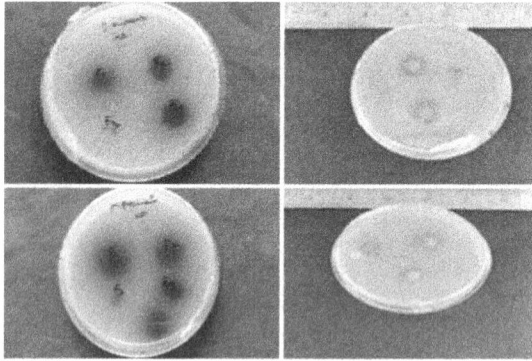

Figure 11.1.   Antibacterial activity against Staphylococcus aurous and Klebsiella pneumoniae.

Source: Author.

3.2. Phytochemical screening of leaf extract

Phytochemical tests; Test for steroidal Triterpenes, alkaloids, saponins, carbohydrates, tannins and glycoside of extracts were done and results were tabulated in Figure 11.1 and Table 11.2.

3.3. Solubility screening of the extract

It was done with water, coconut oil, tween-20 and span-20, peopylene glycol; using a vortex mixer and ultrasonicator. It was found that the extract was soluble in propylene glycol.

3.4. Physical evaluation

The color of all the formulations was found to be brownish and all the organogel showed satisfactory homogeneity and were found odourless. Washability of all the formulation was good (Figure 11.2).

3.5. Spreadability

Spreadability is an important parameter to estimate the behavoiur of the

Table 11.2.   Results for phytochemical screening

Sl.no	Tests	Results
1	Test for steroidal triterpenes	Absent
2	Test for saponins	Present
3	Test for tannins	Present
4	Test for alkaloid	Present
5	Test for glycoside	Absent

Source: Author.

formulation after application on to the skin. The spredability of gels were found in the range of 39.16 g to 46.24 g.cm/s. the obtained result suggested that the formulation having satisfactory spreadability and will spread over the skin easily upon application on the affected area.

3.6. pH

It was done to determine the basicity and acidity level in the samples. Developed formulations showed compatible pH range with the skin pH, which suggested that it will not irritate the skin upon application. The obtained pH ranged from 6.3 to 7.1.

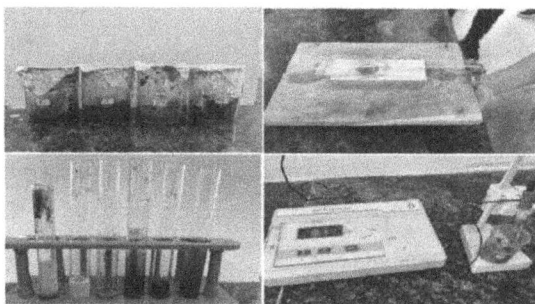

Figure 11.2. Developed herbal gels, Spreadability, Phytochemical screening of leaf extract and pH.

Source: Author.

3.7. Viscosity

The Brookfield Viscometer (DV-I PRIME, USA) was utilized to measure the viscosity of the compositions. The range of the reported viscosity was 8642–9412 cPs. This suggested that the formulation will bind to the skin with good binding ability.

3.8. Stability testing

The prepared leaf extract formulations did not exhibit any physical instability and there was no discernible change in pH before or after the study, showing that the formulations were steady throughout the investigation time frame for stability. The formulas were therefore determined to be stable.

4. Conclusion

Neolamarckia cadamba leaf extractions resulted in extracts with a yield percentage that was as follows. These extracts' initial screening revealed a high concentration of secondary plant metabolites like phenolics, tannins and terpenoids. As a result of their acceptable physical characteristics, the created formulations were skin-compatible. The prepared ointment of Neolamarckia cadamba's methanol-based extract showed substantially greater antibacterial efficacy against both fungi and bacteria in vitro than the created gel. Furthermore, the developed formulations exhibited short-term stability, demonstrating the product's chemical and physical stability. As a result, the developed formulations of Neolamarckia cadamba's methanolic extracts were effective and safe carriers with strong antibacterial properties. The bactericidal activity of the extracts was likewise good. Neolamarckia cadamba is a promising tree, as evidenced by the diverse bioactivities shown by all of its extracts, it may be concluded. Pharmacological and chemical properties of extracts can be used to lead bioassay-guided isolation procedures and data to produce potentially useful pharmaceuticals in the future.

References

[1] Plants, A. T. R. (2010). Neolamarckia cadamba. *Centre for Australian National Biodiversity Research, Canberra.*

[2] Dubey, A., Nayak, S., & Goupale, D. C. (2011). Anthocephalus cadamba: A review. *Pharmacognosy Journal, 2,* 71–76.

[3] Bhandary, M. J., Chandrashekar, K. R., & Kaveriappa, K. M. (1995). Medical ethnobotany of the siddis of Uttara Kannada district, Karnataka, India. *Journal of Ethnopharmacol, 47,* 149–158.

[4] Krisnawati, H., Kallio, M., & Kanninen, M. (2011). Ecology, silviculture and productivity: Anthocephalus cadamba Miq. *CIFOR, Bogor. Indonesia, Center for International Forestry Research (CIFOR),* 11.

[5] Banerji, N. (1977). New saponins from stem bark of Anthocephalus cadamba MIQ. *Indian Journal of Chemistry-B, 15,* 654.

[6] Bhardwaj, S. K., & Laura, J. S. (2008). Antibacterial properties of some plants extract against plant pathogenic bacteria Rathyibacter tritici. *International Journal of Biosciences and Biotechnology Research Asia, 4*(2), 693–698.

[7] Patel, D. A., Darji, V. C., Bariya, A. H., Patel, K. R., & Sonpal, R. N. (2011). Evaluation of antifungal activity of Neolamarckia cadamba (roxb.) bosser leaf and bark extract. *International Research Journal of Pharmacy, 2,* 192–193.

[8] The wealth of India. (1972). A dictionary of Indian raw materials and industrial products.

[9] NISCAIR Press Publishers. (2006). New Delhi, 305–308.

[10] Khare, (2011). *Indian herbal remedies: Rational Western therapy, ayurvedic and other traditional usage, botany.* New York: Springer.

[11] A. Pandey, A. S. Chauhan, D. J. Haware, P. S. Negi. Proximate and mineral composition of Kadamba (Neolamarckia cadamba) fruit and its use in the development of nutraceutical enriched beverage. Journal of Food Science and Technology, 2018, 55:4330–4336.

[12] Pal, I., Majumdar, A., Khaled, K. L., & Datta, S. (2014). Quantitative estimation of some essential minerals in the fruit of Neolamarckia cadamba. *IOSR Journal of Pharmacy and Biological Sciences, 9,* 20–22.

[13] Islam, T., Das, A., Shill, K., Karmakar, P., Islam, S., & Sattar, M. (2015). Evaluation of membrane stabilizing, anthelmintic, antioxidant activity with phytochemical screening of methanolic extract of Neolamarckia cadamba fruits. *Journal of Medicinal Plants Research, 9*(5), 151–158.

[14] AOAC. (1990). *15th Official methods of Analysis.* Association Official Analysis Chemists. Washington D. C. USA, 807–928.

[15] Ranganna, (1986). *Handbook of analysis and quality control for fruit and vegetable products.* Tata McGraw Hill Pub Col. Ltd., New Delhi, India, 1112.

[16] Zhang, D., Hamauzu, Y. (2004). Phenolics, ascorbic acid, carotenoids and antioxidant activity of broccoli and their changes during conventional and microwave cooking. *Food Chemistry, 88,* 503–509.

[17] Danot, M., Nahmias, S., & Zoller, U. (1984). An undergraduate column chromatography experiment. *Chem Educ, 61*(11), 1019.

[18] Gautam, R., & Jachak, S. M. (2009). Recent developments in anti-bacterial natural products. *Medical care and Research Review, 29*(5), 767–820.

[19] Winter, C. A., Risely, E. A., & Nuss, G. W. (1962). Evaluation of anti-microbial activity of some Indian medicinal plants. *Society for Experimental Biology and Medicine, 3,* 554.

[20] Battu G. R., Zeitlin, I. J., & Fray, A. I. (2000). Anti-bacterial activity of

myeloperoxidase inhibitory molecules isolated from resin extracts of Commodore Kua. *British Journal of Pharmacology, 131,* 187.

[21] Barry, A. L. (1976). *The antimicrobial susceptibility test: Principle and Practice, Lea and Fibiger.* Philadelphia, 180.

[22] de Boer, H. J., Kool, A., Broberg, A., Mziray, W. R., Hedberg, I., & Levenfors, J. J. (2005). Anti-fungal and anti-bacterial activity of some herbal remedies from Tanzania. *Journal of Ethnopharmacology, 96*(3), 461–469.

12 Development of in-situ ocular gel containing topiramate to understand its effect in intraocular pressure

Biswajeet Puhan, Aradhana Panigrahi, Guptanjali Sahu, Yashwant Giri, Bikash Ranjan Jena, and Gurudutta Pattnaik[a]

School of Pharmacy and Life Sciences, Centurion University of Technology and Management, Odisha, India

Abstract: The current study aims to create topiramate loaded pH-responsive *in-situ* gels to instill into the eye cavity to evaluate its impact on intraocular pressure (IOP). Different grades of Carbopol, HPMC and Sodium Alginate were selected as polymers for the formulation of ocular *in-situ* gel. The optimization of the formulations was done by evaluating its gelling capacity. The *in-situ* gel compositions were meticulously characterized for different physicochemical and molecular parameters like gelation, FTIR, spectroscopy, viscosity etc. *In-vitro* drug diffusion and microbiological studies were undertaken. The FTIR analysis verifies the compatibility of the polymers with Topiramate. Following a 6-hour *in-vitro* drug diffusion testing, the optimized formulations exhibited drug release in F10 > F5 > F9 > F4. Upon the conclusion of 6 hours, it was noted that formulation F10 exhibited the maximum release among all, whereas F4 had the lowest release among the optimized formulations. No significant alterations in pH and drug content were seen in the accelerated stability studies of the formulations.

Keywords: Topiramate, gelation time, *in-situ* gel, viscosity, ocular deliver

1. Introduction

The current discussion on the design and evaluation of therapeutic goods must specifically address the attributes of the eye and the necessities of ocular delivery systems. The eye may be considered superior to other organs and serves as an ideal structure for evaluating pharmacological efficacy [1]. A primary restriction in ocular administration is achieving and maintaining optimal concentration of the medication at the place of action inside the eye. Diverse ophthalmic formulations, including solutions, ointments, gels and polymeric inserts, have been explored to prolong the ocular residence time of topically administered drugs for the eye [2]. A significant challenge in ocular therapies is achieving an appropriate concentration of medication at the site of action. The formation of tears, brief residence duration and impermeability of the corneal epithelium pose significant challenges, leading to inadequate bioavailability and absorption of ocular dose forms [3]. Innovative methods for ocular drug delivery involve the insertion of solid, lipophilic and hydrophilic devices instilled

[a]gurudutta.pattnaik@cutm.ac.in

DOI: 10.1201/9781003672869-12

into the ophthalmic cul-de-sac, which enhance drug residence time, facilitate prolonged absorption and decrease the frequency of administration for specific drug products [4]. The ocular tear volume is 7 μL, predominantly located in the conjunctival sacs, with 1 μL enveloping the cornea. Commercial eye drops deliver 50 μL. Owing to physical limitations, the eye often eliminates the administered substance within 5–6 minutes, with just a minor fraction (1%–3%) of an eye drop effectively reaching the intraocular tissue [5]. Innovative drug delivery technologies, such as emulsions, ointments, suspensions, aqueous gels, nanoparticles, nanosuspensions and *in situ* gels, have been created to address ocular drug delivery challenges and improve bioavailability [6]. Influenced by temperature, pH and the existence of electrolytes, *in-situ* gels represent a unique formulation that undergoes sol-to-gel transitions [7]. These *in situ* gels are non-Newtonian formulations exhibiting pseudoplastic qualities, characterized by a reduction in viscosity with an increase in shear rate induced by eye movement and blinking. As compared to viscous Newtonian formulations pseudoplasticity demonstrates considerably greater acceptability and presents markedly reduced resistance to blinking [8]. Topiramate is a sulfamate-substituted monosaccharide that inhibits voltage-gated sodium channels, hyperpolarises potassium currents, amplifies postsynaptic GABA receptor activation and suppresses AMPA/kainate receptors. It penetrates the blood brain barrier and get swiftly absorbed following oral administration. It is primarily eliminated with a duration of action of 21 hours via urine [9, 10]. In pediatric patients, it was first authorized in July 1999 as an additional

therapy for those aged two years and older experiencing partial onset seizures. It subsequently received approval for seizures related to Lennox-Gastaut syndrome, generalized tonic-clonic seizures and a primary treatment for partial-onset or main generalized epilepsy. Acute myopia and angle-closure glaucoma are two detrimental consequences linked to topiramate. The primary mechanism of acute myopia and acute angle-closure glaucoma is ciliochoroidal effusion. This causes ciliary body edema, resulting in the relaxation of zonular fibres, lens thickness and anterior displacement of the lens-iris complex. The iris protruding anteriorly obstructs the ocular drainage, inhibiting the outflow of aqueous humor. This finally results in secondary angle-closure glaucoma and myopia. Ciliochoroidal effusion induced by sulphonamides is an idiosyncratic reaction inside the uveal tissue and is independent of dosage [11]. The hapten hypothesis asserts that reactive drug metabolites attach to proteins, resulting in modified proteins identified as foreign entities, triggering immunological responses [11]. A patient must receive a sensitizing dosage before initiating the immunological response with the second dose. The probability of experiencing an adverse reaction to a sulfonamide is 3% [12–15]. In addition to topiramate, several sulfonamides have been documented to induce a comparable clinical condition, such as acetazolamide [16], sulfasalazine [17], hydrochlorothiazide [17] and indapamide [11, 18]. All ocular manifestations are reversible if identified promptly and the medication is discontinued. The study aimed to administer topiramate via the ocular route and assess the enhancement in intraocular pressure (IOP).

2. Material and Methods

Topiramate was procured from Yarrow Chemicals in Mumbai, India. HPMC (K4M), Carbopol (940/934), Potassium Dihydrogen Phosphate, Sodium Chloride, Sodium Bicarbonate, Sodium Hydroxide Pellets, Calcium Chloride and Potassium Chloridewere acquired from Central Drug House, Pvt. Ltd. throughout the study, double-distilled water was utilized.

2.1. Preparation of pH-triggered in-situ *ocular gel*

The required amount of distilled water was taken and to it HPMC K15 was added and stirred until no HPMC lumps were evident with a magnetic stirrer. Carbopol 940/934 was dispersed over the mixture and permitted to sit overnight. A drug solution was formulated and incorporated into the polymeric solution with constant agitation. All created compositions underwent evaluation studies (Table 12.1).

2.2. *Optimization of developed* in-situ *gels*

The gelling capability was assessed by introducing a drop of the formulation into a watch glass containing 2 ml of

Table 12.1. Composition of prepared formulation

Formulation	Drug (mg)	Carbopol-940 (%)	Carbopol-934(%)	HPMC K4M (%)	Distilled Water
F1	50	0.6		0.3	20ml
F2	50	0.6		0.6	20ml
F3	50	0.6		0.9	20ml
F4	50	0.6		1.2	20ml
F5	50	0.6		1.5	20ml
F6	50		0.6	0.3	20ml
F7	50		0.6	0.6	20ml
F8	50		0.6	0.9	20ml
F9	50		0.6	1.2	20ml
F10	50		0.6	1.5	20ml
F11	50		0.4	0.4	20ml
F12	50		0.5	0.4	20ml
F13	50		0.6	0.4	20ml
F14	50		0.7	0.4	20ml
F15	50		0.8	0.4	20ml
F16	50	0.4		0.4	20ml
F17	50	0.5		0.4	20ml
F18	50	0.6		0.4	20ml
F19	50	0.7		0.4	20ml
F20	50	0.8		0.4	20ml

Source: Author.

freshly generated simulated tear fluid, with visual observation of gel formation and concurrent recording of gelling time. *In-situ* gel compositions were optimized based on their gelling capability. Among all the generated *in-situ* gel formulations, four distinct formulations (F4, F5, F9 and F10) exhibiting superior gelling capacity were optimized and subsequently evaluated [19].

2.3. Evaluation of optimized in-situ *gels*

2.3.1. Physical appearance and clarity
The optimized formulations of topiramate ocular *in-situ* gels were evaluated for overall appearance and the presence of suspended particulate matter visually. The formula's clarity was assessed against white and black backgrounds [20].

2.3.2. pH
A digital pH meter was used to measure the topiramate ocular *in-situ* gels' pH right away as they were manufactured. To minimize eye irritation and enhance patient compatibility and tolerance, the gel formulation's pH range should be close to the ocular pH [21].

2.3.3. Viscosity
Using a rotational viscometer the viscosity of the *in-situ* gel was analyzed. Into a beaker the formulated mixture was transferred and the spindle was immersed. The angular velocity of spindle No. 4 was gradually raised from 1 to 10 RPM [22].

2.3.4. Microstructure
The microstructure of the developed *in-situ* gels was evaluated using a bright-field microscope at 40x. On a glass side a small quantity of the gel was applied, covered with a cover slip and examined to analyze the microstructure of the gel.

2.3.5. FTIR
To analyze tiny particles and compounds Fourier transform infrared spectroscopy is frequently employed. This method provides significant insights into the three-dimensional structural data acquired by X-ray diffraction. *In-situ* gels were examined in the 4,000–4,500 cm^{-1} range using an attenuated total reflection infrared (ATR-IR) spectrometer (AlphaE ATR-FTIR, Bruker, USA). This tool is used to examine the chemical interactions between the *in-situ* gels [23].

2.3.6. DSC
The thermal characteristics of the drug and excipients, both separately and together, were examined using the DSC-60 Shimadzu and TA-60 WS collection program. We then ascertained the medication's and the polymer's endothermic and exothermic characteristics [23].

2.3.7. Texture analysis
The constructed nanoemulgels' texture profile was analyzed using a CTX texture analyzer (Ametek Brookfield, US) with a 5.0 g starting trigger force. At a preprogrammed pace of 1 mm/s, a 35 mm glass prob was dipped 10 mm twice in each formulation with an interval of 10 seconds. 30 g of each nanoemulgel was used for this study. The software (Texture Pro) measured firmness, adhesive force, adhesiveness, cohesiveness and springiness [24].

2.3.8. In-vitro drug release
Employing a modified dissolution device, the topiramate release from the prepared formulations was investigated *in vitro*.

As a diffusion medium, the newly prepared buffer (pH 7.4) was used. A specially constructed glass cylinder with an inner diameter of 3.4 cm was attached to one end of a semi-permeable membrane that had been submerged in the diffusion medium for the previous night. The cylinder was open at both ends. The donor chamber was a glass cylinder into which one milliliter of the formulation was carefully pipetted. A stirring rate of 50 RPM was used to maintain the acceptor chamber at 37 ± 2°C. At predetermined intervals, 5 ml of the aliquot was removed and replaced with an equivalent volume of the novel diffusion medium. The sample was analyzed using a UV spectrophotometer at 272 nm.

3. Result and Discussions

3.1. Clarity and visual appearance

All formulations exhibited satisfactory clarity. All the formulations were clear and transparent. No foreign particle was found (Table 12.2).

3.2. pH

Generally, the pH for *in-situ* gel should be near the p^H of the eye to avoid any irritation. All the formulations showed a pH that was compatible with the eye. The pH values of the formulations were ranged from 6.22 to 7.38. Which is compatible with the eye pH.

3.3. Viscosity measurement

The viscosity of the prepared formulations was measured and the determined viscosities were suggested that the formulations were in solution form and showed no changes in viscosity when applied shear stress. The recorded viscosities were found in the range of 86–94 cps. Among all, the F5 showed minimum, while F9 showed the maximum viscosity.

3.4. FTIR

FTIR is used to characterize and analyze different materials and samples. The functional groups of all four formulations were ranges from 4000-600 cm⁻¹. The stretching of the C=O, or carbonyl, group caused some distinctive peaks of the pure drug and formulations to be found at about 1645.48 cm⁻¹, the stretching of the C=N group caused 2391.3 cm⁻¹ and the stretching of the N-H group, or amino group, caused 3376.55 cm⁻¹ (Figure 12.1).

Table 12.2. Physiochemical properties of prepared *in-situ* gel

Formu-lation	Gelling time	Clarity	pH	Viscosity cPs
F4	30.46 min	Clear	6.81	92
F5	20.14 min	Clear	7.00	86
F9	20 min	Clear	7.38	94
F10	30 min	Clear	6.22	92

Source: Author.

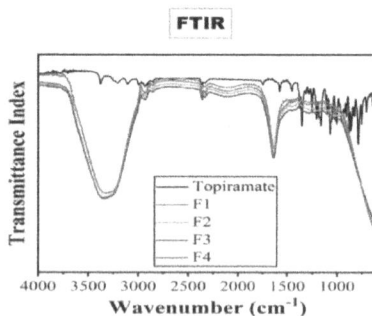

Figure 12.1. FTIR study of pure drug and optimized formulation.

Source: Author.

3.5. DSC

The purpose of the study was to evaluate the change in a sample's physical properties as well as the rise in temperature over time and to verify that the medication changed from a crystalline to an amorphous state when introduced to the formulation. As the melting point of the drug, that is, topiramate, is 125°C, so from the above thermogram, an endothermic peak was observed at around 122.41°C due to the endothermic reaction of the drug caused by melting. No distinct peaks at the melting point region were visible in any of the drug-loaded formulations, indicating that the drug changed from a crystalline to an amorphous state upon addition to the formulation (Figure 12.2).

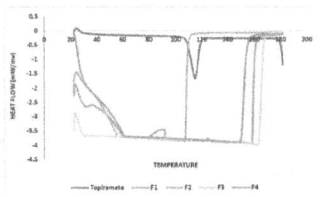

Figure 12.2. DSC study of pure drug and optimized formulation.

Source: Author.

3.6. Texture study

Texture data of the optimized formulations suggested that the formulations will attach in the cul-de-sac cavity with good binding affinity. The detailed recorded values are shown in Figure 12.3 and Table 12.3.

3.8. Microstructure

The microstructure of the developed *in-situ* gels was evaluated using a bright-field microscope at 40x. No gritty particles were found. A homogenous matrix of the polymer was observed.

3.9. In-vitro drug release

Following a 6-hour *in vitro* drug diffusion analysis, the optimized formulations exhibited drug release in the F10 > F5 > F9 > F4 sequence. After 6hours, it was observed the formulation F10 was found to have the maximum release, whereas F4 was found to have the lowest release (Figure 12.4).

4. Conclusion

In-situ gels represent a promising ophthalmic drug delivery method capable of

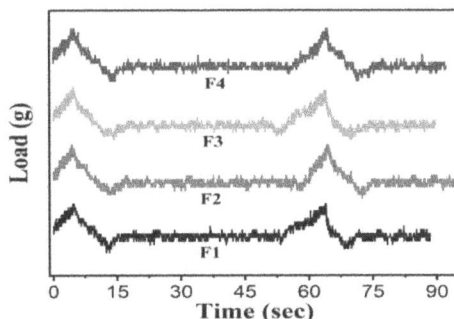

Figure 12.3. Texture study of formulation.

Source: Author.

Table 12.3. Texture study result

Formulations	F5	F9	F10	F11
Hardness Cycle (g)	9	9	9	9
Cohesiveness (mg)	0.96	0.97	0.99	0.2
Springiness (mm)	5.21	4.56	0.78	0.65
Springiness Index	1	0.91	0.96	4.43
Gumminess (g)	10	9	9	0.89
Chewiness (mg)	0.5	0.2	0.3	2.3
Chewiness Index (g)	10	8	9	8

Source: Author.

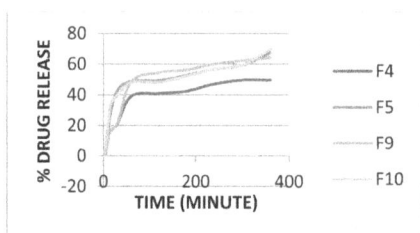

Figure 12.4. Drug release of optimized formulations.

Source: Author.

administering diverse pharmaceuticals to the ocular surface. This review examines the principles, formulation strategies, characterization techniques and applications of *in situ* gels. The current research indicated that the formulated compounds yielded excellent outcomes and this study will be expanded to evaluate the *in-vivo* effects of the medicine on intraocular pressure (IOP). The formulations were assessed for viscosity, pH, gelling capability, rheological properties, *in-vitro* permeation and histological analysis. According to the pH of the prepared gels, the formulations are found to be non-irritant and safer for topical use.

References

[1] Gangadia, B., Modi, D., Patel, G., Bhimani, B., & Patel, U. (2014). Formulation and evaluation of thermo sensitive in-situ gel for local action: A review. *Int J Pharm Res Biosci, 3,* 217–228.

[2] Gratieri, T., Gelfuso, G. M., de Freitas, O. D., Rocha, E. M., & Lopez, R. F. (2011). Enhancing and sustaining the topical ocular delivery of fluconazole using chitosan solution and poloxamer/chitosan in situ forming Gel. *Eur J Pharm Biopharm, 79*(2), 320–327.

[3] Kumar, D., Jain, N., Gulati, N., & Nagaich, U. (2013). Nanoparticles laden in situ gelling system for ocular drug targeting. *J Adv Pharm Technol Res, 4,* 9–17.

[4] Banker, G. S., & Rhodes, T. C. (2002). *Modern Pharmaceutics,* 4th edition. New York, Basel: Marcel Dekker, Inc.

[5] Davies, N. M. (2000). Biopharmaceutical considerations in topical ocular drug delivery. *Clin Exp Pharmacol Physiol, 27,* 558–562.

[6] Patel, A., Cholkar, K., Agrahari, V., Mitra, A. K. (2013). Ocular drug delivery systems: An overview. *World J Pharmacol,2,* 47–64.

[7] Baranowski, P., Karolewiczm, B., Gajda, M., & Pluta, J. (2014). Ophthalmic drug dosage forms: Characterization and research methods. *Sci World J,* 1–14.

[8] Almeida, H., Amaral, M. H., Lobão, P., & Lobo, J. M. (2014). In situ gelling systems: A strategy to improve the bioavailability of ophthalmic

pharmaceutical formulations. *Drug Discov Today*, *19*, 400–412.

[9] Fraunfelder, F. W., Fraunfelder, F. T., & Keates, E. U. (2004). Topiramate-associated acute, bilateral, secondary angle-closure glaucoma. *Ophthalmol*, *111*, 109–111.

[10] Abtahi, M. A., Abtahi, S. H., Fazel, F., Roomizadeh, P., Etemadifar, M., Jenab, K., & Akbari, M. (2012). Topiramate and the vision: A systematic review. *Clin Ophthalmol*, *6*, 117–131.

[11] Senthil, S., Garudadri, C., Rao, H. B., & Maheshwari, R. (2010). Bilateral simultaneous acute angle closure caused by sulphonamide derivatives: A case series. *Indian J Ophthalmol*, *58*, 248–252.

[12] Panday, V. A., & Rhee, D. J. (2007). Review of sulfonamide-induced acute myopia and acute bilateral angle-closure glaucoma. *Compr Ophthalmol Update. 8*, 271–276.

[13] Guier, C. P. (2007). Elevated intraocular pressure and myopic shift linked to topiramate use. *Optom Vis Sci*, *84*, 1070–1073.

[14] Sen, H. A., O'Halloran, H. S., & Lee, W. B. (2001). Case reports and small case series: Topiramate-induced acute myopia and retinal striae. *Arch Ophthalmol*, *119*, 775–777.

[15] Kumar, M., Kesarwani, S., Rao, A., & Garnaik, A. (2012). Macular folds: An unusual association in topiramate toxicity. *Clin Exp Optom*, *95*, 449–452.

[16] Malagola, R., Arrico, L., Giannotti, R., & Pattavina, L. (2013). Acetazolamide-induced cilio choroidal effusion after cataract surgery: Unusual posterior involvement. *Drug Des Devel Ther*, *7*, 33–36.

[17] Lee, G. C., Tam, C. P., Danesh-Meyer, H. V., Myers, J. S., & Katz, L. J.

(2007). Bilateral angle closure glaucoma induced by sulphonamide-derived medications. *Clin Experiment Ophthalmol*, *35*(1), 55–58.

[18] Blain, P., Paques, M., Massin, P., Erginay, A., Santiago, P., & Gaudric, A. (2000). Acute transient myopia induced by indapamide. *Am J Ophthalmol*, *129*, 538–540.

[19] Gill, H. K., & Bhagat, S. (2015). A novel in-situ for sustained ophthalmic delivery of ciprofloxacin hydrochloride and diclofenac sodium: Design and characterization. *World J Pharm Pharm Sci*, *4*, 1347–1356.

[20] Raj, A., Paresh, M., & Rishikesh, C. (2015). Formulation and evaluation of nasal in-situ gel of Bupropion hydrochloride. *WJPPS*, *4*, 595–614.

[21] Panchal, V. S., Chilkwar, R. N., Sabojil, J. K., Patil, S. M., Nanjwade, B. K. (2015). Development and evaluation ophthalmic in-situ gel of Betaxolol HCl by temperature dependent method for treatment of glaucoma. *J Pharm Sci Pharmacol*, *2*, 1–5.

[22] Dasankoppa, F. S., & Swamy, N. G. (2013). Design, development and evaluation of cationic guar and hydroxypropyl guar based in situ gels for ophthalmic drug delivery. *Indian Drugs*, *50*, 30–41.

[23] Behera, B., Sagiri, S. S., Pal, K., & Srivastava, A. (2013). Modulating the physical properties of sunflower oil and sorbitanmonopalmitate-based organogels. *J Appl Polym Sci*, *127*(6), 4910–4917.

[24] Fujimoto, K., et al. (2016). Hardness, cohesiveness, and adhesiveness of oral moisturizers and denture adhesives: Selection criteria for denture wearers. *Dentistry Journal*, *4*(4), 34.

13 Formulation and evaluation of Amphotericin-B loaded topical antifungal organogel

Rozalika Mohanty, Himansu Bhusan Samal[a], Aradhana Panigrahi, Yashwant Giri, and Gurudutta Pattnaik

School of Pharmacy and Life Sciences, Centurion University of Technology and Management, Odisha, India

Abstract: The most prevalent dermatological conditions are fungal infections of the skin. Several systemic and topical treatment strategies are already available but have some limitations. This study aimed to develop and assess an organogel system that could carry Amphotericin-B (AmB), which has known antifungal activities. Excipients were selected based on the literature and preliminary studies. The chosen components were Span-60, Isopropyl Myristate and Propylene Glycol. Six different formulations loaded with Amphotericin-B were developed. Developed organogels were evaluated for various physicochemical and molecular properties. In-vitro drug release and antifungal activity studies were performed. All the organogels showed good rheological behavior and the FTIR study suggested the compatibility between the drug and excipients. At the end of 6 hours, 52.24% drug release from the optimized formulation was observed. Optimized organogels showed good in-vitro antifungal activity against fungal strains responsible for topical fungal infections. Hence, the optimized organogels will be helpful in the management of topical fungal infections.

Keywords: Amphotericin-B, organogels, topical fungal infections, antifungal activity

1. Introduction

Micro and macro-organisms are constantly present in the air, water and food that humans consume and occupy different parts of the human body [1]. Topical fungal diseases are mainly affecting the human body. Fungal diseases are fatal for immune suppressant patients [2]. Numerous fungi can cause primary or secondary infections in the skin and soft tissue. Developing antifungal resistance, especially in Aspergillus, Candida and, to some extent, dermatophyte transmission, may have significant clinical implications even though these infections are on the rise [3]. For fungus to infect people, they must fulfil four criteria: the ability to grow at body temperature, the capacity to bypass or penetrate surface barriers, the capability to lyse and assimilate tissue and resistance to the immune system, even at elevated body temperatures. Fungi navigate host barriers employing morphogenesis, the process by which small, round, detachable and long connected cells fuse. Ironically, by weakening immune systems, modern medical advancements have exposed millions of people to fungal infections for the first time [4]. Various drug compounds are administered through different routes [2]. A

[a]hbsamal@gmail.com

DOI: 10.1201/9781003672869-13

specific route of administration may be used depending on the type, location, urgency and severity of the disease. Routes and drug delivery methods have significant benefits with different drawbacks. The drug Amphotericin B has a non-crystalline structure and is classified under the polyethylene category of antibiotics [5]. Amphotericin B, owing to its complexity and extensive range of efficacy, is used in the treatment of several illnesses, including mucormycosis, extracutaneous sporotrichosis, aspergillosis, blastomycosis, cryptococcosis, histoplasmosis, paracoccidioidomycosis. It can also treat some cases of hyalohyphomycosis and phaeohyphomycosis [6, 7]. A general definition of dermatophytosis is an infection of the nails, hair, or glabrous skin. The keratinophilic fungi Trichophyton spp., Microsporum spp. and Epidermophyton, isolated from symptomatic and asymptomatic people, are the source of infections. Although dermatophytosis is usually not fatal, these infections are among the most prevalent globally and have risen frequently in recent years [8]. Due to several adverse effects of oral antifungals, like hepatotoxicity, drug-drug interaction and high treatment cost, patient compatibility with oral antifungals is less. To overcome these associated adverse effects with oral antifungals and to provide an alternative option in case of drug tolerance of microorganisms, this work relies on the development and characterization of an Amphotericin-B loaded topical antifungal organogel with enhanced permeability and antifungal activity. Organogel is a semisolid transdermal pharmaceutical formulation characterized by the physical interaction between the gelator compounds forms a 3-dimensional structure [9–13].

2. Materials and Methods

2.1. Materials

Amphotericin B was acquired from Yarrow Chemicals Pvt. Ltd. Propylene glycol was acquired from Nice Chemicals (p) LTD. Span-60 and Isopropyl Myristate was procured from Sisco Research Laboratory Pvt. Ltd. Other used chemicals were obtained from local suppliers. Double distilled water was used throughout the research work.

2.2. Solubility of Amphotericin B

Amphotericin B is insoluble in water, ether, dehydrated alcohol, benzene and toluene. It dissolves readily in dimethyl sulfoxide, the solution of dimethyl formamide, propylene glycol and very little methyl alcohol. Amphotericin B is insoluble with water at pH 6–7 but dissolves about 0.1 mg/ml in other pHs.

2.3. Calibration curve

The 10 mg of AmB was precisely weighed in a 10 ml volumetric flask. Amphotericin B was dissolved in a sufficient amount of Propylene glycol, shaken frequently and the volume was brought up with water to 10 ml to create a stock solution with 10000 µg/ml. A 1 ml solution was withdrawn (1000 µg/ml) and diluted up to 10 ml using Propylene glycol blended phosphate buffer. Different working solutions were prepared from this solution, orbances were taken and a standard curve was developed.

2.4. Preparation and optimization of organogel

Organogels based on sorbitan monostearate (span 60) were made using a straightforward technique that called for

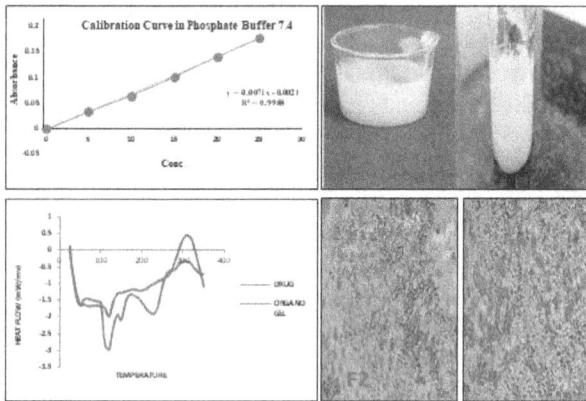

Figure 13.1. Calibration curve of Amphotericin-B, Visual evaluation of formulations, differential scanning calorimetry study of optimized formulation, Microscopic view of optimized organogels.

Source: Author.

dissolving the necessary amount of span 60 in isopropyl myristate in a beaker. A homogenous solution was produced after heating this mixture in a water bath to 60°C. Using the same process, six organogels were prepared. A fixed 0.2% w/w solution of Amphotericin B was dissolved and prepared in six different beakers with 1 ml of propylene glycol, which was then incorporated into the formulated organogels by continuous stirring at ambient temperature. The solution was further cooled and drug-loaded organogel was prepared; a drug diffusion study was performed with all the formulations and F2 and F4 were selected as optimized formulations based on % drug release and further evaluated for different parameters (Table 13.1).

2.5. Evaluation of developed AmB organogels

2.5.1. Physical evaluation
The prepared organogel samples were observed visually for color, homogeneity and consistency. The odor evaluation was done by smelling a small amount of formulations.

2.5.2. Viscosity
A labman rotational viscometer was used to determine the viscosity of each organoge 1.50 gm samples were added in a beaker, a program was constructed and the viscosity was assessed using spindle L2 at room temperature with an angular velocity varying from 5 to 100 RPM.

Table 13.1. Composition of developed formulations

Components	F1	F2	F3	F4	F5	F6
Amphotericin-B	0.2%	0.2%	0.2%	0.2%	0.2%	0.2%
Span-60	20	24	28	32	36	40
Propylene Glycol (ml)	1	1	1	1	1	1
Isopropyl Myristate	Up to 100	Up to 100	Up to 100	Up to 100	Up to100	Up to100

Source: Author.

2.5.3. pH

pH of the prepared formulation was measured at room temperature. By setting an electronic pH probe into direct contact with the samples, the pH was recorded.

2.5.4. Microscopic study

A small amount of prepared organogel was applied on a glass slide and covered with a coverslip. An upright bright-field compound microscope was employed to observe the organogel's microstructure organization.

2.5.5. Spreadability

Prepared formulations were assessed for spreadability via a manual spreadability equipment. The apparatus had two glass plates; 1 g of formulations was placed between glass plates and 100 grams of weight was tied with the upper slide and allowed to pull the plate through a pulley. Spreadability was measured based on the time required to travel the top slide over the fixed lower slide.

2.5.6. FTIR

Using an FTIR spectrometer, the formulations and the pure drug's FTIR spectra were captured. Wave numbers express the signals resulting from different intramolecular stretching and bending vibrations. A comparison was made between the drug's obtained spectrum and the final formulation.

2.5.7. DSC

Differential scanning calorimetry was done to check the transformation of drug molecules from crystalline to amorphous state. Each sample was divided into six milligrams and added to a separate aluminum pan. The temperature ranged from 25°C to 350°C, with a heating scan rate set at 20°C/min. With STARe 15.00 software, data were evaluated.

2.5.8. Stability study

The optimized organogels underwent a stability test. The gels were placed inside a collapsible tube and kept at 25±2°C and 40±2°C with 75% relative humidity (RH) for three months, away from light. The collected samples were examined for visual characteristics, pH level and % of drug concentration.

3. Result and Discussion

3.1. Physical evaluation

The color of all the formulations was found to be canary yellow and all the organogels showed satisfactory homogeneity and were found odorless (Figure 13.1).

3.2. pH

The pH of the optimized formulations F2 and F4 was 6.2 and 6.6, respectively. This suggested that the formulations would not irritate the skin when applied because the observed pH was near the skin pH.

3.3. Viscosity

The viscosity of the prepared formulation for F2 and F4 was found to be 6782cps and 7496cps, respectively. With the increase in polymer concentration, the viscosity of the formulation increased adversely.

3.4. FTIR

The FTIR spectrum was obtained in the 500 to 4000 cm^{-1} region and the sample was compared with the pure drug amphotericin-B. Furthermore, it was found that all the distinctive peaks of amphotericin-B were available in all the formulations. This confirmed no interaction between the drug and excipients (Figure 13.2).

3.5. DSC

Thermal scans were carried out from 30 to 300°C at a heating rate of 10°C/min under dry nitrogen purge (80 mL/min). No sharp peak was observed in any formulations compared to the pure drug. That suggested the transformation of drug molecules from crystalline to amorphous state (Figure 13.1).

3.6. In-vitro antifungal study

The antifungal efficacy of the optimized formulation showed that the developed formulation was sensitive to candida albicans and showed a significant zone of inhibition.

3.7. Microscopic study

The microstructure of the organogel is shown in Figure 13.1. It was visualized that F2 shows a globular particle where the concentration and the globular size increased; thus, F4 shows a larger globular size.

3.8. Stability

The optimized organogels were placed inside a collapsible tube and kept at fixed conditions. The collected samples were examined for visual characteristics, pH level and percentage of drug content. When checked periodically, no significant change was found in the parameters.

4. Conclusion

Long-term therapy for a topical fungal infection necessitates consistent adherence to the recommended course of action. The current study aimed to create

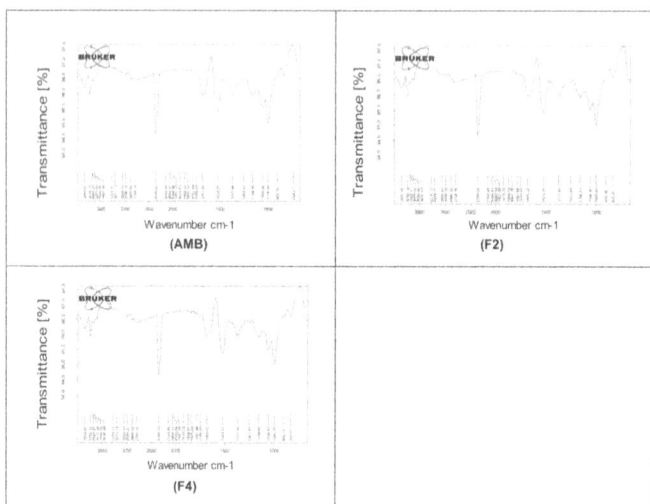

Figure 13.2. FTIR of pure drug Amphotericin B and optimized organogels.

Source: Author.

a novel formulation with a novel combination that could function as a permeation enhancer.

The current study set out to create and assess an antifungal drug delivery system for treating topical fungal infections using an organogel loaded with Amphotericin-B. Amphotericin B was selected as the model drug and formulations containing the keratolytic agent salicylic acid and permeation enhancer span 60 were made. These formulations' pH, adhesion, non-volatile content, drug content, drug diffusion and anti-microbial investigations were then assessed. It was determined from the aforementioned studies that there was compatibility between the drug and the excipients used in the formulations. Every formulation exhibited good penetration and adhesion.

References

[1] Zhao, L., Huang, S., Wang, Z., & Lu, G. (2024). Topical drug delivery strategies for enhancing drug effectiveness by skin barriers, drug delivery systems and individualized dosing. *Frontiers in Pharmacology, 14*, 1333986.

[2] Vallabhaneni, S., Mody, R. K., Walker, T., & Chiller, T. (2016). The global burden of fungal diseases. *Infectious Disease Clinics, 30*(1), 1–11.

[3] Gunaydina, S. D., Arikan-Akdaglib, S., & Akova, M. (2020). Fungal infections of the skin and soft tissue. *Curr Opin Infect Dis, 33*(2), 130–136.

[4] Köhler, J. R., Hube, B., Puccia, R., Casadevall, A., & Perfect, J. R. (2017). Fungi that infect humans. *Microbiology Spectrum, 5*(3), 10–1128.

[5] Nicolaou, K. C., Chakraborty, T. K., Ogawa, Y., Daines, R. A., Simpkins, N. S., & Furst, G. T. (1988). Chemistry of amphotericin B. Degradation studies and preparation of amphoteronolide B. *Journal of the American Chemical Society, 110*(14), 4660–4672.

[6] Lemke, A., Kiderlen, A. F., & Kayser, O. (2005). Amphotericin b. *Applied Microbiology and Biotechnology, 68*, 151–162.

[7] Ellis, D. (2002). Amphotericin B: spectrum and resistance. *Journal of Antimicrobial Chemotherapy, 49*(suppl_1), 7–10.

[8] Kyle, A. A., & Dahl, M. V. (2004). Topical therapy for fungal infections. *American Journal of Clinical Dermatology, 5*, 443–451.

[9] Sangale, P. T., & Manoj, G. (2015). Organogel: A novel approach for transdermal drug delivery system. *World J Pharm Res, 4*(3), 423–42.

[10] Gupta, R., Gupta, M. K., & Sharma, H. K. (2014). A review on organogels and fluid filled method. *International Journal of Research and Review, 3*(2), 274–288.

[11] Kaur, L. P. (2013). Topical gel: A recent approach for novel drug delivery. *Asian Journal of Biomedical and Pharmaceutical Sciences, 3*(17), 1.

[12] Patil, K. D., Bakliwal, S. R., & Pawar, S. P. (2011). Organogel: Topical and transdermal drug delivery system. *Int J Pharm Res Dev, 3*(6), 58–66.

[13] Cerqueira, M. A., Valoppi, F., & Pal, K. (2022). Oleogels and organogels: A promising tool for new functionalities. *Gels, 8*(6), 349.

14 Formulation and evaluation of vildagliptin loaded in-situ gel for ocular drug delivery: An improved approach towards management of diabetic retinopathy

Pritish Kanungo, Subhashree Das, Debaprasad Routray, Guptanjali Sahu, Yashwant Giri, and Gurudutta Pattnaik[a]

School of Pharmacy and Life Sciences, Centurion University of Technology and Management, Odisha, India

Abstract: Most often, diabetic retinopathy (DRP) is caused by diabetes mellitus (DM). DRP's classification as a microvascular illness has long been established. Vildagliptin (VLD) was one of the medications prescribed to treat DRP. In addition to having hypoglycemic properties, VLD lowers eye inflammation and promotes retinal blood flow in individuals with type 2 diabetes. The aim of the experiment focuses on the development of vildagliptin-loaded *in-situ* gels, using polymers Carbopol-940 and HPMC K4M as a pH-triggered gelling agent to improve contact time, control drug release, lower the frequency of administration and boost the drug's therapeutic efficacy. The clarity of all the prepared formulations was found to be satisfactory as the formulations were clear and no particulate matter was present. The FTIR study suggested that there is no interaction between the drug and used excipients. The developed formulations showed excellent gelling capability as they were converted into gel immediately and maintained their consistency for extended periods. The enhanced viscosity due to gel formation increases the drug's contact time at the site of administration. The pH of all formulations was evaluated and found to be between 6.2 and 7.3, indicating that they have no impact on the eyes, as proved by the Draize test. The isotonicity test was performed under a microscope, confirmed that the prepared *in-situ* gels are isotonic with blood and the tear fluid. After 360 minutes of the *in-vitro* drug release, the cumulative drug release of formulations was in the range of 84.01 % to 97.68%. Docking analysis was used to determine all relationship between vildagliptin and its target, dipeptidyl peptidase-4. The study revealed that vildagliptin has a favorable docking score of −6.9 Kcal/mol with dipeptidyl peptidase-4. After six months of accelerated stability study, the study found that no developed formulation deviates significantly from the initial recorded parameters. After completing the research, it was revealed that the optimized formulations showed improved contact time, controlled drug release, lowered frequency of administration and boosted the drug's therapeutic efficacy.

Keywords: Diabetes mellitus, retinopathy, vildagliptin, *in-situ* ocular gel

1. Introduction

Diabetes Mellitus, also known as DM is a chronic condition caused by low insulin synthesis by the pancreas or inadequate insulin usage by the body [1]. Chronic hyperglycemia causes a gradual malfunction of the retinal blood vessels known as diabetic retinopathy (DRP). DRP could be a type 1 or type 2 diabetes conditions [2]. It is unremarkable initially, but if

[a]gurudutta.pattnaik@cutm.ac.in

DOI: 10.1201/9781003672869-14

left untreated, it can result in blindness or loss of vision [3]. By 2030, there will likely be 439 million people with diabetes worldwide, up from the present approximately 285 million. 1.8 million of the 37 million blind people in the globe are caused by DRP [1]. Principal risk factors for DRP are duration of diabetes mellitus and degree of hyperglycemia [4]. Laser therapy, steroid eye implants, anti-VEGF injections (aflibercept, ranibizumab) and vitreoretinal surgery are the main treatments for DRP [5]. A key difficulty in ocular therapeutics is attaining the optimal drug concentration at the site of action, which is limited primarily by precorneal loss, which results in minimal ocular absorption. By adopting *in situ* gel-forming technologies, the effective dose can be changed by prolonging the duration of drug retention in the eye. Drug delivery to the eyes is a really fascinating and difficult endeavor [6, 7]. The eye's physiology, anatomy and biochemistry make it remarkably resistant to outside chemicals. The formulator's task is to get past the eye's defenses without permanently harming the tissue [8, 9]. Ophthalmic ointments prevent nasolacrimal drainage, prolong the duration of contact and reduce dilution by tears to promote optimal medication absorption. The main drawback of the ointment is that it causes blurry vision, which makes it suitable for usage at night or for treating the outer and outer borders of the eyelids. The use of suspension as an ocular delivery method is predicated by the possibility that particles will remain in the corneal spaces. Pre-corneal drugs loss can be minimized by employing a diffusion-controlled, non-erodible polymeric insert to delay drainage. Due to their challenging administration, inserts have a significant drawback in getting accepted by patients

[10, 11]. More effective ocular delivery systems must be developed quickly due to the advent of more sophisticated, sensitive diagnostic methods and treatment substances. The current research and development efforts to design superior therapeutic systems are the primary focus of this research endeavour, as the outmoded ocular solution, suspension and topical dosage forms are evidently insufficient to combat these disorders [12–16].

Vildagliptin (VG) may have extra beneficial effects on ocular blood flow, making it a promising treatment for vision disorders. Recent research has shown that the Vilda Gliptin can reduce ocular inflammation while improve ocular blood flow in people with type 2 diabetes. Vildagliptin is an inhibitor of dipeptidyl peptidase-4 (DPP-4) that is used to manage blood glucose levels for people with type 2 diabetes. Vildagliptin is claimed to successfully regulate inflammation in addition to improving the activity of beta cells. Using topically applied plasticized ocular film formulation, the potential anti-inflammatory properties of vildagliptin for the eyes have been investigated [17, 18]. Using polymers Carbopol-940 and HPMC K4M as a pH-triggered gelling method, vildagliptin *in-situ* gel was invented in the current study to improve contact time and controlled release, lower the frequency of administration and boost the drug's therapeutic efficacy.

2. Materials and Methods

2.1. Materials

Vildagliptin was obtained from Optimus Pharma (Hyderabad, India), HPMC (K4M), Sodium hydroxide pallets and potassium dihydrogen phosphate was purchased from Central Drug House Pvt. Ltd (New Delhi, India). Carbopol-940, Citric acid, Tween-80, Calcium chloride dihydrate, Sodium bicarbonate and Benzalkonium chloride were acquired from Sisco Research Laboratories Pvt. Ltd. (SRL, Maharashtra, India). Distilled water was used in the experimental study.

2.2. Preparation of calibration curve

To prepare the stock solution, ten milligrams of Vildagliptin was mixed in 100 mL of pH 7.4 phosphate buffer solution (Figure 14.1). The stock solution's concentration is 100 µg/ml. Using the stock solution, working concentrations of 1, 2, 3, 4 and 5 µg/ml were produced and measured using a UV spectrophotometer. A calibration curve was plotted using the absorbance measured at 211 nm. The results are presented in Table 14.1.

2.3. Preparation of in-situ gel

In 75 mL of distilled water, buffer salts (Citric acid and dihydrogen phosphate) were dissolved, requisite quantity of HPMC (K4M) was incorporated into the above prepared solution and overhead stirrer was used for homogenous mixing. Required amount of Carbopol 940 was spread on the prepared polymeric

solution and permitted to hydrate for about 12 hours. The hydrated mixture was stirred using a magnetic stirrer. Vildagliptin followed by benzalkonium chloride (BKC) were dissolved in small amount of water. Following that, the drug solutions was incorporated into the polymeric mixture while stirring frequently and the volume was increased to 100 ml with distilled water (Table 14.2).

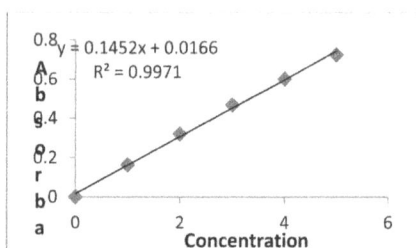

Figure 14.1. Standard curve for vildagliptin.

Source: Author.

Table 14.1. Recorded absorbance against concentration

Concentration (µg/ml)	Wavelength	Absorbance
1	211 nm	0.1638
2	211 nm	0.3203
3	211 nm	0.4676
4	211 nm	0.6021
5	211 nm	0.7241

Source: Author.

Table 14.2. Composition of developed in-situ gels

Formulations	VG	Citric Acid (g)	DHP (g)	HPMC K4M (g)	Carbopol 940 (g)	BKC (g)	Distilled Water (ml)
F1	0.05	0.407	1.125	0.2	0.5	0.002	Upto 100
F2	0.05	0.407	1.125	0.4	0.5	0.002	Upto 100
F3	0.05	0.407	1.125	0.5	0.5	0.002	Upto 100
F4	0.05	0.407	1.125	0.6	0.5	0.002	Upto 100

Source: Author.

2.4. Evaluation in-situ gels those were prepared

2.4.1. Clarity, homogeneity and physical appearance

All the in-situ gels that were prepared visually checked for color, homogeneity and clarity, using clarity test apparatus.

2.4.2. FTIR analysis of ocular in-situ gels

Attenuated total reflection infrared (ATR-IR) spectrometer (Alpha E ATR-FTIR, Bruker, USA) was used for analyzing produced ocular in-situ gels in the 400–4,500 cm^{-1} range.

2.4.3. Estimation of the gelling capacity and gelation period

At room temperature, one drop of each created in-situ gel was added to a separate petridish containing 5 ml freshly made artificial tear fluid. The gelling capacity was determined by monitoring the amount of time needed for the gel to develop and dissolve. This amount of time was computed as follows:

1. Absence of gel formation
2. It takes little time for the formation of gel and few minutes gel and dissolved rapidly
3. Formation of gel was rapid and the gel was stable for some hours
4. Quickly gel was formed and remained stable for several hours.

2.4.4. pH and isotonicity of developed formulations

The pH of the produced mixtures was measured at room temperature with a calibrated pH meter (Spancotek, Microprocessor, India). The electrode of the pH meter was dipped into a 30 ml beaker containing the in-situ gel and the pH was recorded. The experiment was repeated three times to minimize the error. The hemolytic technique is used to determine isotonicity. The produced mixtures were applied to a small amount of blood droplets, which were then examined under a 40X optical microscope. The blood droplets were tested with hypotonic, hypertonic and normal saline solutions.

2.4.5. Viscosity measurement

The rheological features of the generated in-situ gels were determined using a rotational viscometer. (Brookfield, DVE-II). The developed in-situ gel was placed into a 50 ml beaker and the viscosity was measured by dipping the spindle into it (spindle no.64). The angular velocity, also known as shear rate, was then steadily raised from 30 to 100 rpm.

2.4.6. Spreadability study

The synthesized in-situ gel's spreadability was determined using a modified Spreadability equipment. A glass slide of 7.5 cm was attached on to a wooden block (Stationary slide) and adequate amount of in-situ gel was placed over the stationary slide. Another glass slide of same size was placed over the applied in-situ gel and 100gm weight was applied for 5 minutes over the upper slide. The upper slide was secured with 50gm weight through a thread and pulley. The travelling time of the upper slide over the stationary slide was noted till it gets detached.

$$S = m*l/t$$

where,
  S = Spreadability
  M = Weight tied to upper slide (20g)
  l = Glass slide Length (7.5cm)
  t = Time taken in sec.

2.7. Molecular docking study

The examination was conducted using Discovery Studio Visualizer 2021, AutodockVina (version 1.1.2) and the online software MCULE. The sc-PDB source offered the crystal structure for dipeptidyl peptidase-4. The test ligands vildagliptin and dipeptidyl peptidase-4 were docked together using MCULE software. The binding affinities of the target protein's optimum binding configuration were determined in Kcal/mol.

2.8. Stability of developed in-situ gel

In-situ gels were packed in 50 ml tubes and stored in a stability chamber at 40 ± 2°C and 75 ± 5% relative humidity for six months. At regular intervals, 0 samples were obtained and evaluated for several criteria, including physical appearance, pH, drug content and viscosity.

3. Results and Discussion

3.1. Visual appearance, homogeneity and clarity

The clarity of all produced formulations was satisfactory and homogeneous and homogenous without any presence of unwanted particles. All the formulations were found to be milky white in color. The addition of oil to the formulas increased the tone of white.

3.2. Fourier transforms IR spectroscopy analysis

The FTIR analysis was performed to assess the interactions that occur between HPMC (K4M), Carbopol 940, Citric acid, DHP, Tween 80, BKC and VG in the *in-situ* gel. The FTIR spectra of all formulations display broadband in the range of 3,500 cm^{-1} to 1000 cm^{-1}. The FTIR spectra of VG showed signal at 3324.20 cm^{-1} for stretching of N-H, C=O at 1636.97 cm^{-1}, N=O band at 1396 cm^{-1} and nitrite and C=N stretching at 2116.48 cm^{-1} and 1151.01 cm^{-1}, respectively. The peak characteristics of VG are not-changed and showed no difference in their spot. As a result, the medication and polymer do not interact throughout the formulation process (Figure 14.2).

3.3. Estimation of gelation time and gelling capacity

The most significant need for *in-situ* gelling systems is their gelling capacity. Gelation period is important in *in-situ*

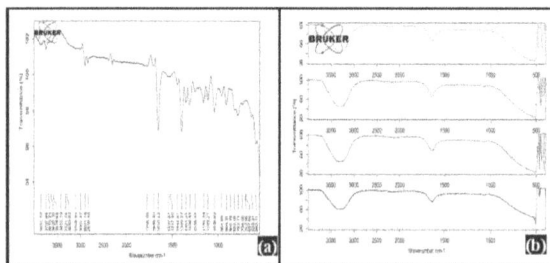

Figure 14.2. FTIR of (a) vildagliptin (b) *in-situ* gel formulations.

Source: Author.

formulations because it influences drug release at the site of action. The gelling time of the formulations varied from around 2 mins to 26 mins roughly. The drug's water binding affinity may be an important variable in the sol dynamic rate of its loaded solution. The introduction of VG (a water-soluble medication) had a significant impact on sol dynamics because it took longer to slide. The produced formulations demonstrated remarkable gelling capabilities, as they were immediately transformed into gel and retained their consistency for an extended period of time. The increased viscosity caused by gel formation extends the drug's contact time at the site of delivery, namely the eye. The converted rheological properties of the formulation offer prevention from washing out of the medication from the cul-de-sac cavity of the eye.

3.4. pH and isotonicity of the prepared gels

The pH of all formulations was evaluated and shown to fall between 5.7 and 6.3, which does not irritate the eyes, as proven by the Draize test. The isotonicity of the generated in-situ gels with lachrymal fluid was validated under a microscope.

3.5. Viscosity measurement

The viscosity of the developed in-situ formulations was estimated to be between 24 and 50 cp. From which F3, is having viscosity around 40 cPs which was similar to the criteria of in-situ gel.

3.6. Spreadability study

Spreadability data indicated that all developed in-situ formulations were easily spreadable. Among these prepared formulations, F3 has a spreadability of 0.28 cm²/s.

3.7. Molecular docking study

Docking analysis was performed to figure out all interactions between vildagliptin and its target, dipeptidyl peptidase-4. The results of the study revealed that vildagliptin has a favorable docking score of -6.9 Kcal/mol with dipeptidyl peptidase-4 (Figure 14.3).

3.8. Stability study

After 6 months of accelerated stability study it was found that no developed formulation deviates from the initial recorded parameters significantly. Detail results of tested parameters are mentioned in the table.

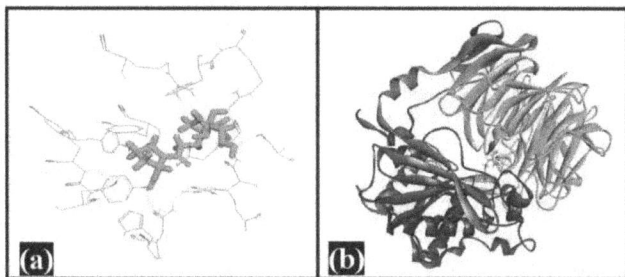

Figure 14.3. Binding of drug with dipeptidyl peptidase-4.

Source: Author.

4. Conclusion

The study focused on developing and evaluating Vildagliptin-loaded *in-situ* gels for ocular drug delivery to treat diabetic retinopathy (DRP), a microvascular disease caused by diabetes mellitus. Vildagliptin (VLD), known for its hypoglycemic and anti-inflammatory properties, was formulated using pH-sensitive polymers Carbopol-940 and HPMC K4M to enhance therapeutic efficacy. The gels demonstrated satisfactory clarity, gelling capability and isotonicity with blood and tears, ensuring safe application. They showed extended retention time in the eye, controlled drug release and improved drug contact time, thus reducing the need for frequent administration. FTIR analysis confirmed no drug-excipient interaction and molecular docking revealed a strong binding affinity between VLD and dipeptidyl peptidase-4, enhancing drug efficacy. The formulations remained stable over six months, making them promising candidates for better management of DRP.

References

[1] Balaji, R., Duraisamy, R., & Kumar, M. P. (2019). Complications of diabetes mellitus: A review. *Drug Invention Today*, *12*(1).

[2] Mbata, O., El-Magd, N. F., & El-Remessy, A. B. (2017). Obesity, metabolic syndrome and diabetic retinopathy: Beyond hyperglycemia. *World Journal of Diabetes*, *8*(7), 317.

[3] Dekhil, O., Naglah, A., Shaban, M., Ghazal, M., Taher, F., & Elbaz, A. (2019). Deep learning based method for computer aided diagnosis of diabetic retinopathy. In *2019 IEEE International Conference on Imaging Systems and Techniques (IST)* (pp. 1–4). IEEE.

[4] Leske, M. C., Wu, S. Y., Hennis, A., Hyman, L., Nemesure, B., Yang, L., Schachat, A. P., Barbados Eye Study Group. (2005). Hyperglycemia, blood pressure, and the 9-year incidence of diabetic retinopathy: The Barbados Eye Studies. *Ophthalmology*, *112*(5), 799–805.

[5] Mansour, S. E., Browning, D. J., Wong, K., Flynn, Jr H. W., & Bhavsar, A. R. (2020). The evolving treatment of diabetic retinopathy. *Clinical Ophthalmology*, 653–678.

[6] Ashim, K. M. (1993). *Ophthalmic drug delivery system* (Vol. 58, pp. 105–110). New York: Marcel Dekker Inc.

[7] Kaur, I. P., Garg, A., Singla, A. K., & Aggarwal, D. (2004). Vesicular systems in ocular drug delivery an overview. *Int J Pharm*, *269*, 1–14.

[8] Singh, S. K., & Bandyopadhyay, P. (2006). Pharmacia Corporation. Ophthalmic formulation with novel gum composition. US 7128928.

[9] Thorsteinn, L., & Tomi, J. (1999). Cyclodextrins in ocular drug delivery. *Adv Drug Del Rev*, *36*, 59–78.

[10] Gokulgandhi, M. R., Parikh, J. R., Megha Barot, M., & Modi, D. M. (2007). A pH triggered in situ gel forming ophthalmic drug delivery system for tropicamide. *Drug Delivery Technology*, *5*, 44–49.

[11] Zhidong, L., Jiawei, L., Shufang, N., Hui, L., Pingtian, D., & Weisan, P. (2006). Study of an alginate/HPMC based in situ gelling ophthalmic delivery system for gatifloxacin. *Int J Pharm*, *315*, 12–17.

[12] Indu, P. K., Manjit, S., & Meenakshi, K. (2000). Formulation and evaluation of ophthalmic preparations of acetazolamide. *Int J Pharm*, *199*, 119–127.

[13] Pandit, D., Bharathi, A., Srinatha, R., & Singh, S. (2007). Long acting ophthalmic formulation of indomethacin:

Evaluation of alginate gel systems. *Indian J Pharm Sci, 69*, 37–40.

[14] Johan, C., Katarina, E., Roger, P., & Katarina, J. (1998). Rheological evaluation of gelrite in situ gel for opthalmic use. *Eur J Pharm Sci, 6*, 113–116.

[15] Katarina, E., Johan, C., & Roger, P. (1998). Rheological evaluation of poloxamer as an in situ gel for ophthalmic use. *Eur J Pharm Sci, 6*, 105–112.

[16] Srividya, B., Cardoza, R. M., & Amin, P. D. (2001). Sustained ophthalmic delivery of ofloxacin from a pH triggered in situ gelling system. *J Control Release, 69*, 379–388.

[17] Berndt-Zipfel, C., Michelson, G., Dworak, M., Mitry, M., Löffler, A., Pfützner, A., & Forst, T. (2013). Vildagliptin in addition to metformin improves retinal blood flow and erythrocyte deformability in patients with type 2 diabetes mellitus–results from an exploratory study. *Cardiovascular Diabetology, 12*, 1–7.

[18] Nandi, S., Ojha, A., Nanda, A., Sahoo, R. N., Swain, R., Pattnaik, K. P., & Mallick, S. (2022). Vildagliptin plasticized hydrogel film in the control of ocular inflammation after topical application: study of hydration and erosion behaviour. *Zeits chrift für Physikalische Chemie, 236*(2), 275–290.

15 Design, optimization and evaluation of telmisartan fast dissolving film employing mango kernel starch as a new natural super disintegrant

Medisetty Gayatri Devi[1,2,a], Santosh Kumar R.[1], and Anusha Kusuma[3]

[1]Department of Pharmaceutics, GITAM School of Pharmacy, GITAM (Deemed to be University), Rushikonda, Visakhapatnam, Andhra Pradesh, India
[1,2]Department of Pharmaceutics, Viswanadha Institute of Pharmaceutical Sciences, Visakhapatnam, Andhra Pradesh, India
[3]Department of Pharmaceutics, Balaji Institute of pharmaceutical Sciences, Warangal, Telangana, India

Abstract: The concept of fast dissolving dosage forms indeed holds potential to decrease the frequency of dose while increasing therapeutic efficacy, bioavailability and stability. The current study used the solvent casting process to manufacture fast dissolving films (FDF) of the anti-hypersensitive medication telmisartan and optimization through 2^3 factorial design employing mango kernel starch as a natural super disintegrant. The concentrations of the various superdisintegrants (maltodextrin, sodium starch glycolate and mango kernel starch) were chosen as the independent variables and disintegration time, percentage drug dissolved in 10 minutse were taken as the depended variables. The optimized formulations (TF2) have impressive attributes with minimized disintegration time (9.23±0.23) and highest percentage drug release in 10 minutes 9.23±0.23.

Keywords: Telmisartan, fast dissolving dosage form, solvent casting method, in-vitro drug release, pharmacokinetic studies, stability studies

1. Introduction

Telmisartan, an angiotensin II receptor antagonist, is a pivotal therapeutic agent for hypertension and cardiovascular diseases [1]. Its efficacy in reducing blood pressure and mitigating cardiovascular risk underscores its significance in clinical practice [2]. To address these limitations, the formulation of telmisartan into fast dissolving films (FDFs) emerges as a promising strategy. FDFs offer several advantages over traditional dosage forms, especially in scenarios requiring rapid drug onset and ease of administration [3]. Nevertheless, the successful formulation of telmisartan into FDFs necessitates overcoming obstacles such as prolonged disintegration times and inadequate drug dissolution rates [4].

Natural super disintegrants, derived from plant or microbial sources, offer an attractive alternative to synthetic counterparts due to their biocompatibility and eco-friendliness [5].

The incorporation of natural super disintegrants in telmisartan FDFs holds

[a]gayatri.minnu@gmail.com

DOI: 10.1201/9781003672869-15

promise in achieving faster onset of action and enhanced drug bioavailability [5, 6].

This research article aims to explore the formulation of telmisartan into FDFs using natural disintegrant, both in single versus in combination, with a focus on their impact on disintegration time and drug dissolution in 10 min [7, 8].

2. Materials

Telmisartan, was obtained from Hyderabad-based Hetero Pvt Ltd. HPMC E15 was obtained from the TM Media Delhi, Mango kernel starch was extracted in the laboratory setting. Sodium starch glycolate, Propylene glycol (PG) was obtained from SD Fine Chemicals in Mumbai. Citric acid and maltodextrin, were acquired from Gattefosse India Private Limited in Mumbai, mannitol was obtained from LobaChemie Pvt Ltd in Maharashtra.

3. Optimization of Telmisartan Fast Dissolving Films using 2^3 Factorial Design

The formulation variables were optimized using a 2^3 factorial design, with the primary factors being the amounts of sodium starch glycolate, maltodextrin and mango kernel starch [8, 9] (Table 15.1).

Table 15.1. Optimization variables and their levels used in Telmisartan fast dissolving films Factorial design

Variables	Low (-1)	High (+1)
Mango kernel starch (%) - A	0	5
Maltodextrin(%) - B	0	5
Sodium starch glycolate (%) - C	0	5

Source: Author.

4. Preparation of Telmisartan Fast Dissolving Films

The formulation process began by dispersing 115 mg of HPMC E15 and super disintegrants in sufficient quantity (70%) of distilled water, which was then stirred for 2 hours at 2000 RPM on a magnetic stirrer at room temperature (Solution A). Separately, a mixture containing 40 mg of telmisartan and 10 mg of citric acid was dispersed in sufficient quantity of distilled water (30%) containing 15% plasticizer (PG) and stirred for 1 hour (Solution B). To the solution B, 10 mg of mannitol was added slowly to the solution A, while manually agitating to ensure uniform distribution. After achieving a clear solution, it was left undisturbed for 6 hours to release any remaining bubbles or air. The solution was then transferred to a petri dish and allowed to dry for 24 hours at 40°C in a hot air oven. The final films were then divided into 2 by 2 cm² pieces and kept in desiccator-safe aluminium sachets [9–11].

5. Evaluation of Prepared Telmisartan Fast Dissolving Films

5.1. Weight variation studies

Weight of three 2 × 2 cm² fast dissolving films of each formulation was noted by using an electronic weighing balance [12, 13].

5.2. Thickness

Three 2 × 2 cm² samples from each formulation were measured for thickness using a screw gauge with a 0–10 mm range and a minimum count of 0.01 mm [14].

5.3. *In vitro disintegration time*

A thermostat shaker with a 50 RPM setting and a petri dish containing 25 mL of 6.8 pH buffer solution was used to maintain a temperature of $37 \pm 0.3°C$. After inserting the 2×2 cm^2 film into the petri dish, the amount of time needed for the film to dissolve was noted [15].

5.4. *Surface pH*

First, the 2×2 cm^2 film was put on a petri dish and moistened with purified water. It was then brought into touch with a pH meter's electrode and the pH was recorded [16].

5.5. *Drug content uniformity*

After cutting the 2×2 cm^2 film, it was put into a 100 mL volumetric flask filled with phosphate buffer and it was agitation using a mechanical shaker to achieve uniformity. After filtration the drug content was subsequently measured spectroscopically at 296 nm [17]

$$\text{Drug content (\%)} = \frac{\text{Actual amount of drug}}{\text{Theoretical amount of drug}} \times 100$$

5.6. *Tensile strength*

A 2×2 cm^2 piece of film was placed vertically in between two clamps. Keeping the bottom clamp still allowed the upper clamp to be pulled at a speed of 100 mm per minute. The weight at which the film breaks is noted. The obtained values were then used in the formula below to calculate the tensile strength [18].

$$\text{TS(\%)} = \frac{\text{Load force at failure}}{\text{Film Thickness}} \times \text{Film width}$$

5.7. *Percentage elongation (%E)*

Each sample (2×2 cm^2) was securely held in a vertical position between two clamps. The films were then subjected to tension at a rate of 100 mm per minute by the top clamp, while the bottom clamp remained fixed. The obtained initial length of the film and length of the film after fracture was used in the below formula to get the percentage elongation of the film [19].

$$\text{Percentage Elongation} = \frac{\text{Increased length of film}}{\text{Initial length of film}} \times 100$$

5.8. *Folding endurance*

The number of times a film could be folded at the same spot before breaking or reaching the folding limit was used to determine the film's folding endurance [20].

5.9. *Percent moisture absorption (uptake) and moisture loss*

First, the weight of the 2×2 cm^2 film is recorded. The films are then stored at ambient temperature and 75% relative humidity for a week. Next, the film's final weight is mentioned. The formula below, formula, is used to compute the percentage of moisture absorption [21].

$$\text{Percentage moisture absorption} = \frac{\text{Final weight of film - Initial weight of film}}{\text{Initial weight of film}}$$

Desiccators containing anhydrous calcium chloride were used to ensure the films maintained their weight. The films were removed and weighed after 3 days. The formula for determining moisture loss and moisture absorption rates is as follows [22].

$$\text{Percentage moisture loss} = \frac{\text{Initial weight of film - Final weight of film}}{\text{Initial weight of film}} \times 100$$

5.10. **In vitro *dissolution studies***

A modified type 5 dissolution apparatus was employed to carry out the in vitro dissolution studies. Each 2×2 cm^2 film, carrying 40 milligrams of telmisartan, was wrapped with nylon wire mesh and set on a watch glass. After that, the assembly was placed in a dissolution flask with 500 mL of artificial saliva (phosphate buffer with a pH of 6.8) as the dissolution media. A 37°C temperature was maintained and a 50 RPM rotation speed was set. A 5 mL sample was taken at predetermined intervals of 5, 10, 15, 20, 25, 30, 45 and 60 minutes. After each withdrawal, an equal volume of fresh dissolution medium was added. Appropriate dilutions were performed on the dissolution medium and the samples were analysed at 296 nm. The measurements were carried out three times [15].

6. Results and Discussion

6.1. Drug excipient compatibility studies

The thermo gram of the telmisartan along with the mango kernel starch exhibited sharp endothermic peak at 263.46°C, which corresponds to pure telmisartan meting point 261.52°C that indicates the prepared mango kernel starch compatible with the telmisartan (Figures 15.1 and 15.2).

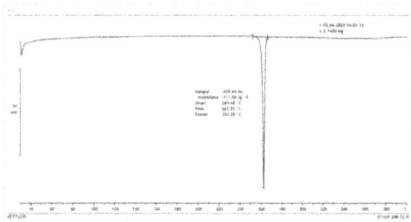

Figure 15.1. Telmisartan's thermo gram revealed an endothermic peak at 261.52°C, which is its melting point.

Source: Author.

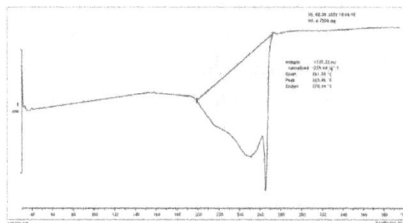

Figure 15.2. At 263.46 degrees Celsius, an endothermic melting peak was visible on the telmisartan + MKS thermo gram.

Source: Author.

6.2. Optimization of the independent variables

Based upon the preliminary studies, the films prepared with 5% concentration of the super disintegrants, it was observed that the formulation with 5% natural super disintegrant showed increase in dissolution time and decrease in disintegration time.

6.3. Impact of formulation parameters on dependent variables: percentage of medication dissolved in 10 minutes and in vitro disintegration time

In vitro Disintegration Time (DT) = $-54.165 + 40.8775A + 33.225B + 38.63C - 36.1275AB - 39.6725AC - 36.32BC + 36.4325ABC$ ($R^2 = 1.000$).

Percentage dissolved in 10 min (PD_{10}) = $+57.75875 + 21.57875A - 10.65875B + 7.94625C - 9.08875AB - 11.22375AC - 7.35625BC + 4.88375ABC$ ($R^2 = 1.000$).

By seeing the above two polynomial equations it can be interpreted that the factor A, B, C and ABC has positive effect on the DT and AB, AC and BC has negative effect on the DT.

The factors A, C and ABC has positive effect on the PD_{10} and B, AB, AC and BC has negative effect on the PD_{10}.

ANOVA of *in vitro* disintegration time and ANOVA of percentage drug dissolved in 10 min results indicate that the individual and combination effects of the Mango kernel starch (MKS), Maltodextrin (MDX) and Sodium starch glycolate (SSG) were significant (P < 0.05).

The impact of several independent factors on the disintegration time and percentage of drug dissolved in 10 minutes is displayed in the contour plot and surface response plot in Figures 15.3 and 15.4.

6.4. Characterization of telmisartan fast dissolving films

No morphological changes were observed in the telmisartan fast dissolving films across all formulations. The prepared telmisartan fast dissolving films were found to weigh between 16.24 ± 0.01 and 28.56 ± 0.03 mg as shown in Figure 15.5. The films exhibited a thickness ranging from 0.196 ± 0.03 mm to 0.282 ± 0.05 mm, as shown in Figure 15.5B. The films demonstrated

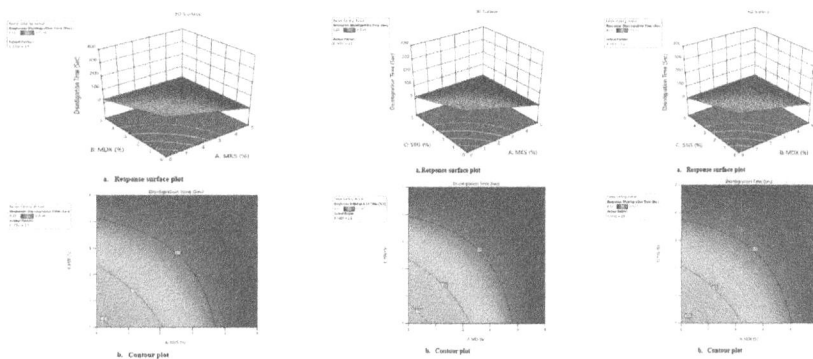

Figure 15.3. Images of 3D surface response plots and contour plots shows the effect of combination of super disintegrants on *in-vitro* disintegration time.

Source: Author.

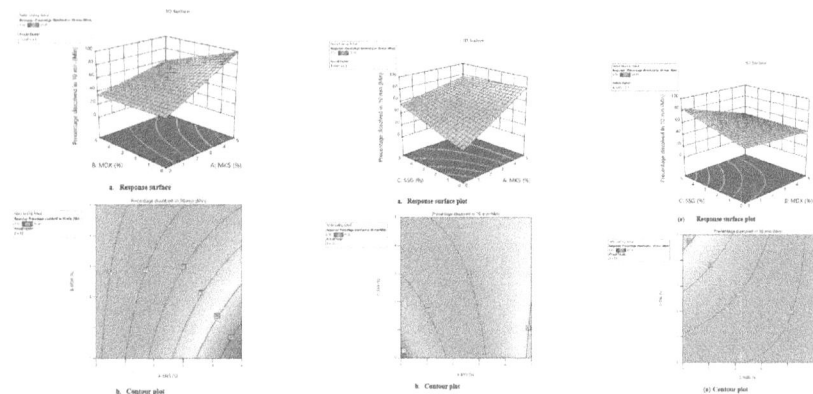

Figure 15.4. Images of 3D surface response plots and contour plots shows the effect of combination of super disintegrants on the percentage drug dissolved in 10 min.

Source: Author.

good uniformity in thickness across the samples. The films' of *in-vitro* disintegration time ranged from 9.23 ± 0.23 to 315.45 ± 0.22 seconds and the formulation TF2 prepared with 5 % mango kernel starch showed the least disintegration time of 9.23±0.23 seconds as shown in Table 15.2. The film surface pH ranged between 6.67±0.52 and 7.03±0.93 as shown in Figure 15.5G proves that the film pH matches the pH of the saliva. The drug content of films ranged from 97.54±0.90 to 99.92±0.73 percent as shown in Figure 15.5D. The films tensile strength ranged from 1.26±0.76 MPa to 3.02± 0.43 MPa as shown in Figure 15.5C. The films percentage elongation was found to be in between 23.16±0.94%

to 10.85±0.59 %as shown in Figure 15.5E. The formulation TF2 prepared with 5% mango kernel starch showed the highest elongation time 23.22±0.94 percentage. The films folding endurance of was found to be in between 77±0.36% to 95±0.80% as shown in Figure 15.5F. The percentage of moisture loss in the films ranged from 1.52±0.66 to 2.4± 0.78 %w/w as shown in Figure 15.5G. The percentage moisture uptake of films ranged from 3.96±0.66 to 4.95±0.44 %w/w as shown in Figure 15.5G. The drug cumulative percentage dissolved from telmisartan fast-dissolving films ranged from 31.52±0.35 to 99.89±0.57% and the cumulative percentage drug dissolved of all the formulation shown in

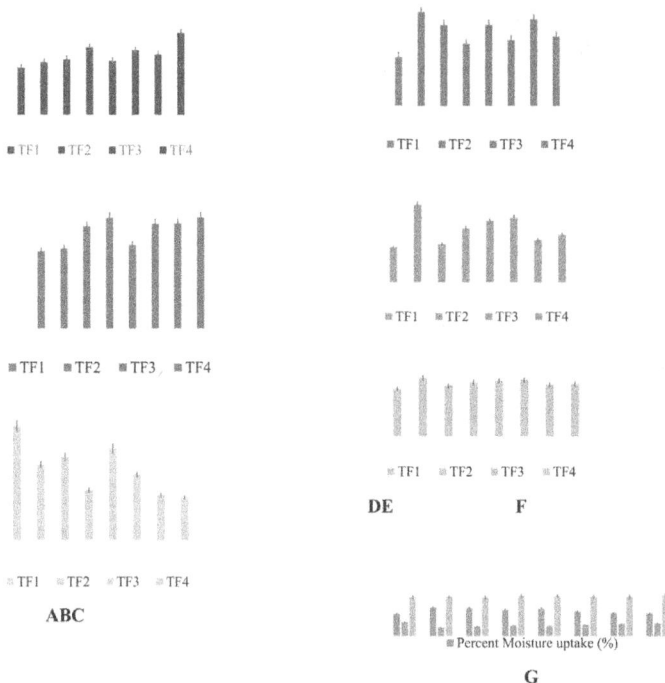

Figure 15.5. Graphical presentation of weight variation (A), Thickness (B), Tensile strength (C), Drug content uniformity (D), Percentage elongation (E), Folding endurance (F), Percentage moisture uptake, percentage moisture loss and surface pH (G) of all batches (TF1–TF8).

Source: Author.

Table 15.2. Evaluation parameters (disintegration time in sec, percentage dissolved in 10 min) of telmisartan fast dissolving films

Formulations	Disintegration time in sec	Percentage dissolved in 10 min
TF1	315.45±0.22	6.34±0.84
TF2	9.23±0.23	99.89±0.57
TF3	31.24±0.91	27.68±0.91
TF4	15.26±0.33	65.34±0.68
TF5	13.34±0.15	69.16±0.43
TF6	11.54±0.84	98.28±0.27
TF7	20.14±0.78	41.54±0.65
TF8	17.12±0.50	53.84±0.51

Source: Author.

Figure 15.6 and percentage drug dissolved in 10 min shown in Table 15.2. The formulation TF2 prepared with 5% concentration of the mango kernel starch showed the highest percentage of drug dissolved in 10 min (99.89±0.57).

7. Conclusion

The produced mango kernel starch had super disintegration property and was fine, free-flowing and amorphous in nature. A 2^3 factorial design and the solvent casting process were used to prepare

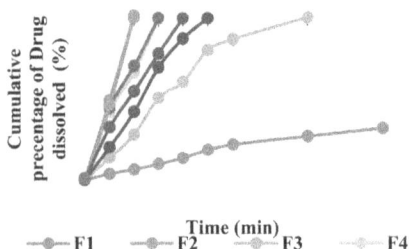

Figure 15.6. Cumulative percentage of drug dissolved from the formulation TF1 to TF8.

Source: Author.

the telmisartan fast dissolving films using mango kernel starch, a new super disintegrant. The optimized formulation of telmisartan (TF2) FDFs with 5% mango kernel starch as super disintegrant exhibited a maximum PD_{10} and least DT in comparison to the other formulations. As a result, it can be recommended as a novel natural super disintegrant in the formulation of poorly soluble drugs in order to further improve in vitro dissolution, bioavailability and therapeutic action.

8. Acknowledgment

We would like to express our sincere gratitude to Santhosh Kumar Rada, Gitam School of Pharmacy, whose support and guidance were crucial for the completion of this study.

References

[1] Yir-Erong, B., Bayor, M. T., Ayensu, I., Gbedema, S. Y., & Boateng, J. S. (2019). Oral thin films as a remedy for noncompliance in pediatric and geriatric patients. *Therapeutic Delivery*, *10*(7), 443–464.

[2] Dixit, R. P., & Puthli, S. P. (2009). Oral strip technology: Overview and future potential. *Journal of Controlled Release: Official Journal of the Controlled Release Society*, *139*(2), 94–107.

[3] Darshan, P. S., & Sudheer, P. (2023). Fast dissolving films—an innovative approach for delivering nutraceuticals. In *Industrial Application of Functional Foods, Ingredients and Nutraceuticals* (pp. 361–396). Elsevier.

[4] Lee, Y., Kim, K., Kim, M., Choi, D. H., & Jeong, S. H. (2017). Orally disintegrating films focusing on formulation, manufacturing process, and characterization. *Journal of Pharmaceutical Investigation*, *47*, 183–201.

[5] Qin, Z. Y., Jia, X. W., Liu, Q., Kong, B. H., & Wang, H. (2019). Fast dissolving oral films for drug delivery prepared from chitosan/pullulan electrospinning nanofibers. *International Journal of Biological Macromolecules*, *137*, 224–231.

[6] Liew, K. B., Odeniyi, M. A., & Peh, K. K. (2016). Application of freeze-drying technology in manufacturing orally disintegrating films. *Pharmaceutical development and technology*, *21*(3), 346–353.

[7] Cilurzo, F., Cupone, I. E., Minghetti, P., Buratti, S., Selmin, F., Gennari, C. G., & Montanari, L. (2010). Nicotine fast dissolving films made of maltodextrins: a feasibility study. *Aaps Pharmscitech*, *11*, 1511–1517.

[8] Drago, E., Campardelli, R., Lagazzo, A., Firpo, G., & Perego, P. (2023). Improvement of natural polymeric films properties by blend formulation for sustainable active food packaging. *Polymers*, *15*(9), 2231.

[9] Darekar, A., Sonawane, S., & Saudagar, R. (2017). Formulation and evaluation of orally fast dissolving wafer by using natural gum: Review article. *International Journal of Current Pharmaceutical Review and Research*, *8*(03).

[10] Lebaka, V. R., Wee, Y. J., Ye, W., & Korivi, M. (2021). Nutritional composition and bioactive compounds in three different parts of mango fruit. *International journal of environmental research and public health*, *18*(2), 741.

[11] Bangar, S. P., Kumar, M., & Whiteside, W. S. (2021). Mango seed starch: A sustainable and eco-friendly alternative to increasing industrial requirements. *International journal of biological macromolecules*, *183*, 1807–1817.

[12] Nisha, R., Dhruv, D., & DN, P. (2022). Recent Trends in developments of Superdisintegrants: An Overview. *Journal of Drug Delivery & Therapeutics*, *12*(1), 163–69.

[13] Badekar, R., Bodke, V., Tekade, B. W., & Phalak, S. D. (2024). An overview on oral thin films–Methodology, characterization and current approach. *International Journal of Pharmacy and Pharmaceutical Sciences*, 1–10.

[14] Kusuma, A., & Santosh Kumar, R. (2024). Optimization of fast-dissolving tablets of carvedilol using 23 factorial design. *International Journal of Applied Pharmaceutics*, 98–107.

[15] Shahrim, N. A., Sarifuddin, N., & Ismail, H. (2018). Extraction and Characterization of Starch from Mango Seeds. *Journal of Physics. Conference Series*, *1082*, 012019.

[16] Noor, F. (2014). Physicochemical properties of flour and extraction of starch from jackfruit seed. *International Journal of Nutrition and Food Sciences*, *3*(4), 347.

[17] Sonthalia, M., & Sikdar, D. C. (n.d.). Production of starch from mango (Mangifera Indica L.) seed kernel and its characterization. Ijtra.com. Accessed September 30, 2024. https://www.ijtra.com/view/production-of-starch-from-mango-mangifera-indical-seed-kernel-and-its-characterization.pdf.18.

[18] Patil, P. D., Gokhale, M. V., & Chavan, N. S. (2014). Mango starch: Its use and future prospects. *Innov. J. Food Sci*, *2*, 29–30.

[19] N.d. Iosrjournals.org. Accessed September 30, 2024b. https://www.iosrjournals.org/iosr-jac/papers/vol3-issue6/D0361623.pdf.

[20] Egharevba, H. O. (2020). Chemical properties of starch and its application in the food industry. In *Chemical Properties of Starch*. IntechOpen.

[21] Ratnayake, W. S., Wassinger. A. B., & Jackson, D. S. (2007). Extraction and characterization of starch from alkaline cooked corn masa. *Cereal Chemistry*, *84*(4), 415–422.

[22] Vemuri, V. D., Nalla, S., Gowri, G., & Mathala, N. (2021). Cramming on potato starch as a novel superdisintegrant for depiction and characterization of candesartan cilexetil fast dissolving tablet. *The Thai Journal of Pharmaceutical Sciences*, *45*(2), 137–147.

16 Deeper and updated insights on breast cancer

Anup Kumar Dash[1] and Sucharita Babu[2,a]

[1]Shri Rawatpura Sarkar College of Pharmacy, Shri Rawatpura Sarkar University Raipur, Chattishgarh, India
[2]Department of Pharmacology, School of Pharmacy and Life Sciences, Centurion University of Technologies and Management, Odisha, India

Abstract: When it comes to malignant tumours breast cancer is the most often type of cancer in women diagnosed global and the primary reason for fatalities coming out of cancer. Globally, the chance of getting breast cancer is rising at an alarming rate. Because of this, even if the disease has been identified and treated to a greater extent, lowering the overall mortality rate, it is still necessary to look for novel therapeutic approaches in addition to predictive and prognostic signs. Many different treatment modalities are used and the molecular subtype determines which one to use. Two parts of the multimodal treatment for breast cancer are systemic therapy and locoregional therapy, which comprise radiation therapy and surgery. Systemic therapies include immunotherapy (which was created more recently), Chemotherapy, anti-HER2 therapy and hormone therapy for disorders that test positive for hormones are available options. Between 15% and 20% of all cases of breast cancer are triple negative. Over 15% of patients have this subtype of breast cancer. Due to the condition's extremely invasive nature and poor response to treatment, it poses a therapeutic challenge and, thus, attracts a great deal of scientific attention.

Keywords: Cancer, breast cancer, women, prevalence, gene therapy, tumor

1. Introduction

Normal cells that have deviated from the rules of life and death are called cancer cells. Some scientists view their independence and self-control as an evolutionary process occurring during cell division. As an organism that goes through a process of evolution Cancer cells advance through the process of selected transformation into malignancy in addition to natural selection and mutation. Since 1960, the global rate of cancer-related mortality has increased by 66%, Research on Cancer International Agency (IARC) claims that. Breast cancer (BC) is currently the second most prevalent cancer globally, after lung cancer. Consequently, almost one in eight women in the US alone are predicted to at some time in their lives develop invasive breast cancer [1]. With an expected in 2020, 2.26 million cases of breast cancer were reported; it is the most often diagnosed illness in the world the leading cause of cancer-related mortality among females. It thus presents a major risk to world health. Though formerly believed to be a disease primarily affecting wealthy countries, in 2020 over half of all diagnoses and in less developed nations, breast cancer claimed the lives of two thirds of victims [2]. As the most prevalent disease diagnosed worldwide, Lung cancer has been eclipsed by breast cancer, accounting caused

[a]sucharitababu2@gmail.com

DOI: 10.1201/9781003672869-16

2.3 million new cases in both sexes and 1 in 8 cancer diagnoses combined [3]. The majority by far typical cancer in women to receive a diagnosis in 2020–25 percent of all cases in females—its frequency has been increasing worldwide, particularly in nations going through transition [1]. In 2020, breast cancer was responsible for an estimated 685,000 women's deaths, or 16% of all cancer deaths among females. The WHO started the Global Breast Cancer Initiative. At a time when public health responses to this development were insufficient, the WHO launched the Global Breast Cancer Initiative [4] (Figure 16.1).

2. Epidemilogy

In the United Kingdom, there are 55,000 new cases of breast cancer each year, making it the most common cancer in the country. Cancer Research UK predicts a 2% increase in this by the year 2035. As women age, their risk of breast cancer increases; in fact, women aged 50 and older account for approximately 80% of cases. Of all cancers, breast cancer was the leading killer in 2020, taking the lives of 685 000 people around the world (World Health Organisation, 2021) [6]. When it comes to malignant tumours affecting women, breast cancer ranks first globally. Up to

Figure 16.1. Cancer statistics [5].

Source: Author.

36% of all cancer diagnoses are breast cancer. In 2018, just over 2.089 million women received a breast cancer diagnosis [7]. The development of this malignant growth is on the rise globally, with developed nations having the highest frequency. Rich countries make up almost half of all cases world wide. This tendency is mostly the result of the so-called Western way of life, which is connected to bad eating habits, diabetes, elevated stress levels and inadequate physical exercise [8]. The proportion of in the previous 30 years, the number of adult (20–49 year old) premenopausal women has almost doubled. SadlyPolish ladies are still don't seem to care all that much about avoidance. They disregard the significance of routine checkups and their breasts. Polish women receive fewer preventative interventions than women in other European countries [9]. Of the women surveyed, 80% in the Netherlands, 71% in England and just 44% in Poland reported receiving free mammography preventive programs. For example, Poland's the 5-year survival rate for breast cancer is 78.5%. Considerably lower than the US's 90% 5-year survival rate [8]. Malignant neoplasms are responsible for 107.8 million disability-adjusted life years (DALYs) for women around the world each year, with 19.6 million of those instances being breast cancer. The World Health Organisation (WHO) provided estimates for these figures [10]. In 2020 [11], there were 2.26 million [95% UI, 2.24–2.79 million] new instances of breast cancer worldwide, making it the most common cancer diagnosis for women worldwide. Projections indicate that breast cancer alone will account for 29% of all new cancer diagnoses among women in the US [10].

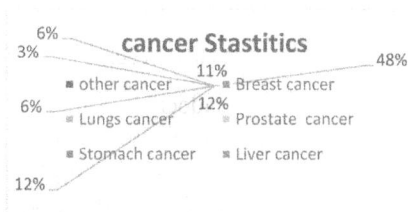

GLOBOCAN data from 2018 [12] show a strong positive correlation between the Human Development Index (HDI) and age-standardized incidence rates (ASIR) of breast cancer. In comparison to extremely high HDI countries (75.6 per 100,000), the ASIR was more than 200% lower in medium and low HDI countries (27.8 per 100,000 and 36.1 per 100,000, respectively) [4]. Not only is breast cancer the most common kind, but it also accounts for the majority of cancer-related deaths among women worldwide. In all, breast cancer contributed to 684,996 deaths worldwide. In 2020, 63% of fatalities worldwide took place in Asia and Africa, despite these regions having the highest incidence rates among developed nations [11]. The majority of breast cancer patients will survive in high-income countries; the prognosis is significantly poorer in low- and middle-income countries [10]. In 2020, the mortality-to-incidence ratio (MIR) for BC was 0.30, which indicated a reasonable five years survival rate [11].

3. Types of Breast Cancer

Tumour size, nodal involvement, metastatic presence and some biomarkers, including progesterone and estrogen receptors and the ERBB2 receptor (formerly known as HER2), all contribute to the determination of the stage of breast cancer [13]. All pathology samples should be analysed for ERBB2 status, progesterone receptors and estrogen receptors following a histologic diagnosis of breast cancer [14, 15]. Breast tumours that display none of these markers are triple-negative malignancies. DCIS, or non-invasive stage 0

breast cancer, is a condition. Stages I, IIa and IIb represent early invasive cancer and stages IIIa, IIIb and IIIc represent locally advanced cancer and these stages of breast cancer are nonmetastatic where stage IV is for metastatic breast cancer [13].

3.1. Non-invasive breast cancer

Breast cancer in stage 0, DCIS, non invasive DCIS, a type of pure, noninvasive cancer, is most frequently detected by breast duct-confined microcalcifications on mammography. If therapy is not received, 40% or more of DCIS cases will eventually develop into aggressive breast cancer.27 Treatment options for DCIS include mastectomy or radiation therapy, with a 2-mm surgical margin as the ideal result [13]. When a mastectomy is performed, a sentinel lymph node (SLN) biopsy is performed to look for the (very uncommon) presence of lymph node involvement. After a mastectomy, it might not be technically possible to do an SLN biopsy. Radiation treatment is provided to individuals undergoing lumpectomies; this mix of treatment is thought to preserve breast tissue. Radiation Counseling could be delayed for individuals who have little, low-grade lesions, so they have a lower chance of recurrence. Radiation therapy is not advised for people receiving mastectomy treatment [13].

3.2. Invasive breast cancer

Chemotherapy, endocrine medications, using monoclonal antibodies for immunotherapy that target tumor-related receptors, radiation therapy and surgery are examples of preoperative and postoperative systemic therapies for nonmetastatic breast cancer. When deciding whether to add chemotherapy to a patient's treatment plan, molecular testing can be helpful. The expression of 21 genes test (Oncotype DX) is the best test to forecast the course of the disease and decide whether to add chemotherapy. It can be used for patients, including males, having node-negative, hormone receptor-positive breast cancer [14]. When postoperative chemotherapy and hormone therapy are given, chemotherapy is always given before hormone therapy [15].

3.3. Breast cancer with metastases

The typical rate of survival for people having breast cancer that has spread increased over the past few decades as a result of breakthroughs in breast cancer therapy. Although there is currently a 24 to 40-month survival rate, metastatic breast cancer is rarely curable [16]. Reduction of symptoms, prolonging life and maintaining quality of life are the main objectives of treatment [17]. Chemotherapy, immunotherapy and endocrine therapy may be administered to target specific subtypes of breast cancer. After receiving systemic therapy, patients whose tumour burden impairs their quality of life may benefit from surgery or radiation. Potential therapies for the Denosumab (Prolia) or bisphosphonates like pamidronate (Aredia) or zoledronic acid (Reclast) treat between sixty and eighty percent of patients with metastatic breast cancer who develop bone metastases. These treatments appears to reduce the risk of fractures and hypercalcemia, two outcomes of bone metastases [13]. In stage IV breast cancer, metastases to the liver, lungs and brain are

frequent. Treatments for symptoms must to be suggested when necessary to offer palliation.

3.4. Risk factors of breast cancer

Cancer develops from normal epithelium in a multi-step process. Numerous factors, including those in combination, can lead to breast cancer the diet, the surroundings and one's ancestry. Normal breast tissue has a balanced balance of both positive and negative growth factors, thus specific functions must either increase or decrease for breast cancer to develop [18].

3.5. Age

Growing older may be associated with a rise in the prevalence of breast cancer as a result of the accumulation of somatic mutations commencing a woman's exposure to ovarian hormones when she delays menopause until later in life and this has been connected to an increased breast cancer risk. Certain evidence suggests that breast cancer may be more aggressive in younger women than in older ones. This would be in line with an illness that manifests clinically earlier and grows more swiftly [18].

3.6. Breast cancer and other cancers in the family medical history

Should one or more of your closest blood kin are affected with breast cancer, it is likely that the illness belongs to the family. There are more cases of breast cancer than one might expect to arise at random in some homes. It could be challenging to tell if a family history of cancer is the result of common lifestyle choices, random chance, inherited genes, or some combination of these [19].

3.7. Sex

Women account for 99 per cent of cases of breast cancer. Approximately 1% of Males are impacted by this cancerous growth; in Poland, 0.4/105 is the standardized incidence rate. Annually, no more than 100 instances are documented [20]. Nevertheless, the prevalence of breast cancer in men is increasing consistently, just like it does in women and it is most certainly linked to obesity and extended survival times [21].

4. Changes within the BRCA gene

Genetic mutation is the term used to describe a changed gene. Specific cancer types may be more likely to occur in people with specific gene abnormalities. Gene-inherited mutations can be passed from one parent to the next—just a tiny portion of the breast. A small percentage of malignancies (around 5%–10%) are caused by genetic alterations the BRCA1 and BRCA2 genes, which are associated with breast cancer, are part of normal human physiology. The reason why these genes are called tumour suppressors is that they seem to have a role in stopping the spread of cancer cells. Gene mutations affecting BRCA1 or BRCA2 may impair their capacity to control the growth of cancer. These alterations are uncommon. Approximately 1 in 500 persons encounter them men and women can inherit a mutant BRCA gene from their father or mother offspring of those who have the another way to inherit it is through gene mutation. A kid

has a 50% risk of inheriting the mutation if it is present in one or both parents' two copies of the BRCA gene. This implies that there is a 50% chance that a child won't inherit the gene mutation [22]. Research indicates that women who are born with mutations in either the BRCA1 or BRCA2 gene are 85% more likely to get breast cancer throughout their lifetime. Furthermore, women who have these hereditary mutations are at a greater risk of risk of early-life breast cancer development. Women with the BRCA gene mutation are more likely to develop bilateral breast cancer. If someone has cancer in one breast, they are more likely to develop cancer in the other. If a woman has a mutation in the BRCA gene, she can get ovarian cancer at any age [23].

4.1. The late menopause

When a woman reaches menopause, her menstrual cycle stops because her Ovaries cease releasing progesterone and estrogen. This decrease in hormone levels results in your cells to be exposed to other substances for a longer period of time, this elevates the likelihood of developing breast cancer in you. In a similar vein, breast tissue is exposed to outside influences for a shorter amount of time when the menopause starts early in life, reducing the likelihood from breast cancer [24].

4.2. Estrogen

There is one correlation between internal and external estrogens as well as the chance of breast cancer. Endogenous estrogen is produced by the ovary spontaneously in premenopausal women and ovarian excision can reduce the risk of breast cancer. As far as exogenous estrogen is concerned, hormone replacement therapy and oral contraceptives are the two most common sources. Oral contraceptives have been in use since the 1960s extensively and changes have been made to their formulations to reduce adverse effects. But still, for female groups that are African American and Iranian, Still, the odd ratio is more significant than 1.5. But in ladies who quit taking oral contraceptives for longer, the risk of breast cancer does not rise for more than ten years. Giving external estrogen or other hormones to females going into menopause or beyond is known as hormone replacement therapy (HRT). Studies have shown a correlation between the usage of HRT and a higher risk of breast cancer. The relative risk for people who have never used HRT and those who do so currently is revealed by the Million Women Study in the UK. After four and eight years of using HRT, hazardous ratios (HRs) of 1.48 and 1.95 were discovered, respectively, in a cohort trial of 22,929 Asian women. Two years after ceasing hormone replacement therapy, there is a considerable reduction in the possibility of developing breast cancer. Among breast cancer survivors, the recurrence rate is notable, with a 3.6 HR for a new breast tumour. On hormone replacement therapy. Since the 2003 results of the WHI randomized controlled study regarding the detrimental effects of HRT in the US of America, the Breast cancer incidence rate has dropped by 7% [25].

4.3. Hormone replacement therapy

Hormone replacement therapy (HRT) raises the risk of breast cancer by roughly 8% compared to oestrogen alone, according to research from the Women's Health Initiative (WHI). Each year, estrogen alone increases the risk by about 1%. Even with highly brief use of combination HRT, the study indicated a greater risk than a placebo. After a few years of not taking HRT, the higher risk seems to have gone away. Among Canadian women aged 50 to 69, the WHI study found that the number of new cases of breast cancer decreased significantly between 2002 and 2004. At the same time, fewer people were using combined HRT. Other nations, like the US, Norway, Australia, the Netherlands and Switzerland, have also noticed this trend. Combination HRT has risks that outweigh its advantages when used for a long time [26].

4.4. Being ponderous

For women who have gone through menopause, obesity increases their chance of breast cancer. Research indicates that ladies who have never had a body mass index of 31.1 or above-received hormone replacement treatment (HRT) have a 2.5-fold higher risk of breast cancer than people with a BMI of 22.6 or lower. More specifically, estrogens generated during ovulation have a substantial impact on breast cancer. Breast tissue absorbs cumulative doses of estrogen over time, which is thought to be the cause of multiple breast cancer risk factors. The bulk of ovaries generate most of the body's estrogen, with adipose tissue producing very little of the hormone after menopause. A higher risk of breast cancer can result from having more adipose tissue since it can raise estrogen levels [27].

5. Diagnostic Tools

5.1. Mammography

Diagnostic mammography uses low radiation levels to produce an image of the breast. It investigates unexpected results from screening mammograms or clinical breast exams. During a biopsy, an aberrant region can also be found using mammography [28].

5.2. Ultrasonography

Using high-frequency sound waves, an ultrasound can provide images of different bodily parts. It's employed to ascertain whether a cyst or solid tumour is the source of the breast lump. Furthermore, medical personnel can use ultrasound to guide them to the biopsy location. For women whose breast cancer has progressed, an ultrasound may be done to check for liver metastases [29].

5.3. Biopsy

Only by a biopsy is it possible to diagnose breast cancer. The process of removing bodily parts or cells from a patient for analysis in a lab is known as a biopsy. Whether or not the sample contained cancerous cells will be disclosed in the pathologist's report. Whether a lump is palpable, feelable, non-palpable, or not feelable will determine what kind of biopsy is performed. The doctor may utilize mammography or ultrasound to find the area that must be checked. Most biopsies are carried out at hospitals and the patient is discharged [30].

5.4. Fine needle aspiration

It uses an excellent needle and a syringe to extract a little tissue sample from a lump. It aids medical professionals in identifying solid tumours from cysts in lumps. Fine needle aspiration (FNA) cannot identify if a malignancy is invasive or non-invasive. A medical practitioner uses ultrasound or palpation to guide the insertion of a tiny needle into the breast lump during the treatment. Cells or fluid are suctioned out of the lump using a syringe fastened to the needle. A pathologist then uses a microscope to study these cells or fluid samples to identify if they are benign, not cancerous or malignant. Important information depends on the kind of breast cancer gleaned by a minimally invasive treatment called FNAC. Analysing the cells' characteristics, helping to identify cancerous cells and directing further diagnostic or therapeutic processes aid in the diagnosis of breast cancer. For a more thorough assessment, however, other procedures, such as a surgical biopsy or core needle biopsy, can be suggested based on the circumstances [31].

6. Management of Breast Cancer

6.1. Surgery

The two most popular types of breast surgery are lumpectomy, which preserves the breast tissue and mastectomy, which removes the entire breast and is usually followed by breast reconstruction. During a lumpectomy, the breast tumour and some of the surrounding healthy tissue are removed. The ideal margins status is 'no ink on the tumour' or the lack of tumour cells at the tissue boundary [32]. Mastectomies and breast-conserving surgery are the two main surgical methods that enable the removal of malignant tissues from the breast (BCS). BCS sometimes referred to as a partial/segmental mastectomy, broad local removal, lumpectomy, or quadrantectomy, permits the excision of malignant tissue while protecting healthy breast tissue. BCS is frequently paired with oncoplasty, a plastic surgical treatment. A mastectomy is the removal of the breast in its entirety, usually accompanied by a brief rebuilding of the breast. The techniques used to remove the Axillary lymph node dissection (ALND) and sentinel lymph node biopsy (SLNB) are the lymph nodes that are impacted. People who have gotten BCS often tend to require a full mastectomy in the future, even though it seems to be significantly more beneficial for patients [10]. However, the main benefits of using BCS include significantly better cosmetic outcomes, a reduction in the psychological burden placed on the patient and a decrease in the number of complications following surgery [27]. Specifies the size of the tumour, surgical viability, clinical characteristics and the patient's desire to keep their breast tissue all influence the therapeutic choice among those with breast cancer in its early stages, according to the European Society for Medical Oncology (ESMO) [33].

6.2. Hormonal therapy

For patients with breast cancer who test positive for either the progesterone or estrogen receptors (ER or PR), hormonal therapy, also known as endocrine therapy, is utilized as an adjuvant or neoadjuvant therapy. In around 70% of invasive breast tumours, ERs and PRs are expressed. This is a ubiquitous expression. Hormonal therapy involves blocking receptors and/or lowering estrogen

levels. While aromatase inhibitors like letrozole, anastrozole and exemestane suppress estrogen levels, drugs like tamoxifen, toremifene and fulvestrant block or degrade ERs. For premenopausal women, leuprolide (Lupron) and goserelin (Zoladex) can also inhibit the ovaries. It has been demonstrated that breast cancer patients' mortality rates can be decreased by combining hormonal therapy with chemotherapy. Hormone therapy lasts for five to ten years. The women's menopausal status influences the medication choice. These can be administered orally or by injection. These medications frequently cause the following adverse effects: headache, joint and bone pain, mood swings, weak bones, hot flashes, vaginal dryness and night sweats [34].

6.3. Chemotherapy

BC chemotherapy includes several cytotoxic drug classes, including as alkylating agents, antimetabolites and tubulin inhibitors. A nitrogen mustard alkylating agent that breaks DNA strands is cyclophosphamide. DNA is involved in the mechanism of action of anthracyclines, which include doxorubicin, daunorubicin, epirubicin and idarubicin. Intercalation, preventing the formation of macromolecules [34]. Certain taxanes, such as paclitaxel and docetaxel, attach to microtubules and stop them from disassembling, which causes cell cycle arrest and death. Hemotherapy can be applied as an adjuvant or neoadjuvant in addition to treating metastatic BC [35].

7. Conclusion

Incidence rate of BC is increasing day by day globally. Though there is so many treatment options available for BC mortality rate is rising due to the adverse effects of drugs and risk factors. It has been seen that modifiable risk factors could play a significant role in reducing the incidence of breast cancer. The present review focuses on risk factors of BC that can be considered for prevention and treating the BC. Providing the best possible prevention is vital since, over the past few decades. Early detection can be pivotal in cancer treatment and increase the survival rate. Mammography is currently the most widely used screening technique that allows for relatively early breast cancer identification. By mitigating the risk factors and reducing the adverse effects may reduce the mortality and increase the survival rate in breast cancer.

References

[1] Zubair, M., Wang, S., & Ali, N. (2021). Advanced approaches to breast cancer classification and diagnosis. *Frontiers in Pharmacology, 11*, 632079.

[2] Sung, H., Ferlay, J., Siegel, R. L., Laversanne, M., Soerjomataram, I., Jemal, A., & Bray, F. (2021). Global cancer statistics 2020: GLOBOCAN estimates of incidence and mortality worldwide for 36 cancers in 185 countries. *CA: A Cancer Journal for Clinicians, 71*(3), 209–249.

[3] Rauniyar, A., Hagos, D. H., Jha, D., Håkegård, J. E., Bagci, U., Rawat, D. B., & Vlassov, V. (2023). Federated learning for medical applications: A taxonomy, current trends, challenges, and future research directions. *IEEE Internet of Things Journal, 11*(5), 7374–7398.

[4] Heer, E., Harper, A., Escandor, N., Sung, H., McCormack, V., & Fidler-Benaoudia, M. M. (2020). Global burden and trends in premenopausal and postmenopausal breast cancer: a

population-based study. *The Lancet Global Health, 8*(8), e1027-e1037.

[5] Anderson, B. O., Ilbawi, A. M., Fidarova, E., Weiderpass, E., Stevens, L., Abdel-Wahab, M., & Mikkelsen, B. (2021). The Global Breast Cancer Initiative: a strategic collaboration to strengthen health care for non-communicable diseases. *The Lancet Oncology, 22*(5), 578–581.

[6] Katsura, C., Ogunmwonyi, I., Kankam, H. K., & Saha, S. (2022). Breast cancer: presentation, investigation and management. *British Journal of Hospital Medicine, 83*(2), 1–7.

[7] Nardin, S., Mora, E., Varughese, F. M., D'Avanzo, F., Vachanaram, A. R., Rossi, V., Saggia, C., Rubinelli, S., & Gennari, A. (2020). Breast cancer survivorship, quality of life, and late toxicities. *Frontiers in Oncology, 10*, 864.

[8] Smolarz, B., Nowak, A. Z., & Romanowicz, H. (2022). Breast cancer—epidemiology, classification, pathogenesis and treatment (review of literature). *Cancers, 14*(10), 2569.

[9] Religioni, U. (2020). Cancer incidence and mortality in Poland. *Clinical Epidemiology and Global Health, 8*(2), 329–334.

[10] Łukasiewicz, S., Czeczelewski, M., Forma, A., Baj, J., Sitarz, R., & Stanisławek, A. (2021). Breast cancer—epidemiology, risk factors, classification, prognostic markers, and current treatment strategies—an updated review. *Cancers, 13*(17), 4287.

[11] Ferlay, J., Ervik, M., Lam, F., Colombet, M., Mery, L., Piñeros, M., Znaor, A., Soerjomataram, I., & Bray, F. (2020). Global cancer observatory: Cancer today. *Lyon: International Agency for Research on Cancer*, 20182020.

[12] Sharma, R. (2021). Global, regional, national burden of breast cancer in 185 countries: evidence from GLOBOCAN 2018. *Breast Cancer Research and Treatment, 187*, 557–567.

[13] Waks, A. G., & Winer, E. P. (2019). Breast cancer treatment: A review. *Jama, 321*(3), 288–300.

[14] National Comprehensive Cancer Network. (2008). NCCN clinical practice guidelines in oncology. *http://www. nccn. org/professionals/physician_gls/ PDF/occult. pdf.*

[15] Allison, K. H., Hammond, M. E. H., Dowsett, M., McKernin, S. E., Carey, L. A., Fitzgibbons, P. L., … & Wolff, A. C. (2020). Estrogen and progesterone receptor testing in breast cancer: ASCO/CAP guideline update. *Journal of Clinical Oncology, 38*(12), 1346–1366.

[16] Caswell-Jin, J. L., Plevritis, S. K., Tian, L., Cadham, C. J., Xu, C., Stout, N. K., Sledge, G. W., Mandelblatt, J. S., & Kurian, A. W. (2018). Change in survival in metastatic breast cancer with treatment advances: Meta-analysis and systematic review. *JNCI Cancer Spectrum, 2*(4), pky062.

[17] O'Sullivan, C. C., Loprinzi, C. L., & Haddad, T. C. (2018, June). Updates in the evaluation and management of breast cancer. In *Mayo Clinic Proceedings* (Vol. 93, No. 6, pp. 794–807). Elsevier.

[18] Cuthrell, K. M., & Tzenios, N. (2023). Breast cancer: updated and deep insights. *International Research Journal of Oncology, 6*(1), 104–118.

[19] Maio, F., Tari, D. U., Granata, V., Fusco, R., Grassi, R., Petrillo, A., & Pinto, F. (2021). Breast cancer screening during COVID-19 emergency: patients and department management in a local experience. *Journal of Personalized Medicine, 11*(5), 380.

[20] Religioni, U. (2020). Cancer incidence and mortality in Poland. *Clinical Epidemiology and Global Health, 8*(2), 329–334.

[21] Lima, S. M., Kehm, R. D., & Terry, M. B. (2021). Global breast cancer incidence and mortality trends by region, age-groups, and fertility patterns. *EClinicalMedicine, 38*.

[22] Li, M. R., Liu, M. Z., Ge, Y. Q., Zhou, Y., & Wei, W. (2021). Assistance by routine CT features combined with 3D texture analysis in the diagnosis of BRCA gene mutation status in advanced epithelial ovarian cancer. *Frontiers in Oncology, 11*, 696780.

[23] Hu, X., Zhang, Q., Xing, W., & Wang, W. (2022). Role of microRNA/lncRNA intertwined with the wnt/β-catenin Axis in regulating the pathogenesis of triple-negative breast cancer. *Frontiers in pharmacology, 13*, 814971.

[24] Vatankhah, H., Khalili, P., Vatanparast, M., Ayoobi, F., Esmaeili-Nadimi, A., & Jamali, Z. (2023). Prevalence of early and late menopause and its determinants in Rafsanjan cohort study. *Scientific Reports, 13*(1), 1847.

[25] Belachew, E. B., & Sewasew, D. T. (2021). Molecular mechanisms of endocrine resistance in estrogen-receptor-positive breast cancer. *Frontiers in Endocrinology, 12*, 599586.

[26] Mills, Z. B., Faull, R. L., & Kwakowsky, A. (2023). Is hormone replacement therapy a risk factor or a therapeutic option for Alzheimer's disease?. *International Journal of Molecular Sciences, 24*(4), 3205.

[27] Kunyahamu, M. S., Daud, A., & Jusoh, N. (2021). Obesity among healthcare workers: which occupations are at higher risk of being obese?. *International Journal of Environmental Research and Public Health, 18*(8), 4381.

[28] Chang, C. C., Ho, T. C., Lien, C. Y., Shen, D. H. Y., Chuang, K. P., Chan, H. P., Yang, M.-H., & Tyan, Y. C. (2022, June). The effects of prior mammography screening on the performance of breast cancer detection in Taiwan. In *Healthcare* (Vol. 10, No. 6, p. 1037). MDPI.

[29] Wang, Y., Chen, H., Li, N., Ren, J., Zhang, K., Dai, M., & He, J. (2019). Ultrasound for breast cancer screening

in high-risk women: results from a population-based cancer screening program in China. *Frontiers in Oncology*, *9*, 286.

[30] van der Poort, E. K., van Ravesteyn, N. T., van den Broek, J. J., & de Koning, H. J. (2022). The early detection of breast cancer using liquid biopsies: model estimates of the benefits, harms, and costs. *Cancers*, *14*(12), 2951.

[31] Wang, M., Kundu, U., & Gong, Y. (2020). Pitfalls of FNA diagnosis of thymic tumors. *Cancer Cytopathology*, *128*(1), 57–67.

[32] Schnitt, S. J., Moran, M. S., & Giuliano, A. E. (2020). Lumpectomy margins for invasive breast cancer and ductal carcinoma in situ: current guideline recommendations, their implications, and impact. *Journal of Clinical Oncology*, *38*(20), 2240–2245.

[33] Cardoso, F., Kyriakides, S., Ohno, S., Penault-Llorca, F., Poortmans, P., Rubio, I. T., Zackrisson, S., & Senkus, E. (2019). Early breast cancer: ESMO Clinical Practice Guidelines for diagnosis, treatment and follow-up. *Annals of Oncology*, *30*(8), 1194–1220.

[34] Rai, R., & Tripathi, V. (2023). An overview of breast cancer epidemiology, risk factors, classification, genetics, diagnosis and treatment. *Vantage J Thematic Analysis*, *4*, 45–67.

[35] Burguin, A., Diorio, C., & Durocher, F. (2021). Breast cancer treatments: updates and new challenges. *Journal of Personalized Medicine*, *11*(8), 808.

17 UV-visible spectrophotometric method for estimation of Tacrolimus in bulk and different pharmaceutical capsules

Babita Adhikary[1,a] and Ladi Alik Kumar[2,b]

[1]Department of Pharmaceutical Analysis and Quality Assurance, Centurion University of Technology and Management, Odisha, India
[2]Department of Pharmaceutics, Centurion University of Technology and Management, Odisha, India

Abstract: This paper aims to compare by analyzing different marketed formulations (capsules) of Tacrolimus using UV Spectrophotometric methods. The marketed capsules used were Rolitrans1, Tacrograf 1.0 and Prograf 1mg. A UV-visible Spectrophotometric method has been developed and validated for measuring Tacrolimus in nanoparticles and it is accurate, reproducible and inexpensive. The correlation coefficient for the data was found to be 0.999, indicating a very high degree of correlation and indicating that the method for measuring the analyte is highly reliable.

The analysis uses the regression equation $Y = 0.0117x + 0.0035$. This analysis had a detection limit of 0.09 mg/L. This is the lowest analyte concentration that can be reliably detected but not necessarily quantified. The quantification limit was 0.2 mg/mL. This represents the lowest analyte concentration that can be detected and quantified with sufficient precision and accuracy. The results of this study highlight the significance of carefully weighing patient considerations and particular product attributes when choosing a Tacrolimus product for use in clinical practice. By considering these parameters, clinicians can ensure that their patients receive the safest and most effective care possible, decreasing the likelihood of side effects and other complications.

Keywords: Tacrolimus, spectrophotometry, capsule, validation

1. Introduction

Among a variety of immunosuppressive drugs, following organ transplantation, Tacrolimus (also referred to as FK-506 or Fujimycin) is frequently recommended to lower the immune system activity of the patient and, as a result, the likelihood of organ rejection [1]. Additionally, following bone marrow transplants, it is applied topically to treat severe refractory uveitis, vitiligo and atopic dermatitis. Lipids and organic solvents dissolve fairly readily in it. It's soluble in saturated hydrocarbons but insoluble in water and it's moderately soluble in saturated hydrocarbons[2]. *Streptomycin tsukubaensis,* a bacterium found in Japanese soil, was discovered in its fermentation broth in 1984. The chemical name for Tacrolimus is macrolide[3]. When it attaches itself to the immunophilin the activity of the peptidyl-prolyl isomerase is reduced in FKBP-12 (FK506 binding protein). Calcineeurin is inhibited by the FKBP12-FK506 complex, which impedes the T-lymphocyte signal transduction pathway and the

[a]babita.adhikary@cutm.ac.in, [b]alikkumar3@gmail.com

DOI: 10.1201/9781003672869-17

generation of IL-2. According to the Biopharmaceutics Classification System, tacrolimus is categorized as a class 2 medication (BCS)[4] (Figure 17.1).

For therapeutic usage, pharmaceutical dosage forms include ointments, injections and capsules. This study has developed and validated a unique Tacrolimus in nanoparticles UV visible Spectrophotometric measuring technique. In accordance with ICH requirements, a procedure for assessing analytical performance parameters has been created [5].

Because of its simplicity, selectivity and sensitivity, UV Spectrophotometric procedures are widely used in industrial laboratories [6]. There are no reports of Tacrolimus as a UV method of determination in the literature. In the current work, we devised a straightforward, sensitie and precise UV Spectrophotometric technique to quantify Tacrolimus in pharmaceutical and pure formulations [7].

2. Instrumentation

Throughout the experiment, a double beam UV visible spectrophotometer (UV-1900i, UV Probe, Shimadzu

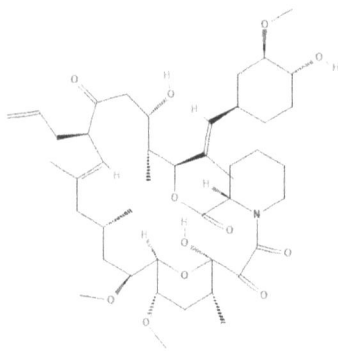

Figure 17.1. Structure of tacrolimus.

Source: Author.

Corporation) with a matching quartz cell with a 1 cm path length was used.

3. Experiment Getting the Standard Solution Ready

A precisely weighed 1 mg tacrolimus sample was added, given time to dissolve and then diluted with chloroform to the appropriate concentration within a volumetric flask with a capacity of 10 ml. A second 10 ml volumetric flask was pipette with about 1 ml of the standard stock solution. Which was then diluted with chloroform to the proper strength to create a working standard solution at a 100 g/ml concentration [8].

A suitable amount of an aliquot was moved to an independent volumetric flask. With a 10ml capacity from a normal Tacrolimus stock solution. Methanol was added to the amount as needed to achieve the desired chloroform concentrations of 2, 4, 6 and 8 mg/ml [9].

4. Material and Method

4.1. Chemicals and reagents

Tacrolimus was received Sarv Bio Labs Pvt. Ltd, Panchkula, Haryana, India. **Sample-1.** Tacrolimus Capsule IP 1.0 mg, Tacrograf (Biocon Biologies), **Sample-2.** Tacrolimus Capsule IP 1.0 mg, Pangraf (Panacea Biotech), **Sample-3.** Tacrolimus capsule IP 1 mg, Cipla (Rolitrans) were purchased from local medicine store (medplus) at Rayagada, Odisha India. Chloroform used was of Nice Chemicals.

4.2. Determination of λ_{max}

Using a UV-Visible spectrophotometer, a calibration curve for the medication

Tacrolimus in methanol (as a solvent) was plotted at 264 nm (max). With a regression coefficient of 0.9999, it demonstrates linearity within the spectrum of concentration of between 0.2 and 0.8 mg/ml. Table 17.1 displays several values and Figure 17.2 displays them graphically [10].

4.3. Calibration curve

Table 17.1. Calibration Curve of Tacrolimus at 264nm

Sl no	Concentration(ug/ml)	Amplitude
1	2	0.0268
2	4	0.0508
3	6	0.0735
4	8	0.0974

Source: Author.

4.4. Preparation of sample solution

Sample1, Sample 2 and Sample 3 each contained about 4 mg of Tacrolimus and the protein was precipitated using 10 milliliters of methanol. Then, Centrifuge for while five minutes at 2000 rpm. The supernatant was suitably diluted and then looked through in the 200–400 nm ultraviolet (uv) range. Tacrolimus concentration at 264 nm was determined using the calibration curve regression equation[11, 12].

4.5. Method of validation

In terms of sensitivity, specificity, linearity, dynamic range, LOD and LOQ, this approach has been demonstrated to be dependable. For every parameter, the relative standard deviations were computed as a percentage. The recommended UV-visible spectrophotometry has been verified in accordance with ICH guidelines[13].

4.5.1. Accuracy

Three distinct concentrations (Low, Medium and High) were added to a sample solution containing a 1 mg/ml concentration in order to assess the precision of the Tacrolimus. Table 17.2 presents the results obtained.

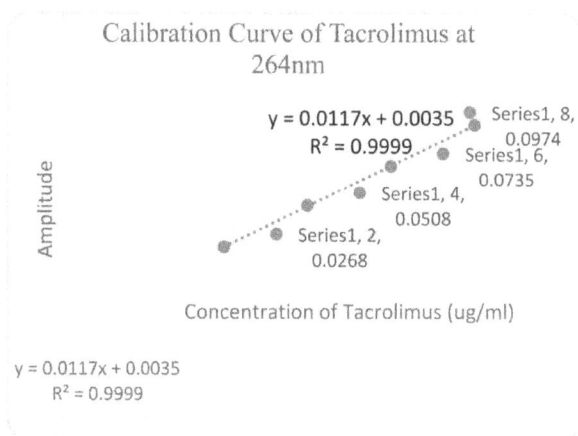

Calibration Curve of Tacrolimus at 264nm

$y = 0.0117x + 0.0035$
$R^2 = 0.9999$

Series1, 8, 0.0974
Series1, 6, 0.0735
Series1, 4, 0.0508
Series1, 2, 0.0268

Amplitude

Concentration of Tacrolimus (ug/ml)

$y = 0.0117x + 0.0035$
$R^2 = 0.9999$

Figure 17.2. Calibration curve of tacrolimus.

Source: Author.

4.5.2. Precision
The degree of consistency among a group of metrics taken during successive controlled samplings of a homogeneous sample defines the precision of an analytical method. The method's weekly, daily and repeatable precision was all expressed as relative standard deviations. The results are shown in Table 17.2 [14].

4.5.3. Repeatability
To assess repeatability, six copies of each concentration of the standard solution were utilized presents the outcomes [15].

4.5.4. Limits of quantification and detection
Small volumes of typical fixes (LOQ) were employed to establish the advanced system's limit of detection (LOD) and limit of quantitation (LOQ) methods.

Table 17.2. Method validation summary for Tacrolimus

Parameters	Tacrolimus
Detection wavelength (nm)	264
(mg/mL) Range of linearity	0.2–0.8
Coefficient of correlation (r2)	0.9999
Regression Equation	Y=0.0117x + 0.0035
The precision, %RSD Intra-day (n=3) Intra-day (n=3) Repeatability of measurement (n=6)	100.65, 0.47% RSD 100.71, 0.69% RSD 101.54, 1.62% RSD
Accuracy (% Recovery, n=3) 50% 100% 150%	99.80 ± 0.34 101.02 ± 1.87 100.48 ± 1.22
LOD (mg/mL)	0.09
LOQ (mg/mL)	0.2

Source: Author.

The outcomes are displayed in Table 17.2 [16, 17] (Figures 17.3–17.5).

UV Spectrophotometric Comparative study of Tacrolimus (Sample 1, Sample 2, Sample 3)

Sample-1 (Biocon Biologics) Tacrolimus Capsule IP 1.0 mg:

Sample-2 (Panacea Biotech) Tacrolimus Capsule IP 1.0 mg:

5. Result and discussion

For this research, a detection wavelength of 264 nanometers (nm) was

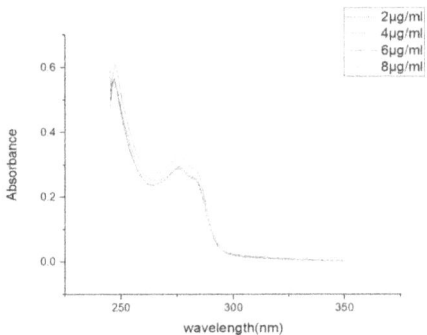

Figure 17.3. UV spectrophotometric graph Sample-1 (Biocon Biologics) Tacrolimus Capsule IP 1.0 mg.

Source: Author.

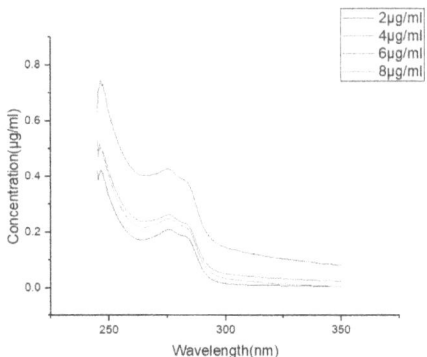

Figure 17.4. UV Spectrophotometric graph Sample-2 (Panacea Biotech) Tacrolimus Capsule IP 1.0 mg.

Source: Author.

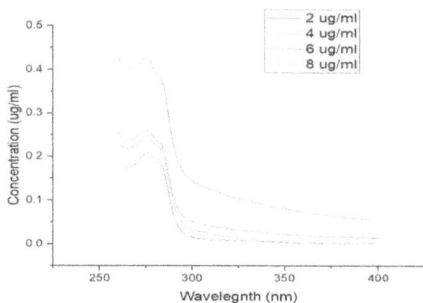

Figure 17.5. UV Spectrophotometric graph Sample-3 (Cipla) Tacrolimus Capsule IP 1mg.

Source: Author.

employed.0.2 milligrams per milliliter the test's linearity range was (mg/mL) to 0.8 mg/mL. This indicates that the technique worked well within this range of concentrations. The data's correlation coefficient, which was determined to be 0.9999, showed a very high degree of correlation and suggested that the analyte measurement technique is quite dependable. The regression equation used for the analysis is $Y = 0.0117x + 0.0035$[18, 19].

When the analysis was repeated the precision was determined to be as follows, three times on the same day:

- 100.65%, with a 0.47% relative standard deviation (RSD).
- 100.71%, with a 0.69% RSD.

When the analysis was repeated six times, the results showed a range of values with a maximum RSD of 1.62%. By running the analysis at various concentrations, the method's accuracy was evaluated:

At 50% concentration, the measured value was 99.80% with a deviation of 0.34%.

- At 100% concentration, the measured value was 101.02% with a deviation of 1.87%.

- At 150% concentration, the measured value was 100.48% with a deviation of 1.22%. These results suggest that the method provides accurate measurements.

The limit of detection for this study was 0.09 mg/mL. Though it may not always be measured, this is the lowest analyte concentration that can be reliably detected. There was a quantitative limit of 0.2 mg/mL. This is the lowest analyte concentration at which it is consistently detectable and quantifiable with reasonable accuracy and precision [20].

These data provide important information about the analytical method's performance, including its accuracy, precision and sensitivity, which are critical for ensuring the reliability of the analytical results in scientific experiments or quality control processes.

6. Conclusion

In conclusion, the comparison of various Tacrolimus products now on the market (Tacrograf, Pangraf and Rolitrans) has provided vital new information about the differences in the efficacy, safety and tolerance of various formulations of this crucial immunosuppressant medication. Tacrolimus is the active ingredient in each of the analyzed medicines, however there are considerable variations in the formulation, delivery method and other elements that can affect the efficacy of the medication and patient outcomes.

According to the results in the comparison analysis, some of the more recent tacrolimus formulations may be superior to previous ones in terms of lowering side effects and enhancing patient compliance. However, more investigation is required to properly comprehend the relative advantages and disadvantages of each product.

The study's findings demonstrate the importance of carefully weighing patient considerations and particular product attributes when choosing a Tacrolimus product for use in clinical practice. Clinicians may ensure that their patients receive the safest and most effective care possible by considering these parameters, hence reducing the likelihood of adverse effects and other problems.

7. Acknowledgement

The Centurion University of Technology and Management, Odisha provided all the facilities required for the research, for which the authors are grateful.

References

[1] Yamashita, K., Nakate, T., Okimoto, K., Ohike, A., Tokunaga, Y., Ibuki, R., Higaki, K., & Kimura, T. (2003). Establishment of new preparation method for solid dispersion formulation of tacrolimus. *Int J Pharm, 267,* 79–91.

[2] Tamura, S., Ohike, A., Ibuki, R., Amidon, G. L., & Yamashita, S. (2002). Tacrolimus is a class ii low-solubility high-permeability drug: The effect of p-glycoprotein efflux on regional permeability of tacrolimus in rats. *J Pharm Sci, 91,* 719–729.

[3] Amidon, G. L., Lennernas, H., Shah, V. P., & Crison, J. R. (1995). A theoretical basis for a biopharmaceutic drug classification: The correlation of in vitro drug product dissolution and in vivo bioavailability. *Pharm Res, 12,* 413–420.

[4] Tamura, S., Ohike, A., Ibuki, R., Amidon, G. L., & Yamashita, S. (2002). Tacrolimus is a class II low-solubility high-permeability drug: the effect of

P-glycoprotein efflux on regional permeability of tacrolimus in rats. *Journal of Pharmaceutical Sciences, 91*(3), 719–729.   https://doi.org/10.1002/jps.10041

[5] Venkataramanan, R., Swaminathan, A., Prasad, T., Jain, A., Zuckerman, S., Warty, V., McMichael, J., Lever, J., Burckart, G., & Starzl, T. (1995). Clinical pharmacokinetics of tacrolimus. *Clinical Pharmacokinetics, 29,* 404–430.

[6] Bowman, L. J., & Brennan, D. C. (2008). The role of tacrolimus in renal transplantation. *Expert Opinion on Pharmacotherapy, 9*(4), 635–643.

[7] Scott, L. J., McKeage, K., Keam, S. J., & Plosker, G. L. (2003). Tacrolimus: A further update of its use in the management of organ transplantation. *Drugs, 63*(12), 1247–1297.

[8] Kremer, J. M., Habros, J. S., Kolba, K. S., Kaine, J. L., Borton, M. A., Mengle-Gaw, L. J., Schwartz, B. D., & Wisemandle, W. (2003). Mekki, for the Tacrolimus-Methotrexate Rheumatoid Arthritis Study Group QA. Tacrolimus in rheumatoid arthritis patients receiving concomitant methotrexate: A six-month, open-label study. *Arthritis & Rheumatism, 48*(10), 2763–2768.

[9] Ruzicka, T., Assmann, T., & Homey, B. (1999). Tacrolimus: The drug for the turn of the millennium?. *Archives of Dermatology, 135*(5), 574–580.

[10] van Gelder, T. (2002). Drug interactions with tacrolimus. *Drug Safety, 25,* 707–712.

[11] Jusko, W. J., Piekoszewski, W., Klintmalm, G. B., Shaefer, M. S., Hebert, M. F., Piergies, A. A., Lee, C. C., Schechter, P., & Mekki, Q. A. (1995). Pharmacokinetics of tacrolimus in liver transplant patients. *Clinical Pharmacology and Therapeutics, 57*(3), 281–290.

[12] Tamura, S., Ohike, A., Ibuki, R., Amidon, G. L., & Yamashita, S. (2002). Tacrolimus is a class II low-solubility high-permeability drug: The effect of

P-glycoprotein efflux on regional permeability of tacrolimus in rats. *Journal of Pharmaceutical Sciences*, *91*(3), 719–729.

[13] Böer, T., Marques, M., & Cardoso, S. G. (2008). Determination of tacrolimus in pharmaceutical formulations by validated spectrophotometric methods. *Revista de Ciências Farmacêuticas Básica e Aplicada, 29*(2).

[14] Camargo, G. A., Lyra, A. M., Barboza, F. M., Fiorin, B. C., Beltrame, F. L., Nadal, J. M., et al. (2021). Validation of analytical methods for Tacrolimus determination in Poly (ε-caprolactone) nanocapsules and identification of drug degradation products. *Journal of Nanoscience and Nanotechnology, 21*(12), 5920–5928.

[15] Sharma, S. M. I. T. A., & Sharma, M. C. (2010). Spectrophotometric and atomic absorption spectrometric determination and validation of azathioprine in API and pharmaceutical dosage form. *J Optoelectron Biomed Mater*, *2*(4), 213–216.

[16] Green, J. M. (1996). Peer reviewed: A practical guide to analytical method validation. *Analytical Chemistry*, *68*(9), 305A–309A.

[17] Gao, L., Li, J., Kasserra, C., Song, Q., Arjomand, A., Hesk, D., & Chowdhury, S. K. (2011). Precision and accuracy in the quantitative analysis of biological samples by accelerator mass spectrometry: application in microdose absolute bioavailability studies. *Analytical Chemistry*, *83*(14), 5607–5616.

[18] Halloran, P. F. (2004). Immunosuppressive drugs for kidney transplantation. *N Engl J Med*, *351*, 2715–2729.

[19] Kino, T., Hatanaka, H., Miyata, S., Inamura, N., Nishiyama, M., Yajima, T., Goto, T., Okuhara, M., Kohsaka, M., Aoki, H., et al. (1987). Fk-506, a novel immunosuppressant isolated from a streptomyces. Ii. Immunosuppressive effect of fk-506 in vitro. *J Antibiot. 40*, 1256–1265.

[20] Joe, J. H., Lee, W. M., Park, Y. J., Joe, K. H., Oh, D. H., Seo, Y. G., Woo, J. S., Yong, C. S., & Choi, H. G. (2010). Effect of the solid-dispersion method on the solubility and crystalline property of tacrolimus. *Int J Pharm Investig, 395*, 161–166.

18 Recent advancement in targeted drug delivery systems for cancer treatment: Updated review

Jyoshna Rani Dash[1,a], Omprakash Pradhan[1], Rajesh Kumar Pradhan[1], Gurudutta Pattnaik[1], and Biswakanth Kar[2]

[1]School of Pharmacy and Life Sciences, Centurion University of Technology and Management, Odisha, India
[2]School of Pharmaceutical Sciences, Siksha O Anusandhan Deemed to be University, Bhubaneswar, India

Abstract: Cancer is a major cause of both death and morbidity. Conventional cancer therapies include targeted therapy, immunotherapy, radiation therapy and chemotherapy. However, obstacles like cytotoxicity, lack of selectivity and multi-drug resistance (MDR) make it difficult to treat cancer effectively. Novel nanoscale targeting technologies that may offer cancer patients new hope have been made possible by protein engineering and materials science developments. For clinical application, several therapeutic nanocarriers have received approval. The development of nanomedicine has become essential for early cancer detection, targeted medication delivery and increasingly individualized cancer treatment strategies. Some drawbacks of conventional chemotherapy management, including low oral bioavailability, chemotherapy resistance, systemic side effects, decreased drug solubility and constrained therapeutic parameters, have been addressed by nanoparticle-based drug delivery systems. Nevertheless, despite the hopes, there are still difficulties with dose and stability in real-world applications of such nanoscale drug delivery systems and care should be taken with the materials' biocompatibility. Considering these factors, academics and small-to-large-scale industry researchers with backgrounds in drug delivery, cancer research and nanotechnology will find this review on nanotechnology drug delivery systems fascinating. Our in-depth analysis leads us to conclude that the most powerful and cutting-edge field of application for nanostructured drug delivery systems is still present today.

Keywords: Multidrug resistance, nanoparticles, chemotherapy, tumor, biocompatible

1. Introduction

Cancer is a potentially fatal disease that causes malignant cells to develop uncontrollably and irregularly. Uncontrolled cell development can cause unwanted growth and reactions that ultimately destroy normal tissues and organs when it invades them [1]. These aberrant cells divide continuously and without control, destroying every tissue they contact. Most malignancies show symptoms after these imperceptible causes have transformed many healthy body cells into malignant ones. These 'rebellion cells' spread throughout the lymphatic system [1]. Approximately 3.4 million fatalities are attributed to cancer globally. Several well-known risk factors for cancer include smoking, which can lead to lung,

[a]jyoshnadash92@gmail.com

DOI: 10.1201/9781003672869-18

breast and ovarian cancer; being over-weight or obese, which can cause different cancers, including breast, colon, lung, prostate, liver, leukemia, ovarian, pancreatic and brain tumors, kidney; radiation exposure, which can cause skin cancer; family history; stress; the environment; and pure luck [2]. Cancer cells can metastasize or create a second tumor by moving throughout the body through lymphatic and blood vessels. Patients are usually given anticancer medicines to kill cancer cells. These medications employ two methods to function: by eliminating the cancer. The cancer cells die through direct chemical agent exposure and the induction of apoptosis [2]. Approximately 200 distinct cancer forms have been documented to date. Surprisingly, viruses such as the human papillomavirus, hepatitis C virus, hepatitis B virus, Epstein-Barr virus, hepatitis C virus and human T-cell lymphotropic virus type 1 also influence the development of several cancer types [3]. In 2018, The World Health Organization reports that, with around 18 million cases and 10 million deaths attributed to the disease, cancer is currently the second greatest cause of death worldwide. Because of how rapidly industrialization is progressing. The death rate from cancer is expected to have nearly doubled by 2040 [4]. Surgical tumor removal, radiation therapy and chemotherapy are examples of conventional cancer therapies that not only kill cancerous cells but also damage good cells in cancer patients. This results in a variety of undesirable side effects, including exhaustion, anemia, internal bleeding and appetite loss. Usually less than 1% of the cells in a tumor are extremely resistant to common therapies.

Cancer stem cells or cancer-initiating cells are these cells [4].

It is generally accepted that gene mutations cause cancer. An estimated 18.1 million new cases of cancer and 9.6 million cancer-related deaths occurred in 2018. According to the Global Cancer Observatory, 30 million people will die from cancer globally by 2030 [5]. Biological therapy, radiation therapy, chemotherapy and surgical resection are common cancer treatments [5]. Malignant solid tumors can be successfully removed with surgery. Particularly in the initial stages of cancer growth. Multiple therapies, including surgery, chemotherapy and radiation, are included in combined therapy. Chemotherapy has been a popular treatment for cancer patients over the years due to its ease of use and convenience [8]. Numerous cancers can be effectively treated with chemotherapy, including lung cancer, Hodgkin's and non-Hodgkin's lymphoma, acute myelogenous leukemia, acute lymphoblastic cancer, germ cell cancer, ovarian cancer and choriocarcinoma [5]. Cancer has a complex etiology that includes both genetic and environmental factors. It is typically possible to identify genetic information modification in cancer cells [6]. For example, Cancer is characterized by widespread epigenome dysregulation, genome instability and mutation. Certain modifications impact how cells function and have a role in cancer development. However, the cancer's phenotype can return to normal by using medication or gene therapy to reverse these changes [6]. Drug delivery systems (DDS) have been employed to administer therapeutic chemicals for treating diseases in preclinical and clinical settings. Oral ingestion or injection are the two administration methods for conventional DDS [7]. Despite the traditional DDS's many benefits, including its simplicity of use, although generally acknowledged by patients, it has significant drawbacks and restrictions.

Limited efficiency: When taken orally, many medications have varied absorption rates. Additionally, certain medications may be broken down before they reach the bloodstream due to the stomach's low pH and digestive enzymes 'action [7]. Lack of selectivity: Because oral drug delivery has poor biodistribution, there are better medication options targeting certain organs. Drug intake may be highly concentrated in detoxifying organs like the liver or kidney, which can lead to toxicity to those organs [7].

The creation of nanoscale drug delivery devices has great potential to lower metabolism since they can dissolve medications in their hydrophilic or hydrophobic compartment and increase the solubility of poorly soluble pharmaceuticals [8]. Furthermore, due to its huge surface-to-volume ratio for drug loading, variable size for modification, prolonged plasma half-life and distinct biodistribution profile, nanomedicine offers an edge over traditional chemotherapy in passive targeting [8]. The second most common cause of mortality globally is cancer. Chemotherapy is a critical component of cancer treatment for both free and undetectable cancer cells. Chemotherapy employs chemicals to either destroy cancer cells or stop them from growing. Chemotherapeutic treatments target fast-growing cells primarily because they are more likely to produce cancer than healthy cells [9]. Nonetheless, the medications also target the rapidly proliferating healthy cells because they can also do so [9]. This unwanted onslaught is the reason why conventional chemotherapy fails. Multidrug resistance (MDR) is yet another important obstacle to effective therapy. When cancer cells become resistant to the cytotoxic medications during or soon after treatment, they may be able to evade the effects of chemotherapy [9]. Thus, the current advancements in targeting nanomaterials for anticancer drug delivery by organelle-specific targeting, tumor microenvironment targeting and cancer cell surface targeting receptors are covered in this study. Given that the delivery of anticancer medications via various channels has revolutionized cancer therapy, special attention has been paid to cancer cell surface and organelle-specific targeting [10]. To do this, active strategies combine molecules that bind to overexpressed antigens or receptors on the target cells with nanocarriers that carry chemotherapy medications [11]. However, even though nanoparticles are very beneficial as drug delivery vehicles, many problems still need to be fixed, such as toxicity, poor tissue dispersion, unstable circulation and low oral bioavailability [11]. This article describes the critical role that nanotechnology plays in cancer research and nanomedicine. The various types and characteristics of nanoparticles, their potential as drug-delivery vehicles to eradicate drug resistance and kill cancer cells and how technological developments in nanoparticles will improve their therapeutic efficacy and functionality in cancer treatments will all be covered in this article [11].

2. Epidemiology

According to WHO the percentage of people affected in various types of cancer worldwide is given below (Figure 18.1).

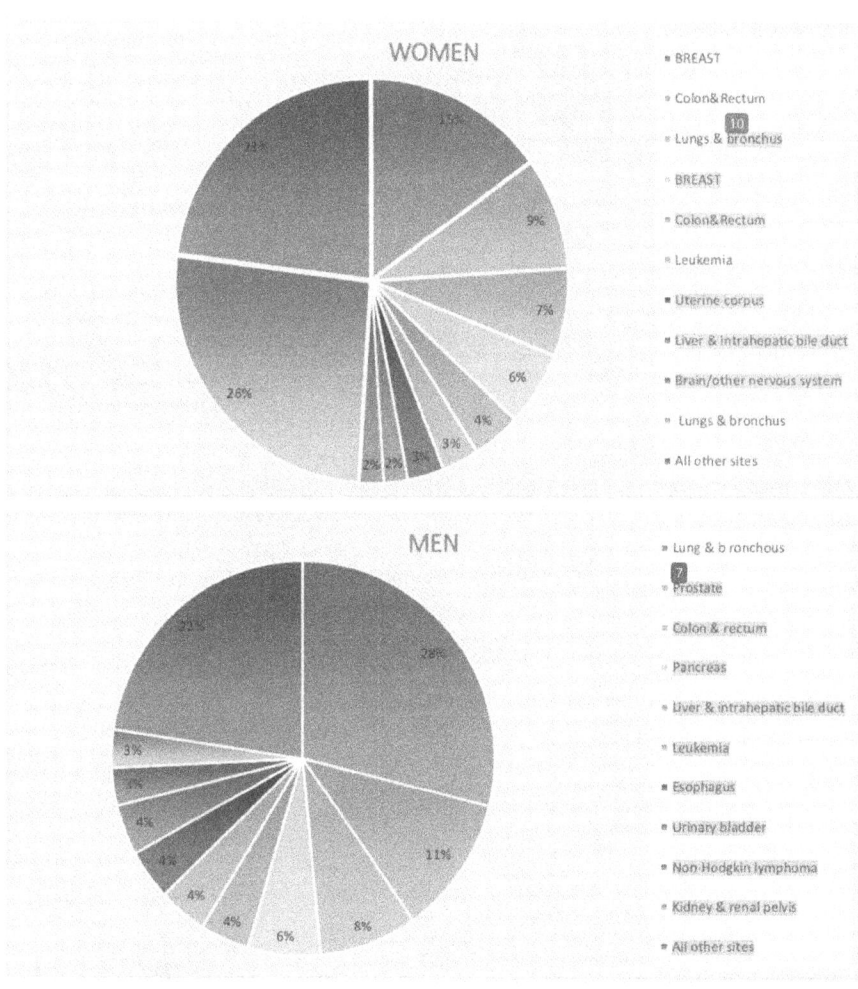

WOMEN

- BREAST
- Colon& Rectum
- Lungs & bronchus
- BREAST
- Colon&Rectum
- Leukemia
- Uterine corpus
- Liver & intrahepatic bile duct
- Brain/other nervous system
- Lungs & bronchus
- All other sites

MEN

- Lung & b ronchous
- Prostate
- Colon & rectum
- Pancreas
- Liver & intrahepatic bile duct
- Leukemia
- Esophagus
- Urinary bladder
- Non-Hodgkin lymphoma
- Kidney & renal pelvis
- All other sites

Figure 18.1. Percentage of affected people in different type of cancer disease.
Source: Author.

3. Pathophysiology

Pathophysiology:

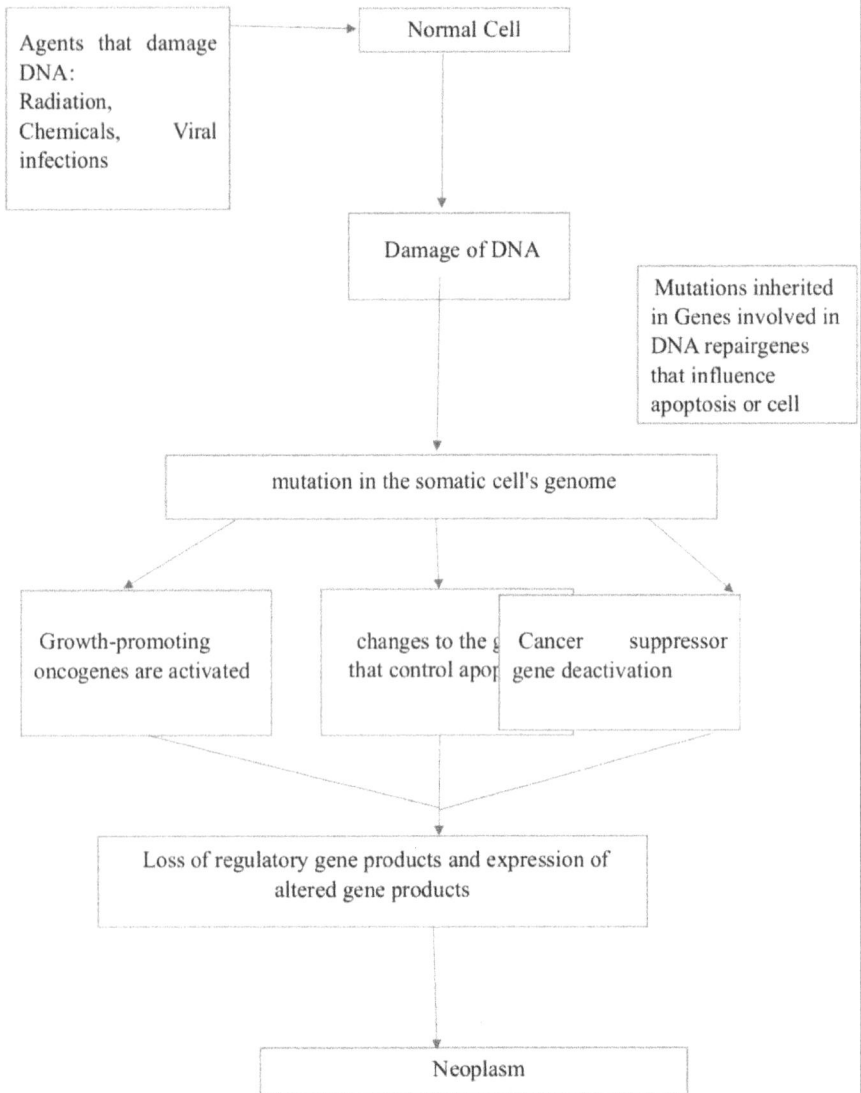

Figure 18.2. General etiology and pathophysiology of cancer.

Source: Author.

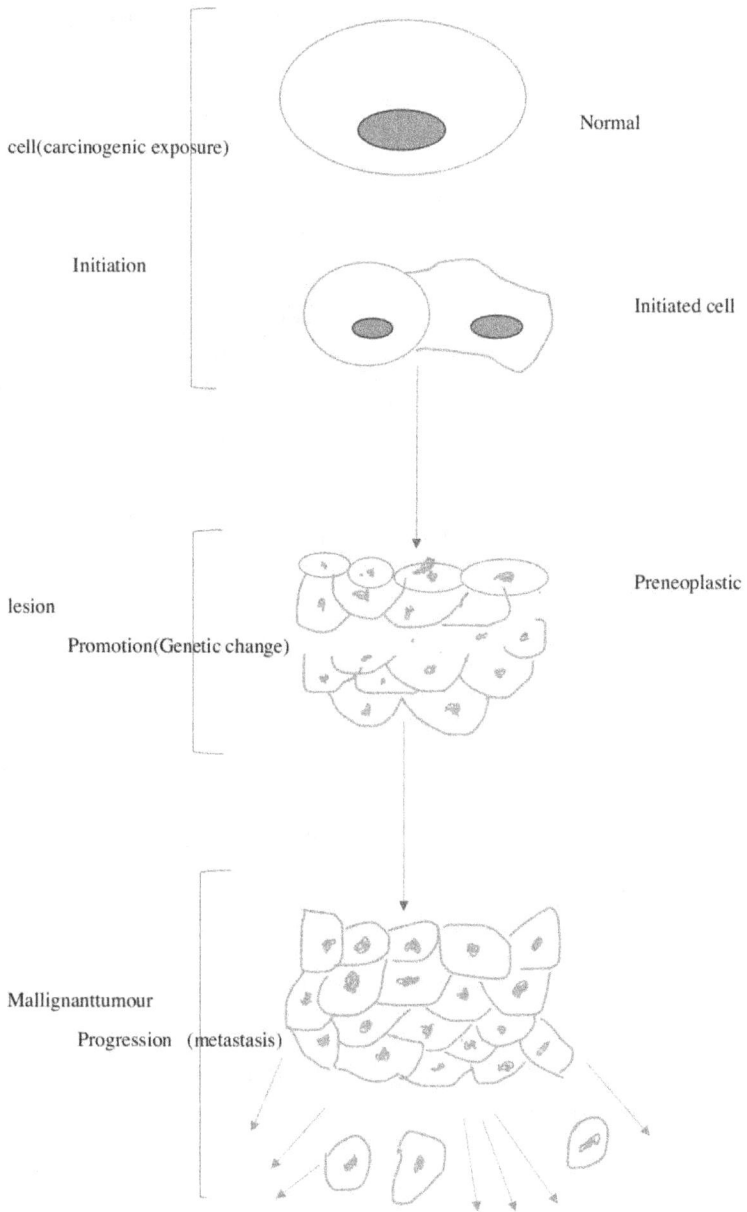

cell(carcinogenic exposure)

Normal

Initiation

Initiated cell

lesion

Preneoplastic

Promotion(Genetic change)

Mallignanttumour

Progression (metastasis)

Figure 18.3. The multistage process of carcinogenesis.

Source: Author.

4. Different Nanomaterials used for Cancer Therapy

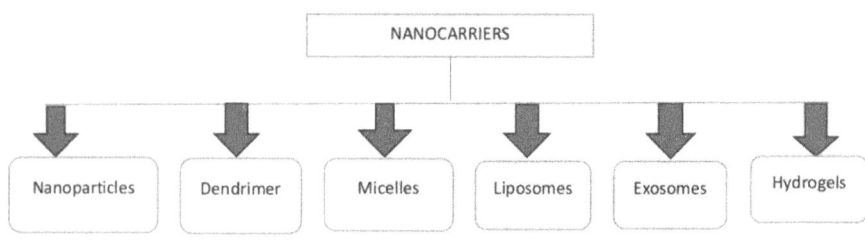

4.1. Nanoparticles used as a carrier

The therapeutic window for anticancer treatments can be expanded by using smart nanoparticles (NPs) to deliver lower medication doses while retaining effective intracellular concentrations. For example, nano polymers have a high drug-loading capacity that minimizes systemic side effects while enabling lesser doses to achieve the same efficacy [12]. Thus, through coordinated targeting mechanisms, smart-NP formulations can improve tumor accumulation and cancer cell selectivity, offering a treatment approach that drastically lowers systemic adverse effects [12, 13]. The term 'nanoparticle' refers to particles having at least one dimension between one and one hundred nanometers. Volume-specific surface area is now used to define nanoparticles. Generally speaking, Particles with a volume-specific surface area (VSSA) of at least 60 m/cm are referred to as nanoparticles [14]. We refer to nanoparticles used as parts of a transport system for other substances as nanocarriers [14]. Particles of nanoscale sizes are known as nanoparticles. Some nanoparticles (NPs) studied extensively are metallic, mAb, polymeric and extracellular vesicles (EVs)

[14]. PNPs are defined as colloidal macromolecules with a size between 10 and 1000 nm, which is less than a micron. PNPs serve as drug carriers, delivering substances to particular malignant sites and ensuring a sustained release. A nanocapsule or nanosphere is created when drugs are encapsulated in or bonded to the surface of nanoparticles [15]. The unique qualities and architectures of these enhanced polymeric nanoparticles provide exceptional benefits. PNPs contribute to the stability of volatile pharmacological drugs. PNPs have a greater loading capacity than free medicines and offer additional delivery options for chemical medications, such as oral and intravenous. Preserving medications from breaking down reduces the amount of harmful substances that can affect healthy tissues [16].

4.1.1. Dendrimer used as a carrier
The hyperbranched specified architecture of dendrimers is a unique class of macromolecules. The most noticeable characteristics of dendrimers are their rapid surface adjustments and extreme branching [17]. The size of these dendrimer polymers is mostly between 1 and 10 nm. Some large dendrimers, however, can have diameters as large as 14 to 15 nm [16–18]. Among other bioscience

domains, dendrimer research has concentrated more on the synthesis and design of biocompatible dendrimers and their application in immunology, drug delivery and the development of vaccines, antimicrobials and antiviral medications [19, 20]. There have been many different types of dendrimers reported; these include dendritic hydrocarbon, porphyrin-based, ionic, silicon-based, phosphorus-based, ionic, carbon/oxygen-based, polypropylene-imine and polylysine dendrimer [21]. The divergent and convergent methods are the dendrimer production techniques that are frequently described. Fritz Vögtle et al. introduced dendrimers in 1978 [22, 23].

4.2. Micelles used as a carrier

Amphiphilic molecules have remarkable self-assembly properties when their hydrophilic and hydrophobic regions are exposed to a solvent [24]. It attracts the polar components of the solvent when the concentration of the hydrophilic solvent exceeds the critical micelle concentration (CMC). According to Shin et al. [24], the hydrophobic regions also stay away from the solvent. Thus, hydrophilic materials form a corona, whereas hydrophobic elements form a core. This arrangement is known as a straight or regular polymeric micelle [25]. Thus, hydrophilic substances create a corona, while hydrophobic substances create a core. According to Letchford et al. [26], this configuration is a straight or regular polymeric micelle. Dialysis or an oil-in-water technique are used instead if the co-polymer is not readily soluble in water [27]. Immature drug release may be revealed by micelles that pass through the CMC [27]. According

to Kohori et al. [28], the equilibrium between the micelle and blood may also be disrupted by the interaction of blood plasma proteins with the absorption of unimers. As previously stated, the problems are usually addressed by forming a disulfide between two polymer chains, which cross-links or connect micellar structures [29]. Different cross-linking techniques include core cross-linked polymer micelles and shell cross-linked polymer micelles [29]. For enhancing the micelle surface, a variety of ligands actively target cancer cells, including folic acid, peptides, carbohydrates, antibodies and aptamers [30]. Cajot state that the anti-cancer drug can be functionalized and released at the appropriate concentration from the micelle's core or corona. As stated by Husseini, et al. [31], oxidation, pH gradients, temperature changes, enzymes and ultrasonography all promote micelle-based SDDSs. Magnetic resonance imaging, ultrasonography and single-photon output are crucial tools for cancer diagnosis and surveillance [32].

4.3. Liposomes used as carriers

Liposomes are round spheres with a hydrophilic core surrounded by a bilayer of different types of amphiphilic lipids, mostly phospholipids [33]. They are still very promising drug transporters and are among the oldest. Generally speaking, the size of liposomes with one or more bilayer membranes can range from 25 nm to 2.5 μm [33]. According to Gregoriadis et al. [34], phospholipids, the main component of the cell membrane, are composed of a hydrophilic head based on phosphate and a hydrophobic tail based on fatty acids. Gregory et al. demonstrated that when in an aqueous solution, phospholipids self-assemble into a bilayered vesicle, with the polar ends facing the water and the non-polar ends forming a bilayer [34]. Drugs that dissolve in water or water may trap the middle of the bilayer [34]. There are two forms of liposomes: uni- and multi-lamellar vesicles, depending on the size and quantity of bilayers. There are two categories of uni-lamellar vesicles: small and large [33]. Numerous methods, such as detergent dialysis, solvent injection, reverse phase evaporation and the thin film hydration method—also referred to as the Bangham method—can be used to create liposomes [35]. The antibody-based method enables liposomes to target tumor tissues actively. To do this, add particular antibodies to the liposomal surface of what are known as immunoliposomes (ILP), which are specific to either the endothelial cells or the cancer cells [36].

4.4. Exosomes used as a carrier

The exosome is an example of a nanosphere with a bilayered membrane made of several proteins and lipids. It is nonimmunogenic since its chemical composition is similar to that of cells in the body. Drug delivery via exosomes is efficient [37]. Because MCF-7 has higher concentrations of different microRNAs, it has been observed that simultaneous and multiplexed detection of microRNAs in a whole exosome is successful [38]. Exosomes containing aflatoxin were produced and studies on drug resistance were carried out. The results showed that medication resistance was elevated when miR-222 was transported by exosomes [39]. By releasing the β-element through microRNAs, drug-resistant exosomes

change the chemoresistance of tumor cells and facilitate information transfer across cells. According to Zhang et al. [40], the anticancer effect of exosomes is ascribed to the transfer of circulating endothelial miRNAs into cancer cells as well as the interaction between the tumor cells and the microenvironment through changes in gene expression. Greater plasmatic miR-503 levels following chemotherapy were discovered to be caused by greater endothelial cell miRNA levels [41]. Phosphatidylethanolamine is the optimal lipid component of U251-MG, which produces fibrosarcoma, breast cancers and the transfer of glioblastoma exosomes into their parent cells. It has also demonstrated comparable efficacy in the targeted delivery of medications [42]. Exosomes laden with docetaxel showed enhanced resistance to treatment because they transported certain miRNA biomolecules [43]. Through the release of exosomes, a drug-resistant tumor cell gives a sensitive BC cell resistance. It was shown that exosomes from drugs like docetaxel and Adriamycin regulate cell death by transferring certain miRNA between cells [44].

4.5. Hydrogels as a carrier

Drug molecules are hidden by a polymeric mesh of hydrophilic polymers dissolved in water to create hydrogels. Because of the expandable nature of these polymeric meshes, the medication can be freed for external dissolution and disintegration [45]. Numerous noteworthy benefits of hydrogels include increased hydrophilicity, high drug-loading capacity, scalability, biocompatibility and simplicity in handling and manufacturing [46]. The aryl hydrocarbon receptor ligands of benzothiazole hydrogels

5F203 and GW 610 showed anticancer action because a double strand in deadly DNA was developed into DNA adducts. Designing prodrugs with high nanomolar potency was proposed by in vitro research and clinical observations [47]. When coupled with cisplatin, thermoreversible chitosan/beta-glycerophosphate hydrogels demonstrated increased anticancer efficacy against human colorectal cancer cells and MCF-7 BC cells [48]. Arg-Gly-Asp (RGD) peptide and estrone produced taxol-loaded self-assembly molecular hydrogels, exhibiting enhanced in vitro anticancer efficacy and targeted administration [49]. Micelles containing docetaxel that functioned as pH-triggered hydrogels showed a greater diffusion rate in intestinal and stomach fluid simulations. Pharmacokinetic findings indicated that docetaxel was present in the micelles. They enhanced the oral administration's bioavailability [50]. To improve medication release and facilitate digitally reversible self-healing with ultrasonic support, ionic cross-linked hydrogels were developed [51]. The results illustrated the advantages of this approach, which permits the planned administration of chemotherapy medications [51].

5. Recent Advancements in Nanoparticle-Based Drug Delivery

Nanoparticle-based drug delivery techniques have shown promise in mitigating the shortcomings of traditional cancer therapies. Nanoparticles can now be used to deliver chemotherapy drugs directly to cancer cells, reducing the negative effects on healthy organs. These advancements can specifically target molecular markers present in cancer cells, increasing therapeutic efficiency while minimizing harm to healthy cells [52]. The prognosis for cancer patients may be considerably improved by these advancements. According to El-Assal [53], drug delivery systems that target cancer cells can increase the effectiveness of therapy and reduce the chance of tumor recurrence. Additionally, reducing harm to healthy cells may improve patients' quality of life both during and after treatment. In addition to medication delivery by nanoparticles, immunotherapies are an intriguing area of research [54]. These therapies may be able to cure pre-existing tumors and stop cancer from returning by making use of the immune system's ability to recognize and destroy malignant cells. By preventing proteins that prevent immune cells from attacking cancer cells, immune checkpoint inhibitors enhance the immune system's capacity to recognize and eliminate cancer cells [55]. Remarkable responses have been seen in some patients with advanced melanoma, with their tumors diminishing or going away entirely. Due to these encouraging outcomes, immune checkpoint inhibitors have been approved for treating several cancers, including kidney, bladder and lung [56]. It is crucial to remember that not every patient responds to these treatments similarly and further investigation is required to determine why some people gain more from these treatments than others [57]. Nevertheless, not every patient benefits from immune checkpoint inhibitors. Sometimes, the treatment is ineffective in stimulating the immune system, which leads to little or no tumor shrinkage. This emphasizes the need for additional study and a deeper comprehension of the variables affecting immunotherapies' efficacy in various patients [58]. It is crucial to remember that immune checkpoint inhibitors may cause adverse effects connected to the immune system in certain people. These adverse effects may range from mild to severe and call for additional medical care. As a result, healthcare providers need to keep a close eye on patients undergoing immunotherapy and make preparations to manage any potential side effects [59].

6. Challenges in Delivering Cancer Nano Drugs

By using the immune system to target and eliminate cancer cells, immunotherapies have transformed the treatment of cancer; nevertheless, further research is required to determine how these treatments may impact a woman's chance of acquiring breast cancer [60]. Researchers want to improve treatment strategies and ensure patient safety by examining the long-term effects of immunotherapies and their impact on breast cancer risk factors. The possibility that immunotherapies may put women at higher risk of breast cancer is one area of worry. Having a thorough understanding of how immunotherapies affect breast cancer risk factors can assist medical providers in selecting the best course of action and delivering individualized care. Furthermore, determining any correlations between immunotherapy and breast cancer risk can aid in creating plans to reduce risks and enhance patient outcomes [61]. An extensive investigation might examine how a particular immunotherapy medication affects breast cancer risk factors over the long run in a sample of individuals with the disease. The study would compare the patients with a control group that did not receive immunotherapy, assess changes in the patient's risk factors and track the patients' development over several years [62]. A subset of individuals who received the immunotherapy medication and showed a decrease in risk variables rather than an increase would be a thorough counterexample. This could be brought on by a person's reaction to the medication, lifestyle modifications, or genetic variations. Before making firm judgments about the drug's impact on breast cancer risk factors, researchers must consider these confounding variables. To ascertain the underlying reasons for some patients' reduced risk factors following immunotherapy, more research should be conducted [63]. Examining potential lifestyle modifications, genetic variations and individual drug reactions can provide a more complete picture of the medicine's impact on breast cancer risk factors. This will enable researchers to make more accurate assessments and recommendations on the use of immunotherapy in the treatment of breast cancer [64]. Giving cancer nanodrugs is made more difficult by the immune system's capacity to either eliminate nanoparticles or collect them in undesirable areas, which reduces their effectiveness. Because of the tumor's abnormal blood vessels and thick extracellular matrix, nanodrugs find it difficult to penetrate and spread across it [65]. Targeting ligands, mainly designed to recognize and bind to cancer cell receptors on nanoparticle surfaces, are incorporated into targeted drug delivery systems to circumvent these challenges. This strategy reduces off-target effects while increasing treatment efficacy. Several targeted ligands have been investigated, including

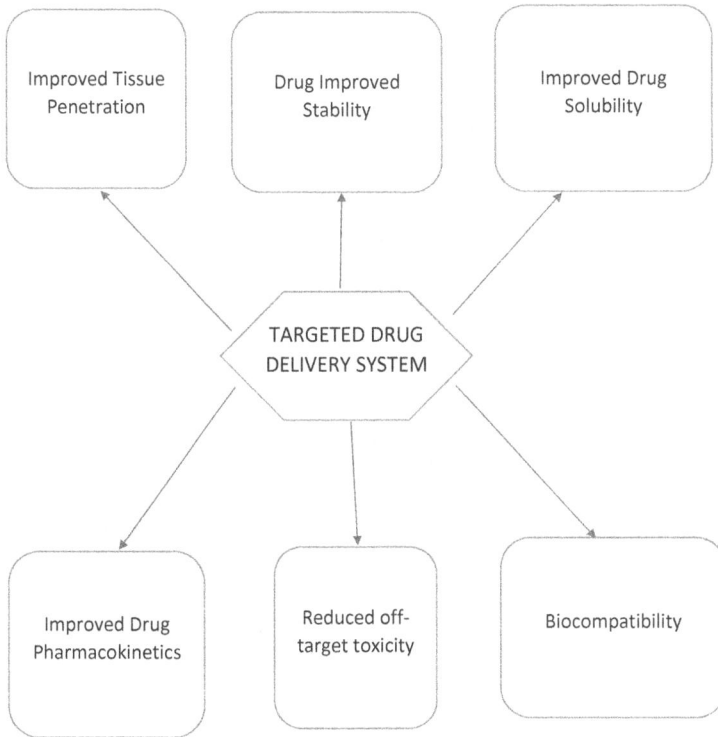

Figure 18.4. Benefits of using a targeted delivery system for drug.

Source: Author.

aptamers, peptides and antibodies [66]. To get over these obstacles, targeting ligands—which are primarily made to identify and attach to cancer cell receptors on the surface of nanoparticles—are added to tailored drug delivery systems. This approach improves therapeutic efficacy while decreasing off-target effects. Aptamers, peptides and antibodies are among the several targeted ligands that have been studied [65, 67–72].

7. Conclusion

Delivering the drug molecule to its intended location within an organism's intricate cellular network is challenging. In the end,

targeted drug delivery is emerging as one of the most cutting-edge medical science techniques for the detection and management of a few deadly illnesses. As a result, there are numerous targets throughout the body and countless ways to deliver medications to those locations. Preclinical data for several disorders have been significantly produced in recent years. As we will see throughout this chapter, several in vitro and in vivo investigations have been carried out to demonstrate the advantages of tailoring pharmaceutical delivery. Recent advancements in Nanoparticle design have enabled drug-delivery systems to be created beyond several physiological and therapeutic obstacles from traditionally administering chemotherapy drugs. By employing multifunctional targeting approaches, smart medicine delivery seeks to pinpoint treatment to tumors in order to decrease cytotoxicity and improve the therapeutic index. In the form of nanoparticle-delivery devices, this quickly developing field presents a fresh approach to the practical application of customized healthcare; yet, success in this attempt requires a greater understanding of the tumor pathophysiology of the patient. By controlling the release of the therapeutic ingredients, these nanoparticles can provide a customized, long-lasting and efficient treatment. Furthermore, as nanotechnology develops, personalized treatment strategies catered to the need of specific patients can become feasible, potentially transforming the way cancer is treated. Readers are encouraged to examine each technique in detail by reading this article to obtain a deeper understanding.

References

[1] Estanqueiro, M., Amaral, M. H., Conceição, J., & Lobo, J. M. S. (2015). Nanotechnological carriers for cancer chemotherapy: the state of the art. *Colloids and surfaces B: Biointerfaces, 126*, 631–648.

[2] Olusanya, T. O., Haj Ahmad, R. R., Ibegbu, D. M., Smith, J. R., & Elkordy, A. A. (2018). Liposomal drug delivery systems and anticancer drugs. *Molecules, 23*(4), 907.

[3] Jabir, N. R., Tabrez, S., Ashraf, G. M., Shakil, S., Damanhouri, G. A., & Kamal, M. A. (2012). Nanotechnology-based approaches in anticancer research. *International Journal of Nanomedicine*, 4391–4408.

[4] Ertas, Y. N., Abedi Dorcheh, K., Akbari, A., & Jabbari, E. (2021). Nanoparticles for targeted drug delivery to cancer stem cells: A review of recent advances. *Nanomaterials, 11*(7), 1755.

[5] Cheng, Z., Li, M., Dey, R., & Chen, Y. (2021). Nanomaterials for cancer therapy: Current progress and perspectives. *Journal of Hematology & Oncology, 14*, 1–27.

[6] Cheng, Y., He, C., Wang, M., Ma, X., Mo, F., Yang, S., Han, J., & Wei, X. (2019). Targeting epigenetic regulators for cancer therapy: Mechanisms and advances in clinical trials. *Signal Transduction and Targeted Therapy, 4*(1), 62.

[7] Dang, Y., & Guan, J. (2020). Nanoparticle-based drug delivery systems for cancer therapy. *Smart Materials in Medicine, 1*, 10–19.

[8] Tang, L., Li, J., Zhao, Q., Pan, T., Zhong, H., & Wang, W. (2021). Advanced and innovative nano-systems for anticancer targeted drug delivery. *Pharmaceutics, 13*(8), 1151.

[9] Hossen, S., Hossain, M. K., Basher, M. K., Mia, M. N. H., Rahman, M. T., & Uddin, M. J. (2019). Smart nanocarrier-based drug delivery systems for cancer therapy and toxicity studies: A review. *Journal of Advanced Research, 15*, 1–18.

[10] Dutta, B., Barick, K. C., & Hassan, P. A. (2021). Recent advances in active targeting of nanomaterials for anticancer drug delivery. *Advances in Colloid and Interface Science, 296*, 102509.

[11] Dadwal, A., Baldi, A., & Kumar Narang, R. (2018). Nanoparticles as carriers for drug delivery in cancer. *Artificial Cells, Nanomedicine, and Biotechnology, 46*(sup2), 295–305.

[12] Kalimuthu, K., Lubin, B. C., Bazylevich, A., Gellerman, G., Shpilberg, O., Luboshits, G., & Firer, M. A. (2018). Gold nanoparticles stabilize peptide-drug-conjugates for sustained targeted drug delivery to cancer cells. *Journal of Nanobiotechnology, 16*, 1–13.

[13] Cheng, Z., Al Zaki, A., Hui, J. Z., Muzykantov, V. R., & Tsourkas, A.

(2012). Multifunctional nanoparticles: Cost versus benefit of adding targeting and imaging capabilities. *Science, 338*(6109), 903–910.

[14] Kreyling, W. G., Semmler-Behnke, M., & Chaudhry, Q. (2010). A complementary definition of nanomaterial. *Nano Today, 5*(3), 165–168.

[15] Masood, F. (2016). Polymeric nanoparticles for targeted drug delivery system for cancer therapy. *Materials Science and Engineering: C, 60,* 569–578.

[16] Martín-Saldaña, S., Palao-Suay, R., Aguilar, M. R., Ramírez-Camacho, R., & San Román, J. (2017). Polymeric nanoparticles loaded with dexamethasone or α-tocopheryl succinate to prevent cisplatin-induced ototoxicity. *Acta Biomaterialia, 53,* 199–210.

[17] Baker Jr, J. R. (2009). Dendrimer-based nanoparticles for cancer therapy. *ASH Education Program Book, 2009*(1), 708–719.

[18] Lim, J., Kostiainen, M., Maly, J., Da Costa, V. C., Annunziata, O., Pavan, G. M., & Simanek, E. E. (2013). Synthesis of large dendrimers with the dimensions of small viruses. *Journal of the American Chemical Society, 135*(12), 4660–4663.

[19] Gillies, E. R., & Frechet, J. M. (2005). Dendrimers and dendritic polymers in drug delivery. *Drug Discovery Today, 10*(1), 35–43.

[20] Menjoge, A. R., Kannan, R. M., & Tomalia, D. A. (2010). Dendrimer-based drug and imaging conjugates: design considerations for nanomedical applications. *Drug Discovery Today, 15*(5–6), 171–185.

[21] Majoros, I. J., Williams, C. R., Tomalia, D. A., & Baker Jr, J. R. (2008). New dendrimers: Synthesis and characterization of POPAM– PAMAM hybrid dendrimers. *Macromolecules, 41*(22), 8372–8379.

[22] Caminade, A. M. (2017). Phosphorus dendrimers for nanomedicine. *Chemical Communications, 53*(71), 9830–9838.

[23] Vögtle, F., Richardt, G., & Werner, N. (2009). Types of Dendrimers and their Syntheses. *Dendrimer Chemistry: Concepts, Syntheses, Properties, Applications,* 81–167.

[24] Shin, D. H., Tam, Y. T., & Kwon, G. S. (2016). Polymeric micelle nanocarriers in cancer research. *Frontiers of Chemical Science and Engineering, 10,* 348–359.

[25] Cagel, M., Tesan, F. C., Bernabeu, E., Salgueiro, M. J., Zubillaga, M. B., Moretton, M. A., & Chiappetta, D. A. (2017). Polymeric mixed micelles as nanomedicines: Achievements and perspectives. *European Journal of Pharmaceutics and Biopharmaceutics, 113,* 211–228.

[26] Letchford, K., & Burt, H. (2007). A review of the formation and classification of amphiphilic block copolymer nanoparticulate structures: Micelles, nanospheres, nanocapsules and polymersomes. *European Journal of Pharmaceutics and Biopharmaceutics, 65*(3), 259–269.

[27] Liu, J., Xiao, Y., & Allen, C. (2004). Polymer–drug compatibility: A guide to the development of delivery systems for the anticancer agent, ellipticine. *Journal of Pharmaceutical Sciences, 93*(1), 132–143.

[28] Kohori, F., Yokoyama, M., Sakai, K., & Okano, T. (2002). Process design for efficient and controlled drug incorporation into polymeric micelle carrier systems. *Journal of Controlled Release, 78*(1–3), 155–163.

[29] Sutton, D., Nasongkla, N., Blanco, E., & Gao, J. (2007). Functionalized micellar systems for cancer targeted drug delivery. *Pharmaceutical Research, 24,* 1029–1046.

[30] Cajot, S., Schol, D., Danhier, F., Préat, V., Gillet De Pauw, M. C., & Jérôme, C. (2013). In Vitro Investigations of Smart Drug Delivery Systems Based on Redox-S ensitive Cross-L inked Micelles. *Macromolecular Bioscience, 13*(12), 1661–1670.

[31] Husseini, G. A., Runyan, C. M., & Pitt, W. G. (2002). Investigating the mechanism of acoustically activated uptake of drugs from Pluronic micelles. *Bmc Cancer, 2*, 1–6.

[32] Blanco, E., Kessinger, C. W., Sumer, B. D., & Gao, J. (2009). Multifunctional micellar nanomedicine for cancer therapy. *Experimental Biology and Medicine, 234*(2), 123–131.

[33] Akbarzadeh, A., Rezaei-Sadabady, R., Davaran, S., Joo, S. W., Zarghami, N., Hanifehpour, Y., Samiei, M., Kouhi, M., & Nejati-Koshki, K. (2013). Liposome: Classification, preparation, and applications. *Nanoscale Research Letters, 8*, 1–9.

[34] Gregoriadis, G. (1973). Drug entrapment in liposomes. *FEBS Letters, 36*(3), 292–296.

[35] Zumbuehl, O., & Weder, H. G. (1981). Liposomes of controllable size in the range of 40 to 180 nm by defined dialysis of lipid/detergent mixed micelles. *Biochimica et Biophysica Acta (BBA)-Biomembranes, 640*(1), 252–262.

[36] Kunjachan, S., Ehling, J., Storm, G., Kiessling, F., & Lammers, T. (2015). Noninvasive imaging of nanomedicines and nanotheranostics: Principles, progress, and prospects. *Chemical Reviews, 115*(19), 10907–10937.

[37] Ha, D., Yang, N., & Nadithe, V. (2016). Exosomes as therapeutic drug carriers and delivery vehicles across biological membranes: Current perspectives and future challenges. *Acta Pharmaceutica Sinica B, 6*(4), 287–296.

[38] Lee, J. H., Kim, J. A., Jeong, S., & Rhee, W. J. (2016). Simultaneous and multiplexed detection of exosome microRNAs using molecular beacons. *Biosensors and Bioelectronics, 86*, 202–210.

[39] Yu, D. D., Wu, Y., Zhang, X. H., Lv, M. M., Chen, W. X., Chen, X., Yang, S. J., Shen, H., Zhong, S. L., Tang, J. H., & Zhao, J. H. (2016). Exosomes from adriamycin-resistant breast cancer cells transmit drug resistance partly by delivering miR-222. *Tumor Biology, 37*, 3227–3235.

[40] Zhang, J., Zhang, H. D., Yao, Y. F., Zhong, S. L., Zhao, J. H., & Tang, J. H. (2015). β-elemene reverses chemoresistance of breast cancer cells by reducing resistance transmission via exosomes. *Cellular Physiology and Biochemistry, 36*(6), 2274–2286.

[41] Bovy, N., Blomme, B., Frères, P., Dederen, S., Nivelles, O., Lion, M., Carnet, O., Martial, J. A., Noël, A., Thiry, M., & Jérusalem, G. (2015). Endothelial exosomes contribute to the antitumor response during breast cancer neoadjuvant chemotherapy via microRNA transfer. *Oncotarget, 6*(12), 10253.

[42] Singh, R., Pochampally, R., Watabe, K., Lu, Z., & Mo, Y.Y. (2014). Exosome-mediated transfer of miR-10b promotes cell invasion in breast cancer. *Molecular Cancer, 13*, 1–11.

[43] Chen, W. X., Cai, Y. Q., Lv, M. M., Chen, L., Zhong, S. L., Ma, T. F., Zhao, J. H., & Tang, J. H. (2014). Exosomes from docetaxel-resistant breast cancer cells alter chemosensitivity by delivering microRNAs. *Tumor Biology, 35*, 9649–9659.

[44] Chen, W. X., Liu, X. M., Lv, M. M., Chen, L., Zhao, J. H., Zhong, S. L., Ji, M. H., Hu, Q., Luo, Z., Wu, J. Z., & Tang, J. H. (2014). Exosomes from drug-resistant breast cancer cells transmit chemoresistance by a horizontal transfer of microRNAs. *PLoS One, 9*(4), e95240.

[45] Bawarski, W. E., Chidlowsky, E., Bharali, D. J., & Mousa, S. A. (2008). Emerging nanopharmaceuticals. *Nanomedicine: Nanotechnology, Biology and Medicine, 4*(4), 273–282.

[46] Caló, E., & Khutoryanskiy, V. V. (2015). Biomedical applications of hydrogels: A review of patents and commercial products. *European Polymer Journal, 65*, 252–267.

[47] Stone, E. L., Citossi, F., Singh, R., Kaur, B., Gaskell, M., Farmer, P. B., Monks, A., Hose, C., Stevens, M. F., Leong, C. O., & Stocks, M. (2015). Antitumour benzothiazoles. Part 32: DNA adducts and double strand breaks correlate with activity; synthesis of 5F203 hydrogels for local delivery. *Bioorganic & Medicinal Chemistry, 23*(21), 6891–6899.

[48] Abdel-Bar, H. M., Abdel-Reheem, A. Y., Osman, R., Awad, G. A., & Mortada, N. (2014). Defining cisplatin incorporation properties in thermosensitive injectable biodegradable hydrogel for sustained delivery and enhanced cytotoxicity. *International Journal of Pharmaceutics, 477*(1–2), 623–630.

[49] Shu, C., Li, R., Yin, Y., Yin, D., Gu, Y., Ding, L., & Zhong, W. (2014). Synergistic dual-targeting hydrogel improves targeting and anticancer effect of Taxol in vitro and in vivo. *Chemical Communications, 50*(97), 15423–15426.

[50] Wang, Y., Chen, L., Tan, L., Zhao, Q., Luo, F., Wei, Y., & Qian, Z. (2014). PEG–PCL based micelle hydrogels as oral docetaxel delivery systems for breast cancer therapy. *Biomaterials, 35*(25), 6972–6985.

[51] Segovia, N., Pont, M., Oliva, N., Ramos, V., Borrós, S., & Artzi, N. (2015). Hydrogel doped with nanoparticles for local sustained release of siRNA in breast cancer. *Advanced Healthcare Materials, 4*(2), 271–280.

[52] Shi, S., Zhang, Z., Luo, Z., Yu, J., Liang, R., Li, X., & Chen, H. (2015). Chitosan grafted methoxy poly (ethylene glycol)-poly (ε-caprolactone) nanosuspension for ocular delivery of hydrophobic diclofenac. *Scientific Reports, 5*(1), 11337.

[53] El-Assal, M. I. A. (2017). Acyclovir loaded solid lipid nanoparticle based cream: A novel drug delivery system. *International Journal of Drug Delivery Technology, 7*(1), 52–62.

[54] Rodriguez-Aller, M., Guinchard, S., Guillarme, D., Pupier, M., Jeannerat,

D., Rivara-Minten, E., Veuthey, J. L., & Gurny, R. (2015). New prostaglandin analog formulation for glaucoma treatment containing cyclodextrins for improved stability, solubility and ocular tolerance. *European Journal of Pharmaceutics and Biopharmaceutics, 95*, 203–214.

[55] Wang, G., & Uludag, H. (2008). Recent developments in nanoparticle-based drug delivery and targeting systems with emphasis on protein-based nanoparticles. *Expert Opinion on Drug Delivery, 5*(5), 499–515.

[56] Salama, H. A., Ghorab, M., Mahmoud, A. A., & Abdel Hady, M. (2017). PLGA nanoparticles as subconjunctival injection for management of glaucoma. *Aaps Pharmscitech, 18*, 2517–2528.

[57] Nie, Y., Fu, G., & Leng, Y. (2023). Nuclear delivery of nanoparticle-based drug delivery systems by nuclear localization signals. *Cells, 12*(12), 1637.

[58] Abo-zeid, Y., & Garnett, M. C. (2020). Polymer nanoparticle as a delivery system for ribavirin: Do nanoparticle avoid uptake by Red Blood Cells?. *Journal of Drug Delivery Science and Technology, 56*, 101552.

[59] Bal-Öztürk, A., Tietilu, S. D., Yücel, O., Erol, T., Akgüner, Z. P., Darıcı, H., Alarcin, E., & Emik, S. (2023). Hyperbranched polymer-based nanoparticle drug delivery platform for the nucleus-targeting in cancer therapy. *Journal of Drug Delivery Science and Technology, 81*, 104195.

[60] Khan, I. U., Serra, C. A., Anton, N., & Vandamme, T. F. (2015). Production of nanoparticle drug delivery systems with microfluidics tools. *Expert Opinion on Drug Delivery, 12*(4), 547–562.

[61] Ahn, J., Ko, J., Lee, S., Yu, J., Kim, Y., & Jeon, N. L. (2018). Microfluidics in nanoparticle drug delivery; from synthesis to pre-clinical screening. *Advanced Drug Delivery Reviews, 128*, 29–53.

[62] Al-Obaidi, H., & Florence, A. T. (2015). Nanoparticle delivery and particle

diffusion in confined and complex environments. *Journal of Drug Delivery Science and Technology*, *30*, 266–277.

[63] Yhee, J. Y., Im, J., & Nho, R. S. (2016). Advanced therapeutic strategies for chronic lung disease using nanoparticle-based drug delivery. *Journal of Clinical Medicine*, *5*(9), 82.

[64] Hallaj-Nezhadi, S., & Hassan, M. (2015). Nanoliposome-based antibacterial drug delivery. *Drug Delivery*, *22*(5), 581–589.

[65] Elumalai, K., Srinivasan, S., & Shanmugam, A. Biomedical Technology.

[66] Le Saux, S., Aubert-Pouëssel, A., Mohamed, K. E., Martineau, P., Guglielmi, L., Devoisselle, J. M., Legrand, P., Chopineau, J., & Morille, M. (2021). Interest of extracellular vesicles in regards to lipid nanoparticle based systems for intracellular protein delivery. *Advanced Drug Delivery Reviews*, *176*, 113837.

[67] Zazo, H., Colino, C. I., Gutiérrez-Millán, C., Cordero, A. A., Bartneck, M., & Lanao, J. M. (2022). Physiologically based pharmacokinetic (PBPK) model of gold nanoparticle-based drug delivery system for stavudine biodistribution. *Pharmaceutics*, *14*(2), 406.

[68] Alqaheem, Y., & Alomair, A. A. (2020). Microscopy and spectroscopy techniques for characterization of polymeric membranes. *Membranes*, *10*(2), 33.

[69] Baiee, R. M. (2019). *Generation of ultra-fine nanoparticles by laser ablation in liquid*. The University of Manchester (United Kingdom).

[70] Kharisov, B. I., Kharissova, O. V., Rasika Dias, H. V., Ortiz Méndez, U., Gómez De La Fuente, I., Peña, Y., & Vázquez Dimas, A. (2016). *Iron-based nanomaterials in the catalysis* (pp. 35–68). Rijeka, Croatia: InTech.

[71] Chalayon, P., & Tanwongsan, C. (2021). Antibacterial effects of copper microparticles/copper nanoparticles/copper (II) oxide nanoparticles and copper microparticles/copper nanoparticles/copper (I) oxide nanoparticles from ultrasono-electrochemical with hydrothermal assisted synthesis method. *Engineering Journal*, *25*(6), 55–66.

[72] Wijaya, C. J., Ismadji, S., & Gunawan, S. (2021). A review of lignocellulosic-derived nanoparticles for drug delivery applications: Lignin nanoparticles, xylan nanoparticles, and cellulose nanocrystals. *Molecules*, *26*(3), 676.

19 IONPs as strategic nanotools for cancer therapy: Concise updates on facile synthetic methods and stabilization

Jnanranjan Panda[1,a], Sagar Rout[2,b], Gurudutta Pattnaik[3,c], Himansu Bhusan Samal[3,d], Bikash Ranjan Jena[3,e], and Bhabani Sankar Satapathy[4,f]

[1]Faculty of Science, Sri Sri University, Cuttack, Odisha, India
[2]School of Pharmaceutical Sciences, Siksha O Anusandhan University, Bhubaneswar, Odisha, India
[3]School of Pharmacy and Life Sciences, Centurion University of Technology and Management, Bhubaneswar, Odisha, India
[4]GITAM School of Pharmacy, GITAM Deemed to be University, Hyderabad campus, Telangana, India

Abstract: Iron oxide-based nanoparticles (IONPs) have recently garnered interest among formulation scientists as facile, alternative, economical, scalable drug delivery platform for various diseases including cancer. As compared to lipoidal or polymeric core, IONPs are more stable, durable with easy synthetic steps. Over the years, IONPs have been emerged as influential theranostic tool in cancer. IONPs, out of various structural ad morphological variants, superparamagnetic type (SPIONPs) is considered suitable for pre-clinical applications. Further, to render biocompatibility, SPIONP's's surface can be modified with various hydrophilic organic polymers like poly ethylene glycol, dextran, chitosan etc. Such surface manipulation in fact renders superior *in vivo* stability at systemic circulation as well as active sites for loading of therapeutic molecules. The drug/diagnostic agents may be encapsulated inside molecular pockets of the larger three-dimensional branched polymeric structure or can be conjugated *via* some chemical reaction through suitable cross-linking agent. Present review discusses on various facile, scalable synthetic methods adapted for the fabrication of SPIONPs with a special focus on various stabilization strategies necessary for their effective biological applications.

Keywords: Iron oxide nanoparticles, synthetic method, stabilization, cancer applications

1. Introduction

Cancer has been recognized as one of the deadliest diseases in recent years, with a staggering death rate worldwide. According to global cancer statistics, approximately 9.7 million deaths related to cancer were reported in 2022, with more than 20 million new cancer cases diagnosed [6]. Among all types of cancer, breast cancer ranks highest in terms of morbidity and mortality among women in Western countries. In 2020, out of 2.3 million new breast cancer cases reported worldwide, 684,996 deaths were attributed due to this cancer

[a]jnanpanda@gmail.com, [b]routsagar5577@gmail.com, [c]gurudutta.pattnaik@cutm.ac.in, [d]himansubhusan.samal@cutm.ac.in, [e]bsatapat@gitam.edu, [f]bbhabanisatapathy@yahoo.com

DOI: 10.1201/9781003672869-19

only [6]. Apart from breast cancer, other cancers including prostate, lung and colon cancer are also been identified as some of the leading deadly diseases, with concerning high death rates across the globe. As a result, advancements in cancer theranostics have attracted significant attention, especially in the post-COVID era.

To date, the primary methods for reducing cancer mortality have included early diagnosis, invasive surgery, immunotherapy, radiation therapy, targeted drug delivery and other advanced treatments. However, once cancer cells begin to metastasize, chemotherapy often remains the only available option. Unfortunately, conventional chemotherapy has significant drawbacks, primarily due to the serious adverse effects it causes during clinical application on patients. Therefore, a major challenge in cancer treatment is achieving cancer cell-specific anticancer drugs delivery while minimizing the adverse effects. Similarly, traditional imaging modalities such as X-ray computed tomography (CT), fluorescence imaging (FI) and magnetic resonance imaging (MRI) also have inherent limitations and cannot locate cancer cells for targeted cancer therapy. These include low sensitivity, limited tissue penetration depth, high costs and the use of radioactive substances. These limitations highlight the urgent need for the development of more effective and innovative theranostic strategies using highly sensitive probe.

In recent times, nanomaterials and nanotechnology has garnered substantial attention in biomedical research and have a significantly influence on global economy [18, 23, 29]. Nanoparticles are ultrafine particles with sizes less than 100 nm, lying in the transition region between molecular and microscopic structures. Nanomedicine has emerged at the forefront of cancer research due to its ability to provide a robust theranostic approach, addressing many of the significant limitations associated with conventional treatment modalities [26, 53, 54]. In nanomedicine, various types of nanoparticles, *viz.*, polymeric, magnetic, carbon-based systems, silica, gold etc. are investigated in which chemotherapeutic drugs are either encapsulated or incorporated into biocompatible polymeric shell [9, 26, 45, 52]. This creates multifunctional nanoformulations designed for integrated therapy and imaging applications. Further, these formulations have the potential to enhance the water solubility of the anticancer drug and extend the blood circulation profile of drug and suppress or eliminate faster renal excretion. This nanoformulation enhances cell-specific drug accumulation and enables controlled activation of the delivered drug at the tumor site, thereby reducing the harmful side effects typically associated with anticancer drugs [9, 45, 49, 52].

In this context, one of the most extensively explored classes of magnetic nanosystems for drug delivery is superparamagnetic iron oxides based nanostructured materials (SPIONPs) [23, 26]. Compared to other nanoparticle-based drug delivery systems, SPIONPs-based materials particularly magnetite (Fe_3O_4) and meghamite (γ-Fe_2O_3) offer dual benefits: they enhance drug delivery efficacy and facilitate imaging and monitoring of treatment outcomes because of their favourable magnetic properties, low toxicity, biocompatibility and high chemical stability [18, 26]. The superparamagnetic behaviour of IONPs was observed at room temperature at sizes smaller than 30nm. This property is characterized by the absence of hysteresis (no remanence or coercivity) when the external magnetic field is removed. Under the influence of a magnetic field, they demonstrate strong paramagnetic behaviour with high saturation magnetization and susceptibility [26].

SPIONPs can be engineered to deliver a wide variety of cytotoxic anticancer drugs specifically to tumor sites using an externally applied magnetic field, minimizing toxic effects on healthy tissues [13, 50, 69]. Such strategies can improve anticancer efficacy while reducing the side effects associated with drugs like doxorubicin and docetaxel, which have narrow therapeutic windows and are highly toxic to healthy cells. Furthermore, SPIONPs can also function as contrast agents for MRI a widely used medical imaging technique.

Another application of SPIONPs as potent anticancer modality lies in their hyperthermia property, which enables them to induce localized heating in tumor regions after guided to the desired location using magnetic fields [26]. Once localized, they induce controlled heating in tumor regions, offering a targeted approach for cancer therapy. Such heating can trigger the release of a loaded drug or cause cell death via temperature-induced apoptosis. The above benefits of SPIONPs make them highly suitable for use as advanced theranostic agent. However, it is noteworthy to mention that without surface engineering the bare SPIONPs can be easily oxidized in the presence of air by hampering both the magnetic property and colloidal stability [14]. In addition, SPIONPs are prone to forming aggregates due to their strong magnetic dipole interactions which further limiting their application

in biomedical field. Therefore, to prevent aggregation and enhance colloidal stability, SPIONPs must be combined with natural or synthetic polymers. Upon exposure to biological systems their biodegradation is facilitated. SPIONPs (surface modified with biocompatible polymers) have the potential to load a larger dose of anticancer drugs, achieving higher local concentrations while minimizing the harmful side effects to healthy tissues.

This review covers facile synthetic methods employed for the development of scalable, high quality SPIONPs for diverse biomedical applications. Furthermore, various stabilization strategies of SPIONPs, especially with aid of biocompatible polymers have also been covered in nutshell.

2. Facile synthesis methods of SPIONPs

Over the years, significant efforts have been dedicated for synthesizing SPIONPs with desirable physicochemical properties for applications in the biomedical field. The primary challenge lies in preparing highly stable, phase-pure monodispersed SPIONPs with an appropriate size and a high degree of crystallinity, which are crucial characteristics for achieving optimal pharmacokinetic (PK) behaviour. On the other hand, MRI contrast enhancement effects are directly influenced by the phase purity and crystallinity of the material. The second challenge is to develop a simplified purification process as an alternative to methods like magnetic filtration or ultracentrifugation process. With significant progress over recent years, various techniques have been reported that efficiently synthesize high quality, water-dispersible and biocompatible SPIONPs (Figure 19.1). The most common methods for synthesizing SPIONPs include chemical co-precipitation, hydrothermal/solvothermal, microemulsion and thermal decomposition techniques, among others. A comprehensive description along with the features of these methods is briefly summarized in Table 19.1.

2.1. Chemical co-precipitation method

This synthesis method is among the simplest techniques for producing SPIONPs. This is a reduction reaction of ferrous (Fe^{2+}) and ferric (Fe^{3+}) salts in a 1:2 stoichiometric ratio at an elevated temperature between 70–90°C under alkaline conditions followed by addition of a base, resulting in the formation of black-colored Fe_3O_4 nanoparticles [14, 41]. The pH of the solution during the synthesis process typically remains between 9 and 14. Massart et al. were the first scientists to employ this synthesis method [43]. The involved chemical reaction can be represented as follows:

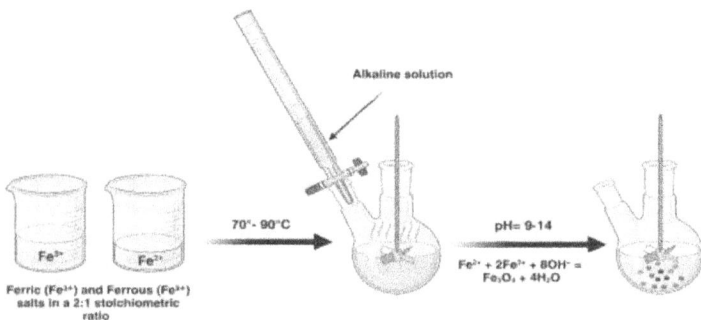

$$Fe^{2+} + 2Fe^{3+} + 8OH- = Fe_3O_4 + 4H_2O$$

The size and morphology of the Fe_3O_4 NPs produced depend on (1) used salt type, such as nitrates, chlorides, sulphates and others (2) strochiometric ratio of Fe^{3+} and Fe^{2+}, (3) the reaction of the temperature as well as (4) pH of solution.

However, the challenges associated with this route include the potential oxidation or reduction of iron salts during the synthesis, which can influence the physicochemical characteristics of the prepared nanoparticles. Fe_3O_4 NPs prepared using this method were highly unstable and exhibited significant polydispersity, resulting in a broad particle size distribution [14, 41, 43]. Large particle size distribution and absence of crystallinity result in a lower saturation magnetization level (30–50 emu/g) for Fe_3O_4 NPs compared to their bulk counterparts (92 emu/g) [41]. It is well established that a brief nucleation burst followed by slow, controlled growth is crucial for producing monodisperse particles. In this context, the use of organic additives as surface-stabilizing agents enables the synthesis of monodispersed Fe_3O_4 nanoparticles of various sizes [20]. Cheng synthesized SPIONPs with

Alkaline solution

70°- 90°C

pH= 9-14

$Fe^{2+} + 2Fe^{3+} + 8OH^- = Fe_3O_4 + 4H_2O$

Ferric (Fe^{3+}) and Ferrous (Fe^{2+}) salts in a 2:1 stoichiometric ratio

Figure 19.1. Schematic representation of chemical co-precipiation method for the synthesis of IONPs.

Source: Author.

Table 19.1. A comparative analysis of important facile methods adopted for synthesis of IONPs

Synthesis Method	Solvent	Reaction Condition	Duration	Size Distribution	Shape Control	Dispersity profile	Yield
Coprecipitation	Water	20–90°C, with/ without a nitrogen atmosphere	Minutes	Relatively narrow	Not good	Polydisperse	High/ scalable
Hydrothermal	Water-Ethanol	High Temperature and High Pressure 200°C	Hours	Very narrow	Very Good	Monodisperse	Medium
Solvothermal	Organic Solvent	High Temperature and High Pressure	Hours	Very narrow	Very Good	Monodisperse	Medium
Microemulsion	Organic Solvent	20–50	Hours	Relatively narrow	Good	Relatively monodisperse	Low
Thermal Decomposition	Organic Solvent	High Temperature	Hours–days	Very narrow	Very good	Monodisperse	High/ scalable

Source: Hu, Y., et al., 2018 and Lu, A., et al., 2007.

a uniform size of 9 nm using tetramethylammonium hydroxide [10].

2.2. Solvothermal/Hydrothermal method

Unlike the co-precipitation method, the solvothermal synthesis technique is scarcely investigated for the formulation of SPIONPs, despite its ability to yield high-quality particles. High-quality Fe_3O_4 NPs can be prepared either in aqueous solutions (referred to as the hydrothermal method) or in organic solvents (referred to as the solvothermal method) using a Teflon-lined stainless-steel autoclave at high temperatures (130–250°C) for 8–72 hours under pressures ranging from 0.3 to 4 MPa [26, 41]. Wang et al. synthesized monodispersed, hydrophilic Fe_3O_4 NPs with tunable size in the diameters ranging from 200 to 800 nm using this synthesis route [63]. In this method, a mixture of $FeCl_3$, sodium acetate and polyethylene glycol (PEG) was dissolved in ethylene glycol and stirred vigorously for 30 minutes to form a clear solution.

The solution was then sealed in a Teflon-lined stainless-steel autoclave and heated at 200°C for 8–72 hours. In this process, $FeCl_3$ served as a precursor, ethylene glycol as the solvent, sodium acetate as a stabilizer for electrostatics and PEG as a surfactant to prevent particle aggregation. Ge et al. synthesized Fe_3O_4 NPs with a tunable diameter ranging from 15 to 31 nm by oxidizing ferrous chloride in a simple aqueous solution under higher pressure of approximately 2 bar and temperature of 134°C [19]. Here, the use of a multicomponent approach proves to be a powerful strategy for preparing monodispersed particles (Figure 19.2).

2.3. Microemulsion method

An isotropic and thermodynamically stable dispersion of a pair of immiscible liquids (oil and water) with presence of suitable surface-stabilizing agents is known as a microemulsion [14, 41]. Oil-in-water (O/W) and Water-in-oil (W/O) microemulsions are two types of systems used to synthesize various types

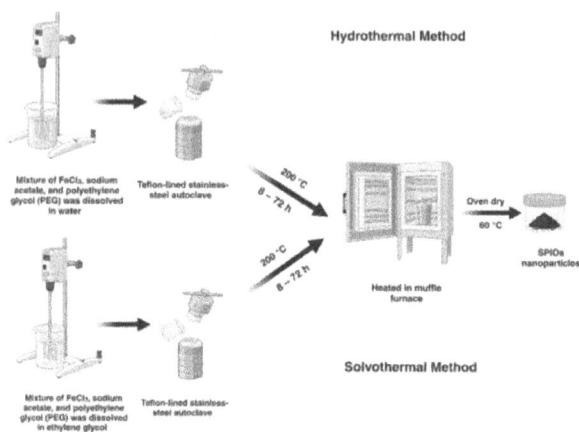

Figure 19.2. Schematic representation of Hydrothermal/solvothermal method for the synthesis of IONPs.

Source: Author.

of magnetic NPs. As per the literatures, sodium dodecyl sulfate (SDS), cetyl-trimethylammonium bromide (CTAB) and sodium bis (2 ethylhexylsulfosuc-cinate) (AOT) are commonly used as surface-stabilizing agents for the fabrication of SPIONPs. The hydrophobic and hydrophilic components of surface coatings have a significant role in stabilizing the NPs and controlling their physicochemical characteristics. The water-in-oil type microemulsion method is extensively utilized for synthesizing SPIONPs with a narrow size range and desired physical properties. The shape and size of the NPs can be precisely guided by adjusting the concentrations of iron precursor, base, surfactant and solvents. Gupta and co-workers synthesized SPIONPs with a particle size of approximately 15 nm using AOT/n-hexane in a nitrogen (N_2) environment [22]. Similarly, Okoli et al. tailored the size of SPIONPs using CTAB [48]. Although this method offers relatively good control over particle shape and size distribution, probable toxicity of the used surfactants favour it's non-biocompatible. Furthermore, the synthesized

SPIONPs need additional stabilization processes to address issues related to aggregation and colloidal instability (Figure 19.3).

2.4. Thermal decomposition method

Higher-quality, monodispersed SPIONPs with smaller sizes and high crystallinity can be achieved using this technique, outperforming other synthesis methods [41]. The process involves the transformation of various organometallic compounds in high-boiling-point organic solvents containing stabilizing surfactants like fatty acid, hexadecylamine, oleic acid and oleyl amine. The common organometallic precursors reported in the literature include Fe(acac)₃ (acac = acetylacetonate), Fe(cup)₃ (cup = N-nitrosophenylhydroxylamine) and Fe(CO)₅ (CO = carbonyl). The ratios and volumes of the starting reagents as well as the reaction conditions are critical factors for precisely controlling the size and morphology of Fe_3O_4 NPs. Yang et al. successfully synthesized monodispersed Fe_3O_4 nanocubes with size of the

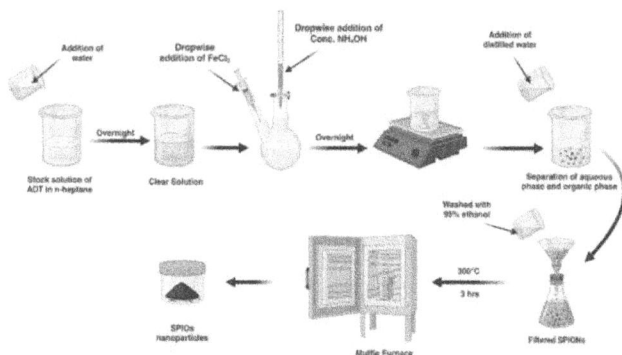

Figure 19.3. Diagrammetic representation of Microemulsion method for the synthesis of IONPs.

Source: Author.

particles ranging between 6.5 to 30 nm. This was achieved through the thermal reaction of ferric acetate in a mixture of oleic acid, 1,2-hexadecanediol, benzyl ether and oleylamine at a reflux temperature of 200°C [65]. Similarly, Kim and co-workers used this strategy to prepare monodispersed Fe_3O_4 nanocubes utilizing thermal reaction of $Fe(acac)_3$ along with the mixture of benzyl ether and oleic acid at a reaction temperature 290°C, thus, yielding particles with size ranging from 20–160 nm [32]. The size and morphology of the SPIONPs were controlled by varying the reaction time and varying the amount of benzyl ether. However, this technique typically requires relatively high temperatures and toxic reagents and the water insolubility of the prepared NPs further limits its applicability in biomedical filed. Therefore, significant efforts have been devoted to transforming hydrophobic NPs into water-soluble ones. In this context, Sun et al. synthesized hydrophobic SPIONPs with sizes ranging from 3 to 20 nm via the thermal decomposition of Ferric acetate in the presence of oleic acid and oleylamine as stabilizing agents [59]. These hexane-dispersed SPIONPs were subsequently converted into hydrophilic ones by adding a bipolar surfactant (Figure 19.4).

3. Stabilization of SPIONPs

Stability is a critical requirement for nearly all type of biomedical applications of SPIONPs. Bare NPs have the tendency to form clusters owing to agglomeration thus, minimize their high surface area-volume ratio, which reduces their colloidal stability and functionality. However, introducing coating layer on the surfaces of Fe_3O_4 NPs not only provides colloidal stability but also creates a protective shell against oxidation. Additionally, selecting an appropriate type of coating enables further functionalization of NPs, facilitating the attachment of specific molecules or groups to tailor their properties for diverse applications. The coating of SPIONPs can be achieved using either natural or synthetic polymers [14, 27] such as chitosan, dextran, polyvinyl alcohol (PVA), polyethylene glycol (PEG), or non-polymeric/

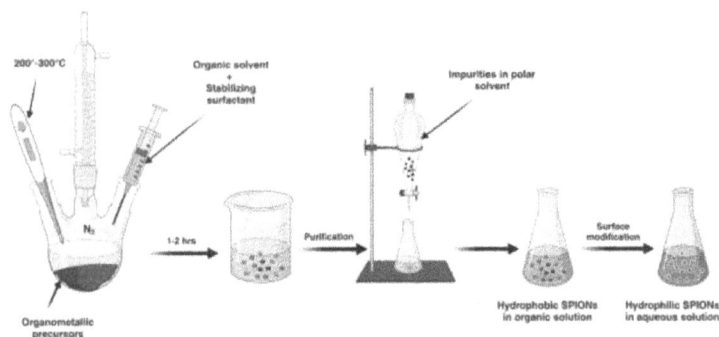

Figure 19.4. Schematic representation of Thermal decomposition method for the formulation of IONPs.

Source: Author.

inorganic materials, including silica [1, 35], carbon [4, 37] and precious metals such as gold [16], silver [57]. Each type of coating material offers unique advantages, such as enhanced biocompatibility, stability and the ability to functionalize the surface for specific applications.

3.1. Stabilization of SPIONPs with polymer coating

3.1.1. Dextran

Dextran is a biocompatible, neutral polysaccharide ($C_6H_{10}O_5$)- commonly used for coating SPIONPs and it has been widely employed in applications such as MRI imaging of the liver and cancer treatment [14, 44]. It stabilizes colloidal solutions and increases the blood circulation time of SPIONPs, making them suitable for in vivo applications [21]. The formation of Fe3O4 NPs in the presence of dextran was first reported by Molday and MacKenzie [47]. Josephson et al. synthesized dextran-coated monocrystalline magnetite to enhance intracellular magnetic labelling of the targeted cells [30]. Pardoe et al. studied the structural and magnetic properties of SPIONPs in presence of dextran and observed that the presence of the polymer helped in reducing the particle size as compared to uncoated particles [51]. While dextran coating shows numerous advantages, it also presents some challenges. Dextran is susceptible to hydrolysis and enzymatic degradation under physiological conditions, which can compromise the long-term stability of the coating [25, 28]. Although dextran provides hydroxyl groups for functionalization, its functional sites are fewer than those of some synthetic polymers, limiting its ability to carry multiple therapeutic or targeting agents. Additionally, the thick dextran

coating can dilute the magnetic properties of SPIONPs, potentially reducing their effectiveness in applications such as magnetic hyperthermia or MRI. Jiang et al. reported that the crystallinity and magnetization of SPIONPs were hindered when coated with carboxy methyl dextran [28].

3.1.2. Chitosan

Chitosan is a hydrophilic, biocompatible and non-toxic copolymer (poly-aminosaccharide) composed of 2-amino-2-deoxy-β-D-glucan units linked via glycosidic bonds [14, 35]. The presence of hydroxyl and amine groups, chitosan is widely utilized in pharmaceutical applications. Its coating properties can prevent the agglomeration of NPs and functionalize their surfaces, enabling the binding of biological entities such as drugs and proteins, making it highly favorable for drug delivery applications [5]. Castelló et al. synthesized SPIONPs with a 12 nm diameter via a controlled coprecipitation method using chitosan as a coating material and demonstrated its suitability for biomedical applications [8]. However, chitosan is highly sensitive to pH and degrades under acidic conditions. The process of coating SPIONPs with chitosan is often challenging due to the inherent variability in chitosan's molecular weight and viscosity. This variability can lead to uneven coatings, affecting the stability and reproducibility of NPs, thereby reducing their effectiveness in biomedical applications. While chitosan is a promising material for SPIONPs coating due to its biocompatibility and biodegradability, its limitations—such as instability, challenges in achieving uniform coatings and potential reduction in magnetic properties at high doses—need to be addressed [3]. Additionally, the potential for toxicity at higher concentrations highlights the need for careful optimization to ensure its effective use in biomedical applications.

3.1.3. Polyvinyl alcohol (PVA)

PVA is a hydrophilic, biocompatible and biodegradable polymer widely used for the functionalization of NPs [35, 55]. PVA coating enhances the colloidal stability of NPs, preventing aggregation and ensuring uniform dispersion in aqueous and other solutions [14, 35]. The hydroxyl groups in PVA facilitate further functionalization, enabling the attachment of biomolecules, drugs, or targeting agents, thereby making it highly suitable for promising biomedical applications. Additionally, the PVA coating acts as a protective layer, reducing the direct exposure of cells to the iron oxide core and minimizing cytotoxic effects. Lee et al. reported the modification of NP surfaces using PVA by precipitating iron salts in an aqueous PVA solution [36]. Similarly, Kayal et al. demonstrated the potential of doxorubicin-loaded, PVA-functionalized SPIONPs in targeted drug delivery applications [31]. However, the PVA layer may slightly reduce the magnetic responsiveness of SPIONPs by forming a non-magnetic barrier around the core NPs. Moreover, achieving a uniform and controlled thickness of the PVA coating can be challenging, potentially affecting the physicochemical properties of the SPIONPs.

3.1.4. Polyethylene glycol (PEG)

PEG is a biodegradable polymer that has been approved by the United States Food and Drug Administration as an outer coating material, with widespread use in biomedical applications [33]. Due to its hydrophilic, biocompatible and non-toxic nature, PEG is highly beneficial for NPs functionalization. The residence of PEG on the surface of NPs imparts stealth properties and hydrophilicity, which can help maintain a prolonged blood circulation profile [21, 50]. PEG coating also enhances the ability of NPs to cross cell membranes, as PEG is soluble in both polar and non-polar solvents and exhibits its high permeability through cell membranes. In addition to its superior stability and solubility in aqueous solutions, PEG retains its stability in biological saline solutions. Anbarasu et al. synthesized PEG-coated SPIONPs *via* a chemical precipitation route using ferric and ferrous sulfates as precursors [2]. Kumargai et al. developed a simple technique for preparing PEG-coated IONPs by hydrolyzing ferric chloride in water, followed by treatment with a PEG-poly(aspartic acid) block copolymer [34]. A. Masoudi et al. proposed an innovative two-step method for PEG-coated IONPs preparation. First, bare IONPs were synthesized via the coprecipitation technique, followed by coating of PEG which was achieved by stirring PEG solution mixed with iron oxide suspension for about 24 hours [42]. However, the primary drawback of PEG coating is it is non-biodegradable in body physiology. Its accumulation over time may pose long-term risks, particularly in applications requiring repeated or high-dose administration. Additionally, the pathways for PEG clearance from the body are not fully understood, raising concerns about its safety and the potential for tissue accumulation.

3.2. Stabilization of SPIONPs within nonpolymeric coating

Polymer-stabilized SPIONPs lose their stability in presence of air and are readily affected by the acidic solutions, which lowers their magnetization values. Therefore, non-polymeric or inorganic coatings on the surface of SPIONPs are often preferred over materials coated with polymers to achieve higher colloidal stability, improved functionality as well as biocompatibility. This approach involves forming a protective layer of inorganic material around the NPs core. Such coatings can prevent aggregation, reduce surface oxidation and provide a versatile platform for further functionalization. As reported in the literatures previously, among different non-polymeric coating materials, silica, carbon and gold have been the most extensively studied and utilized, owing to their distinct advantages and established effectiveness.

3.2.1. Silica coating

Silica coating on the surface of SPIONPs is a well-established technique for improving their stability, functionality and biocompatibility [7, 64]. The silica layer protects against oxidation, prevents particle aggregation and provides a chemically versatile surface for further modifications. However, owing to its natural hydrophilic structure, silica interacts with various types of biological substrates. Silica coating stabilizes SPIONPs by shielding the magnetic dipole interactions through the silica shell. Since silica NPs are negatively charged, their coating also enhances the coulombic repulsion between SPIOs, improving colloidal stability [15, 35]. The Stöber method is the most commonly used technique for silica coating of NPs, where silica is formed through the hydrolysis and condensation of a sol-gel precursor [58]. The thickness of the silica coating can be adjusted by varying the concentration of ammonium hydroxide and the ratio of tetraethyl orthosilicate (TEOS) to water. Silica-coated magnetic NPs have shown great promise in bio-labelling and bio-separation applications. For instance: Santra et al. utilized a water-in-oil type microemulsion technique to fabricate uniform-sized silica-coated SPIONPs and demonstrated their efficacy as magnetic probes for separating target oligonucleotides [56]. Tago et al. synthesized silica-coated Fe_3O_4 NPs using the water-in-oil microemulsion technique [60]. Yang et al. prepared cetyltrimethylammonium bromide-coated SPIONPs using a thermal decomposition technique, followed by silica stabilization using TEOS [66]. Zhao and coworkers developed uniform magnetic nanocomposite spheres (with mesoporous silica shell and a magnetic core) with a particle diameter of 270 nm [68]. However, the primary limitation of this technique is the difficulty in achieving a uniform silica shell thickness over the surface of NPs. An uneven silica coating can create irregular magnetic fields, potentially causing inconsistent heating in magnetic hyperthermia applications.

3.2.2. Carbon coating

Carbon-protected SPIONPs have recently garnered significant attention due to their excellent chemical and thermal stability, as well as their inherently higher electrical conductivity [41]. Carbon-coated SPIONPs act as oxidation barriers, preventing corrosion of the magnetic iron oxide core. Additionally, hydrophilic carbon coatings enhance dispersibility and stability compared to bare SPIONPs. Since they are typically in the metallic state, they exhibit a greater magnetic moment than their oxide-based counterparts. The unique properties of Fe_3O_4-based graphene composites, such as a large surface-to-volume ratio, combined with Fe_3O_4 NPs' high magnetic moment, biocompatibility and cost-effectiveness, offer a promising pathway for preparing highly stable multifunctional magnetic NPs.

Numerous studies have focused on coating or encapsulating magnetic NPs with carbon-based materials. Liu et al. prepared a superparamagnetic Fe3O4-reduced graphene oxide composite (-Fe3O4@rGO) by employing a solvothermal reaction of ferric acetate, GO in ethylenediamine and water [40]. Cong et al. prepared magnetite functionalized rGO sheets and evaluated their potentiality for MRI application [12]. Fan et al. prepared a Fe3O4-graphene composite based nanocarrier system and reported higher drug loading capability (0.35 mg mg) for the anticancer drug 5-fluorouracil [17]. The MTT assay and cellular uptake studies of the nanocarrier system suggested for practical application in the biomedical field. Venkatesha et al. synthesized graphene oxide-Fe3O4 nanocomposite and investigated its potential application as a contrast agent in MRI [61]. Wang et al. reported manganese ferrite-based GO nanocomposite and examined their drug loading capacity along with drug release profile [24]. While carbon coatings provide protection and functionality to SPIONPs, their adverse effects – such as challenges in functionalization, altered magnetic properties, toxicity and aggregation – need to be carefully addressed. Modifications like introducing hydrophilic functional groups or optimizing the coating thickness can help mitigate these drawbacks.

3.2.3. Gold coating

Gold is another inorganic coating material considered highly suitable for enhancing the functionality and stability of SPIONPs in aqueous solutions. Gold coating is particularly attractive because the gold surface can be easily functionalized with thiol groups, allowing the attachment of functional ligands. This enables the NPs to be tailored for various applications, including optical and catalytic uses [11]. For example, Mohammad et al. synthesized gold-coated Fe3O4 NPs and evaluated their potential for hyperthermia applications [46]. Similarly, Liang et al. prepared gold-coated Fe3O4 NPs using the co-precipitation method and demonstrated their effectiveness as a biosensing system for detecting human alpha-thrombin [38].

Gold coating on SPIONPs offers several advantages, such as improved stability, biocompatibility and the ability to functionalize the surface with various biomolecules or ligands. However, there are several drawbacks associated with gold coating on Fe_3O_4 NPs. The gold layer on Fe_3O_4 NPs can attenuate the magnetic properties of the underlying iron oxide core. Since gold is non-magnetic, the overall magnetic behavior of the composite may be hindered compared to bare Fe_3O_4 NPs, which can limit their performance in applications such as magnetic hyperthermia, MRI and targeted drug delivery [39]. Additionally, maintaining a stable and uniform gold coating can be challenging due to the differences in surface properties between gold and Fe_3O_4. This can hinder the long-term stability and uniformity of the gold shell [62, 67]. Achieving consistent, high-quality gold coatings on Fe_3O_4 NPs can be difficult to scale up for commercial applications. Factors such as temperature control, pH and reagent concentrations must be carefully monitored to ensure uniform coatings.

4. Conclusion and Future Perspective

In summary, this review highlights the synthesis, surface stabilization of SPIONPs which are emerging as versatile nano platforms for target specific drug delivery and multi-modal tumor imaging applications. The presence of active functional groups on SPIONPs enables the rational conjugation of biological and therapeutic molecules, enhancing their versatility. Recent advancements in the formulation and surface modification of SPIONPs has further expanded their potential for effective theranostic applications. Moreover, when combined with other treatment modalities, SPIONPs hold great promise for imaging-guided cancer therapy and magnetic hyperthermia therapy-enhanced treatment approaches.

Although significant progress has been made in integrating SPIONPs with other treatment modalities, many approaches remain at the proof-of-concept stage. Numerous challenges must be addressed before SPIONPs can be effectively utilized as probes for diagnostic and therapeutic applications in clinical settings. Clinical trials require a thorough understanding of SPIONPs' *in vivo* behaviour as well as their PK, bio-distribution and clearance processes. Additionally, the challenges associated to their enduring stability and the possibility of organ accumulation is too critical to ensure potential side effects on healthy tissues and enhancing their clinical utility. Potential healthy tissue toxicity associated with SPIONs, especially when administered at high concentrations for prolonged periods, necessitates careful evaluation and optimization. It is highly required that SPIONPs should be eliminated from the body without undesirable tissue accumulation or any adverse effects. However, some SPIONP based formulations show insufficient renal clearance and the risk of side effects from iron overload. The long-term impacts of IONPs on biological systems are being actively addressed by current research and developments, which aim to go past these challenges and completely understand the enormous potential of SPIONPs-based technology. Once these challenges are addressed, multifunctional SPIONPs could be further developed for theranostics of various diseases other than cancer, paving the way for their application in translational medicine in the near future.

5. Acknowledgement

We are highly thankful to the institute for providing necessary support and encouragement.

References

[1] Ali, Z., Andreassen, J.-P., & Bandyopadhyay, S. (2023). Fine-Tuning of Particle Size and Morphology of Silica Coated Iron Oxide Nanoparticles. *Industrial & Engineering Chemistry Research*, 62(12), 4831–4839. https://doi.org/10.1021/acs.iecr.2c03338

[2] Anbarasu, M., Anandan, M., Chinnasamy, E., Gopinath, V., & Balamurugan, K. (2015). Synthesis and characterization of polyethylene glycol (PEG) coated Fe3O4 nanoparticles by chemical co-precipitation method for biomedical applications. *Spectrochimica Acta Part A: Molecular and Biomolecular Spectroscopy*, 135, 536–539. https://doi.org/10.1016/j.saa.2014.07.059

[3] Azmana, M., Mahmood, S., Hilles, A. R., Rahman, A., Arifin, M. A. B.,

& Ahmed, S. (2021). A review on chitosan and chitosan-based bion-anocomposites: Promising material for combatting global issues and its applications. *International Journal of Biological Macromolecules*, *185*, 832–848. https://doi.org/10.1016/j.ijbiomac.2021.07.023

[4] Bae, H., Ahmad, T., Rhee, I., Chang, Y., Jin, S.-U., & Hong, S. (2012). Carbon-coated iron oxide nanoparticles as contrast agents in magnetic resonance imaging. *Nanoscale Research Letters*, *7*(1), 44. https://doi.org/10.1186/1556-276X-7-44

[5] Belessi, V., Zboril, R., Tucek, J., Mashlan, M., Tzitzios, V., & Petridis, D. (2008). Ferrofluids from Magnetic–Chitosan Hybrids. *Chemistry of Materials*, *20*(10), 3298–3305. https://doi.org/10.1021/cm702990t

[6] Bray, F., Laversanne, M., Sung, H., Ferlay, J., Siegel, R. L., Soerjomataram, I., & Jemal, A. (2024). Global cancer statistics 2022: GLOBOCAN estimates of incidence and mortality worldwide for 36 cancers in 185 countries. *CA: A Cancer Journal for Clinicians*, *74*(3), 229–263. https://doi.org/10.3322/caac.21834

[7] Bruce, I. J., & Sen, T. (2005). Surface Modification of Magnetic Nanoparticles with Alkoxysilanes and Their Application in Magnetic Bioseparations. *Langmuir*, *21*(15), 7029–7035. https://doi.org/10.1021/la050553t

[8] Castelló, J., Gallardo, M., Busquets, M. A., & Estelrich, J. (2015). Chitosan (or alginate)-coated iron oxide nanoparticles: A comparative study. *Colloids and Surfaces A: Physicochemical and Engineering Aspects*, *468*, 151–158. https://doi.org/10.1016/j.colsurfa.2014.12.031

[9] Caster, J. M., Patel, A. N., Zhang, T., & Wang, A. (2017). Investigational nanomedicines in 2016: A review of nanotherapeutics currently undergoing clinical trials. *WIREs Nanomedicine and Nanobiotechnology*, *9*(1), e1416. https://doi.org/10.1002/wnan.1416

[10] Cheng, F. (2005). Characterization of aqueous dispersions of Fe3O4 nanoparticles and their biomedical applications. *Biomaterials*, *26*(7), 729–738. https://doi.org/10.1016/j.biomaterials.2004.03.016

[11] Colvin, V. L., Goldstein, A. N., & Alivisatos, A. P. (1992). Semiconductor nanocrystals covalently bound to metal surfaces with self-assembled monolayers. *Journal of the American Chemical Society*, *114*(13), 5221–5230. https://doi.org/10.1021/ja00039a038

[12] Cong, H., He, J., Lu, Y., & Yu, S. (2010). Water-Soluble Magnetic-Functionalized Reduced Graphene Oxide Sheets: In situ Synthesis and Magnetic Resonance Imaging Applications. *Small*, *6*(2), 169–173. https://doi.org/10.1002/smll.200901360

[13] Cui, Y., Zhang, M., Zeng, F., Jin, H., Xu, Q., & Huang, Y. (2016). Dual-Targeting Magnetic PLGA Nanoparticles for Codelivery of Paclitaxel and Curcumin for Brain Tumor Therapy. *ACS Applied Materials & Interfaces*, *8*(47), 32159–32169. https://doi.org/10.1021/acsami.6b10175

[14] Demirer, G. S., Okur, A. C., & Kizilel, S. (2015). Synthesis and design of biologically inspired biocompatible iron oxide nanoparticles for biomedical applications. *Journal of Materials Chemistry B*, *3*(40), 7831–7849. https://doi.org/10.1039/C5TB00931F

[15] Ding, H. L., Zhang, Y. X., Wang, S., Xu, J. M., Xu, S. C., & Li, G. H. (2012). Fe_3O_4 @SiO_2 Core/Shell Nanoparticles: The Silica Coating Regulations with a Single Core for Different Core Sizes and Shell Thicknesses. *Chemistry of Materials*, *24*(23), 4572–4580. https://doi.org/10.1021/cm302828d

[16] Fadeev, M., Kozlovskiy, A., Korolkov, I., Egizbek, K., Nazarova, A., Chudoba, D., Rusakov, V., & Zdorovets, M.

(2020). Iron oxide @ gold nanoparticles: Synthesis, properties and potential use as anode materials for lithium-ion batteries. *Colloids and Surfaces A: Physicochemical and Engineering Aspects*, *603*, 125178. https://doi.org/10.1016/j.colsurfa.2020.125178

[17] Fan, X., Jiao, G., Zhao, W., Jin, P., & Li, X. (2013). Magnetic Fe3O4–graphene composites as targeted drug nanocarriers for pH-activated release. *Nanoscale*, *5*(3), 1143. https://doi.org/10.1039/c2nr33158f

[18] Gao, J., Gu, H., & Xu, B. (2009). Multifunctional Magnetic Nanoparticles: Design, Synthesis, and Biomedical Applications. *Accounts of Chemical Research*, *42*(8), 1097–1107. https://doi.org/10.1021/ar9000026

[19] Ge, S., Shi, X., Sun, K., Li, C., Uher, C., Baker, J. R., Banaszak Holl, M. M., & Orr, B. G. (2009). Facile Hydrothermal Synthesis of Iron Oxide Nanoparticles with Tunable Magnetic Properties. *The Journal of Physical Chemistry C*, *113*(31), 13593–13599. https://doi.org/10.1021/jp902953t

[20] Gupta, A. K., & Curtis, A. S. G. (2004). Lactoferrin and ceruloplasmin derivatized superparamagnetic iron oxide nanoparticles for targeting cell surface receptors. *Biomaterials*, *25*(15), 3029–3040. https://doi.org/10.1016/j.biomaterials.2003.09.095

[21] Gupta, A. K., & Gupta, M. (2005). Synthesis and surface engineering of iron oxide nanoparticles for biomedical applications. *Biomaterials*, *26*(18), 3995–4021. https://doi.org/10.1016/j.biomaterials.2004.10.012

[22] Gupta, A. K., & Wells, S. (2004). Surface-Modified Superparamagnetic Nanoparticles for Drug Delivery: Preparation, Characterization, and Cytotoxicity Studies. *IEEE Transactions on Nanobioscience*, *3*(1), 66–73. https://doi.org/10.1109/TNB.2003.820277

[23] Hao, R., Xing, R., Xu, Z., Hou, Y., Gao, S., & Sun, S. (2010). Synthesis, Functionalization, and Biomedical Applications of Multifunctional Magnetic Nanoparticles. *Advanced Materials*, *22*(25), 2729–2742. https://doi.org/10.1002/adma.201000260

[24] Hoskins, C., Min, Y., Gueorguieva, M., McDougall, C., Volovick, A., Prentice, P., Wang, Z., Melzer, A., Cuschieri, A., & Wang, L. (2012). Hybrid gold-iron oxide nanoparticles as a multifunctional platform for biomedical application. *Journal of Nanobiotechnology*, *10*(1), 27. https://doi.org/10.1186/1477-3155-10-27

[25] Hu, Q., Lu, Y., & Luo, Y. (2021). Recent advances in dextran-based drug delivery systems: From fabrication strategies to applications. *Carbohydrate Polymers*, *264*, 117999. https://doi.org/10.1016/j.carbpol.2021.117999

[26] Hu, Y., Mignani, S., Majoral, J.-P., Shen, M., & Shi, X. (2018). Construction of iron oxide nanoparticle-based hybrid platforms for tumor imaging and therapy. *Chemical Society Reviews*, *47*(5), 1874–1900. https://doi.org/10.1039/C7CS00657H

[27] Jeong, U., Teng, X., Wang, Y., Yang, H., & Xia, Y. (2007). Superparamagnetic Colloids: Controlled Synthesis and Niche Applications. *Advanced Materials*, *19*(1), 33–60. https://doi.org/10.1002/adma.200600674

[28] Jiang, S., Eltoukhy, A. A., Love, K. T., Langer, R., & Anderson, D. G. (2013). Lipidoid-Coated Iron Oxide Nanoparticles for Efficient DNA and siRNA delivery. *Nano Letters*, *13*(3), 1059–1064. https://doi.org/10.1021/nl304287a

[29] Jokerst, J. V., & Gambhir, S. S. (2011). Molecular Imaging with Theranostic Nanoparticles. *Accounts of Chemical Research*, *44*(10), 1050–1060. https://doi.org/10.1021/ar200106e

[30] Josephson, L., Tung, C.-H., Moore, A., & Weissleder, R. (1999). High-Efficiency Intracellular Magnetic Labeling with Novel Superparamagnetic-Tat

Peptide Conjugates. *Bioconjugate Chemistry, 10*(2), 186–191. https://doi.org/10.1021/bc980125h

[31] Kayal, S., & Ramanujan, R. V. (2010). Doxorubicin loaded PVA coated iron oxide nanoparticles for targeted drug delivery. *Materials Science and Engineering: C, 30*(3), 484–490. https://doi.org/10.1016/j.msec.2010.01.006

[32] Kim, D., Lee, N., Park, M., Kim, B. H., An, K., & Hyeon, T. (2009). Synthesis of Uniform Ferrimagnetic Magnetite Nanocubes. *Journal of the American Chemical Society, 131*(2), 454–455. https://doi.org/10.1021/ja8086906

[33] Kizilel, S., Nazli, C., Ergenc, Acar, & Yar. (2012). RGDS-functionalized polyethylene glycol hydrogel-coated magnetic iron oxide nanoparticles enhance specific intracellular uptake by HeLa cells. *International Journal of Nanomedicine*, 1903. https://doi.org/10.2147/IJN.S29442

[34] Kumagai, M., Imai, Y., Nakamura, T., Yamasaki, Y., Sekino, M., Ueno, S., Hanaoka, K., Kikuchi, K., Nagano, T., Kaneko, E., Shimokado, K., & Kataoka, K. (2007). Iron hydroxide nanoparticles coated with poly(ethylene glycol)-poly(aspartic acid) block copolymer as novel magnetic resonance contrast agents for in vivo cancer imaging. *Colloids and Surfaces B: Biointerfaces, 56*(1–2), 174–181. https://doi.org/10.1016/j.colsurfb.2006.12.019

[35] Laurent, S., Forge, D., Port, M., Roch, A., Robic, C., Vander Elst, L., & Muller, R. N. (2008). Magnetic Iron Oxide Nanoparticles: Synthesis, Stabilization, Vectorization, Physicochemical Characterizations, and Biological Applications. *Chemical Reviews, 108*(6), 2064–2110. https://doi.org/10.1021/cr068445e

[36] Lee, J., Isobe, T., & Senna, M. (1996). Magnetic properties of ultrafine magnetite particles and their slurries prepared via in-situ precipitation. *Colloids and Surfaces A:*

Physicochemical and Engineering Aspects, 109, 121–127. https://doi.org/10.1016/0927-7757(95)03479-X

[37] Lewińska, A., Radoń, A., Gil, K., Błoniarz, D., Ciuraszkiewicz, A., Kubacki, J., Kądziołka-Gaweł, M., Łukowiec, D., Gębara, P., Krogul-Sobczak, A., Piotrowski, P., Fijałkowska, O., Wybraniec, S., Szmatoła, T., Kolano-Burian, A., & Wnuk, M. (2024). Carbon-Coated Iron Oxide Nanoparticles Promote Reductive Stress-Mediated Cytotoxic Autophagy in Drug-Induced Senescent Breast Cancer Cells. *ACS Applied Materials & Interfaces, 16*(12), 15457–15478. https://doi.org/10.1021/acsami.3c17418

[38] Liang, G., Cai, S., Zhang, P., Peng, Y., Chen, H., Zhang, S., & Kong, J. (2011). Magnetic relaxation switch and colorimetric detection of thrombin using aptamer-functionalized gold-coated iron oxide nanoparticles. *Analytica Chimica Acta, 689*(2), 243–249. https://doi.org/10.1016/j.aca.2011.01.046

[39] Lin, J., Zhou, W., Kumbhar, A., Wiemann, J., Fang, J., Carpenter, E. E., & O'Connor, C. J. (2001). Gold-Coated Iron (Fe@Au) Nanoparticles: Synthesis, Characterization, and Magnetic Field-Induced Self-Assembly. *Journal of Solid State Chemistry, 159*(1), 26–31. https://doi.org/10.1006/jssc.2001.9117

[40] Liu, Y.-W., Guan, M.-X., Feng, L., Deng, S.-L., Bao, J.-F., Xie, S.-Y., Chen, Z., Huang, R.-B., & Zheng, L.-S. (2013). Facile and straightforward synthesis of superparamagnetic reduced graphene oxide–Fe$_3$O$_4$ hybrid composite by a solvothermal reaction. *Nanotechnology, 24*(2), 025604. https://doi.org/10.1088/0957-4484/24/2/025604

[41] Lu, A., Salabas, E. L., & Schüth, F. (2007). Magnetic Nanoparticles: Synthesis, Protection, Functionalization, and Application. *Angewandte Chemie International Edition, 46*(8), 1222–1244. https://doi.org/10.1002/anie.200602866

[42] Masoudi, A., Madaah Hosseini, H. R., Shokrgozar, M. A., Ahmadi, R., & Oghabian, M. A. (2012). The effect of poly(ethylene glycol) coating on colloidal stability of superparamagnetic iron oxide nanoparticles as potential MRI contrast agent. *International Journal of Pharmaceutics*, *433*(1–2), 129–141. https://doi.org/10.1016/j.ijpharm.2012.04.080

[43] Massart, R. (1981). Preparation of aqueous magnetic liquids in alkaline and acidic media. *IEEE Transactions on Magnetics*, *17*(2), 1247–1248. https://doi.org/10.1109/TMAG.1981.1061188

[44] Mccarthy, J., & Weissleder, R. (2008). Multifunctional magnetic nanoparticles for targeted imaging and therapy*. *Advanced Drug Delivery Reviews*, *60*(11), 1241–1251. https://doi.org/10.1016/j.addr.2008.03.014

[45] Mitchell, M. J., Billingsley, M. M., Haley, R. M., Wechsler, M. E., Peppas, N. A., & Langer, R. (2021). Engineering precision nanoparticles for drug delivery. *Nature Reviews Drug Discovery*, *20*(2), 101–124. https://doi.org/10.1038/s41573-020-0090-8

[46] Mohammad, F., Balaji, G., Weber, A., Uppu, R. M., & Kumar, C. S. S. R. (2010). Influence of Gold Nanoshell on Hyperthermia of Superparamagnetic Iron Oxide Nanoparticles. *The Journal of Physical Chemistry C*, *114*(45), 19194–19201. https://doi.org/10.1021/jp105807r

[47] Molday, R. S., & Mackenzie, D. (1982). Immunospecific ferromagnetic iron-dextran reagents for the labeling and magnetic separation of cells. *Journal of Immunological Methods*, *52*(3), 353–367. https://doi.org/10.1016/0022-1759(82)90007-2

[48] Okoli, C., Boutonnet, M., Mariey, L., Järås, S., & Rajarao, G. (2011). Application of magnetic iron oxide nanoparticles prepared from microemulsions for protein purification. *Journal of Chemical Technology &*

Biotechnology, *86*(11), 1386–1393. https://doi.org/10.1002/jctb.2704

[49] Panda, J., Sankar Satapathy, B., Mishra, A., Biswal, B., & Kumar Sahoo, P. (2023). Potential of Ferrite-Based Nanoparticles for Improved Cancer Therapy: Recent Progress and Challenges Ahead. In M. Khan (Ed.), *Applications of Ferrites*. IntechOpen. https://doi.org/10.5772/intechopen.1002346

[50] Panda, J., Satapathy, B. S., Majumder, S., Sarkar, R., Mukherjee, B., & Tudu, B. (2019). Engineered polymeric iron oxide nanoparticles as potential drug carrier for targeted delivery of docetaxel to breast cancer cells. *Journal of Magnetism and Magnetic Materials*, *485*, 165–173. https://doi.org/10.1016/j.jmmm.2019.04.058

[51] Pardoe, H., Chua-anusorn, W., St. Pierre, T. G., & Dobson, J. (2001). Structural and magnetic properties of nanoscale iron oxide particles synthesized in the presence of dextran or polyvinyl alcohol. *Journal of Magnetism and Magnetic Materials*, *225*(1–2), 41–46. https://doi.org/10.1016/S0304-8853(00)01226-9

[52] Patra, J. K., Das, G., Fraceto, L. F., Campos, E. V. R., Rodriguez-Torres, M. D. P., Acosta-Torres, L. S., Diaz-Torres, L. A., Grillo, R., Swamy, M. K., Sharma, S., Habtemariam, S., & Shin, H.-S. (2018). Nano based drug delivery systems: Recent developments and future prospects. *Journal of Nanobiotechnology*, *16*(1), 71. https://doi.org/10.1186/s12951-018-0392-8

[53] Poller, J., Zaloga, J., Schreiber, E., Unterweger, H., Janko, C., Radon, P., Eberbeck, D., Trahms, L., Alexiou, C., & Friedrich, R. (2017). Selection of potential iron oxide nanoparticles for breast cancer treatment based on in vitro cytotoxicity and cellular uptake. *International Journal of Nanomedicine*, *Volume 12*, 3207–3220. https://doi.org/10.2147/IJN.S132369

[54] Revia, R. A., & Zhang, M. (2016). Magnetite nanoparticles for cancer

diagnosis, treatment, and treatment monitoring: Recent advances. *Materials Today*, *19*(3), 157–168. https://doi.org/10.1016/j.mattod.2015.08.022

[55] Sairam, M., Naidu, B. V. K., Nataraj, S. K., Sreedhar, B., & Aminabhavi, T. M. (2006). Poly(vinyl alcohol)-iron oxide nanocomposite membranes for pervaporation dehydration of isopropanol, 1,4-dioxane and tetrahydrofuran. *Journal of Membrane Science*, *283*(1–2), 65–73. https://doi.org/10.1016/j.memsci.2006.06.013

[56] Santra, S., Tapec, R., Theodoropoulou, N., Dobson, J., Hebard, A., & Tan, W. (2001). Synthesis and Characterization of Silica-Coated Iron Oxide Nanoparticles in Microemulsion: The Effect of Nonionic Surfactants. *Langmuir*, *17*(10), 2900–2906. https://doi.org/10.1021/la0008636

[57] Sobal, N. S., Hilgendorff, M., Möhwald, H., Giersig, M., Spasova, M., Radetic, T., & Farle, M. (2002). Synthesis and Structure of Colloidal Bimetallic Nanocrystals: The Non-Alloying System Ag/Co. *Nano Letters*, *2*(6), 621–624. https://doi.org/10.1021/nl025533f

[58] Stöber, W., Fink, A., & Bohn, E. (1968). Controlled growth of monodisperse silica spheres in the micron size range. *Journal of Colloid and Interface Science*, *26*(1), 62–69. https://doi.org/10.1016/0021-9797(68)90272-5

[59] Sun, S., Zeng, H., Robinson, D. B., Raoux, S., Rice, P. M., Wang, S. X., & Li, G. (2004). Monodisperse MFe$_2$O$_4$ (M = Fe, Co, Mn) Nanoparticles. *Journal of the American Chemical Society*, *126*(1), 273–279. https://doi.org/10.1021/ja0380852

[60] Tago, T., Hatsuta, T., Miyajima, K., Kishida, M., Tashiro, S., & Wakabayashi, K. (2002). Novel Synthesis of Silica-Coated Ferrite Nanoparticles Prepared Using Water-in-Oil Microemulsion. *Journal of the American Ceramic Society*, *85*(9), 2188–2194. https://doi.org/10.1111/j.1151-2916.2002.tb00433.x

[61] Wang, G., Ma, Y., Zhang, L., Mu, J., Zhang, Z., Zhang, X., Che, H., Bai, Y., & Hou, J. (2016). Facile synthesis of manganese ferrite/graphene oxide nanocomposites for controlled targeted drug delivery. *Journal of Magnetism and Magnetic Materials*, *401*, 647–650. https://doi.org/10.1016/j.jmmm.2015.10.096

[62] Wang, L., Luo, J., Maye, M. M., Fan, Q., Rendeng, Q., Engelhard, M. H., Wang, C., Lin, Y., & Zhong, C.-J. (2005). Iron oxide–gold core–shell nanoparticles and thin film assembly. *Journal of Materials Chemistry*, *15*(18), 1821. https://doi.org/10.1039/b501375e

[63] Wang, X., Zhuang, J., Peng, Q., & Li, Y. (2005). A general strategy for nanocrystal synthesis. *Nature*, *437*(7055), 121–124. https://doi.org/10.1038/nature03968

[64] Wu, W., He, Q., & Jiang, C. (2008). Magnetic Iron Oxide Nanoparticles: Synthesis and Surface Functionalization Strategies. *Nanoscale Research Letters*, *3*(11), 397. https://doi.org/10.1007/s11671-008-9174-9

[65] Yang, H., Ogawa, T., Hasegawa, D., & Takahashi, M. (2008). Synthesis and magnetic properties of monodisperse magnetite nanocubes. *Journal of Applied Physics*, *103*(7), 07D526. https://doi.org/10.1063/1.2833820

[66] Yang, H.-H., Zhang, S.-Q., Chen, X.-L., Zhuang, Z.-X., Xu, J.-G., & Wang, X.-R. (2004). Magnetite-Containing Spherical Silica Nanoparticles for Biocatalysis and Bioseparations. *Analytical Chemistry*, *76*(5), 1316–1321. https://doi.org/10.1021/ac034920m

[67] Yu, H., Chen, M., Rice, P. M., Wang, S. X., White, R. L., & Sun, S. (2005). Dumbbell-like Bifunctional Au−Fe$_3$O$_4$ Nanoparticles. *Nano Letters*, *5*(2), 379–382. https://doi.org/10.1021/nl047955q

[68] Zhao, W., Gu, J., Zhang, L., Chen, H., & Shi, J. (2005). Fabrication of Uniform Magnetic Nanocomposite Spheres with a Magnetic Core/Mesoporous Silica Shell Structure. *Journal of the American Chemical Society*, *127*(25), 8916–8917. https://doi.org/10.1021/ja051113r

[69] Zhu, L., Zhou, Z., Mao, H., & Yang, L. (2017). Magnetic Nanoparticles for Precision Oncology: Theranostic Magnetic Iron Oxide Nanoparticles for Image-Guided and Targeted Cancer Therapy. *Nanomedicine*, *12*(1), 73–87. https://doi.org/10.2217/nnm-2016-0316

20 Development of stability indicating RP-HPLC method for simultaneous estimation of sacubitril and valsartan in pure and pharmaceutical dosage form

Kumar Dhiraj[1,a] and Himansu Bhusan Samal[2,b]

[1]Associate ProfessorDepartment of Pharmacy, Institute of Technology and Management, GIDA, Gorakhpur, Uttar Pradesh, India
[2]School of Pharmacy and Life Sciences, Centurion University of Technology and Management, Ramchandrapur, Jatni, Khurda, Odisha, India

Abstract: An HPLC method of reverse phase was developed for the simultaneous detection of sacubitril and valsartan in pure form and dosage form. The chromatography procedure was carried out using the isocratic reverse phase HPLC. For the separation procedure C18 column with a particle size of 1.9 microns was placed inside a 2.1 × 50 mm stainless steel symmetry column. Acetonitrile and phosphate buffer (30:70) with a pH of 3.0 were used in the mobile phase. At 254 nm, the effluent's flow rate was measured to be 1.0 ml/min.

The relative retention times for both medications were 2.56 and 2.88 minutes. For sacubitril and valsartan, the standard curve produced an coefficient of correlation is 0.998 and 0.999, respectively, using a linear distribution ranging from 10–50.1 µg/ml and 49 µg/ml, respectively. The absolute recovery rates for valsartan and sacubitril were 100.2% and 100.5%, respectively. Because degradation products did not interfere while estmation of sacubitril and valsartan, the method is called stability-indicating.

Keywords: Sacubitril, Valsartan, RP-HPLC, dihydrogen phosphate buffer, Acetonitrile

1. Introduction

The drug Valsratan is used for the treatment of several types of heart problem. It is an angiotensin receptor blocker. It mechanism of action is based on the blocking of RAAS (renin-angiotensin-aldosterone) system [1].

2. Drug Profile

An ARB valsartan specifically prevents angiotensin II from binding to AT1, which is present in a variety of organs, including the adrenal glands and vascular smooth muscle [1] (Figure 20.1). Sacubitril forms Sacubitrilat an active form of the drug by the process of de-eth-ylation. Sacubitril is an antihypertensive medication (Figure 20.2). It inhibits the neprilysin enzyme, which breaks down atrial and brain natriuretic peptide. The mentioned two peptide reduces blood pressure by decreasing volume of blood and acting on bradykinin. Bradykinin acts as a strong vasodilator [2, 3].

[a]dhirajkumar5707@gmail.com, [b]hbsamal@gmail.com

DOI: 10.1201/9781003672869-20

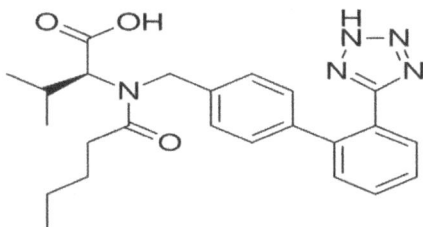

Figure 20.1. Valsartan.

Source: Author.

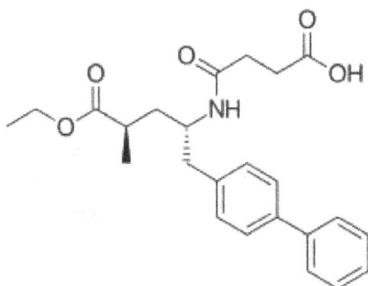

Figure 20.2. Sacubitril.

Source: Author.

3. Experimental Materials and Methods

3.1. Drugs

From Dr. Reddy's, a pure pharmaceutical sample of Sacubitril and Valsartan was acquired.

3.2. Chemicals used

Sl.No	Chemical	Brand names
1	Sacubitril	Yucca Pharma
2	Valsartan	Gift sample (Dr.Reddys Lab)
3	HPLC grade Methanol	SD FineChem
4	Acetonitrile.	SD FineChem
5	Triethyl amine(TEA)	SD FineChem
6	Ortho phosphoric acid(OPA)	SD FineChem

3.3. Instrument

Liquid chromatographic system from LC 20 AD Shimadzu equipped with UV detector and LC solution software.

3.4. Buffer solution (PH-3)

Transferred approximately 1.36 grams of potassium dihydrogen phosphate into 1000 milliliters of water, then use orthophosphoric acid to get the pH down to 3.0 [4].

3.5. Preparation of mobile phase

The mobile phase used in this method are in the ratio of 30:70 (v/v) prepared by mixing of 10 mM dihydrogen phosphate buffer and acetonitrile [5].

3.6. Standard stock preparation

In a volumetric flask of 100 ml 49 mg of sacubitril and 51 mg of valsartan transferred. Initially, each drug was dissolved in 50 milliliters of diluents and the solvent was then added until the volume reached 100 milliliters. The concentrations of sacubitril and valsartan were 490 µg/ml and 510 µg/ml, respectively. 50 ml of the solution transferred in the another flask and marked as C-2 solution. The concentration of the drug was 51 µg/ml for valsartan and 49 µg/ml for sacubitril. After that, solutions with different concentrations were made using a dilution process and calculation [6, 7].

3.7. Preparation of sample stock solution

A mortar and pestle were used to crush ten precisely weighed tablets (Table 20.1). A 100 ml volumetric flask was filled with 100 mg equivalent weight of powdered medication that contained both VAL and STL (marketed formulation CIDMUS; the dose of VAL is 51 mg and the dose of STL is 49 mg in combo tablet). The volume was adjusted with solvent to reach the desired level. (A stock solution). The medication concentrations for sacubitril and valsartan were 49 µg/ml and 51 µg/ml, respectively.

4. Method Development

4.1. Apparatus and chromatographic conditions

A 50 × 2.1 mm Thermo C18 with a particle size of 1.9 µm. The mobile phase consists of a 30:70 (v/v) combination of acetonitrile and phosphate buffer. Prior to use, the mobile phase was degassed by sonication for 30 minutes after passing through 0.45 mm Millipore membrane filters. For detection of component UV detector was used and wavelength was adjusted at 254 nm. In the entire process a constant flow rate was maintained and kept at 1 mL min.

4.2. Optimized method parameters

Table 20.1. System suitability parameters

Instrument used	LC 20 AD Shimadzu
Temperature	30° C
Column	Thermo C18 (50 × 2.1 mm), 1.9 µm
Buffer:	1.5ml of Triethyl amine dissolve in Potassium dihydrogen ortho phosphate dissolved into 1000ml water (pH 3.0)
pH	3
Eluent	Phosphate. B.uffer and A.cetonitrile (30:70)
Solvent flow speed	1ml/ min
Detection Wavlengt.h	254 nm
Injection volume	10 l
Total Analysis Duration.	10 min

Source: Author.

4.3. Running the standard solution of Sacubitril and Valsartan

From the stock solution C-2 two milliliters were poured in a 10-milliliter flask (volumetric) and the solvent was added until the flask's mark of 10µg/ml was reached. The HPLC system was filled with the solution. Figure 20.3 displays the acquired chromatogram.

Optimization of the method: Chromatographic separation was conducted using a solution of buffer and acetonitrile in a 30:70 proportion, with the pH kept at 3. The process was carried out on a 2.1 × 50 mm stainless steel ODS column (C18) containing 1.9 microns particle. A UV detection system used with 1.0 ml/min of flow rate [8–11].

5. Results and Discussion

Linearity: Across the investigated calibration ranges—10.2–50.1 µg/ml for Valsartan and 9.8–49 µg/ml for Sacubitril—the assay method generated a linear calibration plot. Sacubitril and valsartan were shown to have correlation values of 0.998 and 0.999, respectively. Tables 20.2 and 20.3 demonstrate the results, which reveal a substantial correlation between the analyte's peak area and concentration.

5.1. Construction of calibration graph

Precisely measured volumes of 2 mL, 4 mL, 6 mL, 8 mL and 10 mL of the stock solution, which was derived from the C-2 solution for linearity, were transferred into a 10 mL volumetric flask. The solvent was then added until the flask reached the calibration mark. Each concentration level was injected into

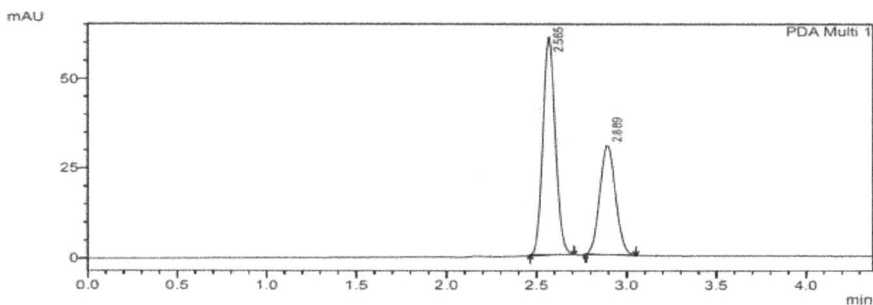

Figure 20.3. Sacubitril and Valsartan standard drug chromatogram.

Source: Author.

Entry. No.	Analyte.	RT index	Peak Area Value.	Peak Lemgth	USP Resolution	T.F. (USP)	Plate. count
1	Sacubitril	2.565	353249	189524	3.6	1.09	5003
2	Valsartan	2.889	1021987	169547	5.2	1.3	3541

the chromatographic system and the peak area was recorded. A graph was then plotted with concentration on the X-axis and peak area on the Y-axis to determine the correlation coefficient [11] (Figures 20.4 and 20.5).

Table 20.2. Linearity results (for Valsartan)

VAL Concentration (ppm)	Area of Peak,
0	0
10.2	510994
20.5	1021999
30.6	1559948
40.8	2064215
50.1	2545141

Source: Author.

Figure 20.4. Calibration curve of valsartan.

Source: Author.

5.2. Precission

The parameter of precission was evaluated through repeatability, interday and intraday precision, studies (Table 20.4).

Table 20.3. Linearity results (for Sacubitril)

STL Concentration (ppm)	Area of Peak,
0	0
4.9	139876
9.8	243277
19.6	489754
29.4	735241
39.2	954728

Source: Author.

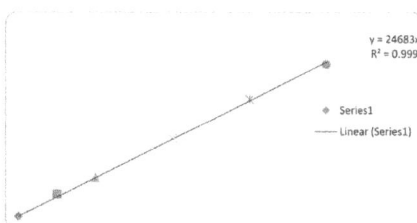

Figure 20.5. Calibration curve of sacubitril.

Source: Author.

5.3. Method Procedure Preparation of standard solution

Standard solution was prepared by diluting two ml C2 standard stock solution to ten ml. As a result concentrations of 9.8 μg/mL for sacubitril and 10.2 μg/mL for valsartan was attained (Table 20.5).

Recovery and accuracy: The correctness of the procedure is confirmed by Tables 20.6 and 20.7, which shows the recovery of VAL and STL in sample bulk drug, ranging from 99.4% to 99.6%.

5.4. Accuracy results

5.4.1. Preparation of stock solution A
Tablet Powder equivalent to 10mg was taken and dissolved in 100ml of diluents.

It is then sonicated for 25 min. This stock solution A contains VAL 51 μg/ml and STL 49 μg/ml.

5.4.2. Preparation of sample solution
1. 50% level: Two milliliters of stock A solution mixed with 0.5 ml of a standard drug solution containing 100 μg/ml. The mixture was diluted to 10 ml.
2. 100% level: Two milliliters of stock A solution mixed with 1 ml of a standard drug solution containing 100 μg/ml. The mixture was diluted to 10 ml.
3. 150% level: Two milliliters of stock A solution mixed with 1.5 ml of a standard drug solution containing 100 μg/ml. The mixture was diluted to 10 ml.

Table 20.4. Precession table (Sacubitril)

Entry. No.	Analyte.	RT index	PeakAreaValue	T.F. (USP)	Plate. count
1	Sacubitril	2.569	355697	1.0	3802
2	Sacubitril	2.567	356745	1.1	3546
3	Sacubitril	2.595	358541	1.4	4633
4	Sacubitril	2.566	357845	1.1	4812
5	Sacubitril	2.570	355241	1.0	3802
	Mean.		**356813**.8		
	Std. Dev.		1393.93		
	% RSD.		1.1		

Source: Author.

Table 20.5. Precession table (Valsartan)

Entry. No.	Analyte.	RT index	PeakAreaValue	T.F. (USP)	Plate. count
1	Valsartan	2.889	1021456	1.2	4759
2	Valsartan	2.889	1029214	1.1	3695
3	Valsartan	2.885	1024254	1.1	4741
4	Valsartan	2.895	1032547	1.2	3793
5	Valsartan	2.888	1025785	1.1	4741
	Mean		1026651		
	Std. Dev.		4324.79		
	% RSD		0.47		

Source: Author.

Table 20.6. Sacubitril accuracy results

Sample Strength (%)	Peak Area Value	Spiked Quantity (mg)	Measured Quantity (mg)	Recovery. efficiency%	Mean. Recovery.
50%	547421.5	4.9	4.9	100 %	100.5%
100%	734365	9.8	9.9	101.0%	
150%	914224.5	13.7	13.8	100.7%	

Source: Author.

Table 20.7. Accuracy results for Valsartan

Sample Strength (%)	Peak Area Value	Spiked Quantity (mg)	Measured Quantity (mg)	Recovery. efficiency%	Mean. Recovery.
50%	1599907	5.1	5.2	101.3%	100.2%
100%	2131690	10.2	10.2	100 %	
150%	2664212	15.3	15.2	99.3%	

Source: Author.

LOD and LOQ: For determination of LOD and LOQ S/N ratio was calculated at lest sample concentration in context to baseline (Table 20.8(A) and 20.8(B)).

5.5. Robustness

By examining the impact of minor modifications to the chromatographic conditions on resolution and peak area, the robustness of the suggested approach was assessed. It was discovered that the technique was resistant to minor variations in the experimental parameters, such as the flow rate (0.9,1.0,1.1) and the composition of the mobile phase (buffer: ACN – 25:75,35:65,40;60) (Tables 20.9 and 20.10).

Table 20.8(A). LOD table

Drug. Name.	Baseline Noise (µV.)	Signal (µV.)	S/N Ratio.
Sacubitril	51	158	3.01
Valsartan	51	155	2.93

Source: Author.

Table 20.8(B). LOQ table

Drug. Name.	Baseline Noise (µV.)	Signal (µV.)	S/N Ratio.
Sacubitril	53	541	9.9
Valsartan	51	519	9.9

Source: Author.

Table 20.9. Flow variation result

Entry. no	Analyte	Solvent Flow. (ml/min)	Peak Area Value	Peak Length	T.F. (USP)	Plate count (USP)
1	Sacubitril	Less (0.9)	345874	189779	4479	0.9
		Actual(1)	365690	191456	4759	0.9
		More (1.1)	369585	180921	3072	0.9
2	Valsartan	Less (0.9)	1096585	206420	4508	0.9
		Actual (1)	1066605	175740	3695	0.9
		More (1.1)	1035252	180921	3072	1.0

Source: Author.

Table 20.10. Mobile phase composition variation in

Entry. no	Drug.	Solvent Flow. (ml/min)	Peak Area Value	Peak Lemgth	T.F. (USP)	Plate. count
1.	Sacubitril	Less (65%)	335894	137954	2028	0.9
		Actual	365697	191456	4759	0.9
		More (75%)	354154	181579	3002	1.0
2.	Valsartan	Less (65%)	1011444	146804	3035	1.0
		Actual	1066604	175740	3695	0.9
		More (75%) (85%)	1069604	181569	3002	1.0

Source: Author.

6. Stability Studies

The goal of the developed technique is to guarantee the efficient separation of both drugs as well as the formulation constituents' degradation peaks. Studies on forced deterioration were conducted to assess the methodologies' specificity and stability-indicating qualities.

6.1. Acid hydrolysis

To attain concentrations of 10.2 μg/mL for Val and 9.8 μg/mL for STL, 2 mL of the stock solution was pippetted from the C-2 solution and poured to a 10 mL flask. With help of diluents volume was adjusted (Figure 20.6).

6.2. Basic hydrolysis

Alkaline hydrolysis was carried out at room temperature for 48 hours. After adding two milliliters of 0.1N sodium hydroxide to two milliliters of stock solution, the liquid was refluxed at 60 degrees Celsius (Figure 20.7).

6.3. Dry heat degradation

A thin layer of heated sample was placed in a petridish at 105C to assess solid state stability. Desired concentration was achieved by diluting with mobile phase to get 10.2 μg/ml Val and 9.8 μg/ml STL (Figure 20.8).

6.4. Photo degradation

Solid samples were exposed to sunlight for seven days in order to conduct photo stability degradation experiments (Figure 20.9).

6.5. Oxidation With (3%) HO

Studies on oxidative stress were conducted at room temperature using 3%

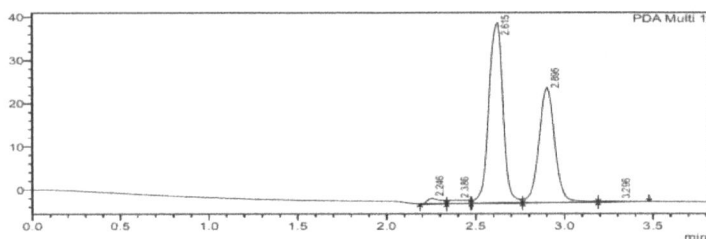

Figure 20.6. Acid hydrolysis chromatogram.

Source: Author.

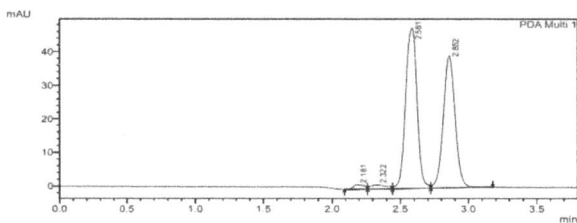

Figure 20.7. Basic hydrlysis chromatogram.

Source: Author.

Entry. No.	Analyte	RT index	Peak Area Value	Peak Length	T.F. (USP)	Plate. count
1	Valsartan	2.56	1441033.47	198574	1.0	3694
2	Sacubitril	2.85	1171571	187452	1.1	4658

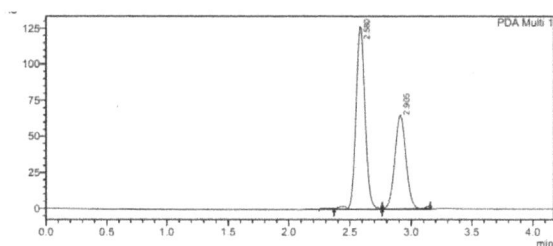

Figure 20.8. Thermal degradation chromatogram.

Source: Author.

Entry. No.	Analyte.	RT index	Peak Area Value	Peak Length	T.F. (USP)	Plate. count
1	Valsartan	2.56	1466445	169587	1.4	3365
2	Sacubitril	2.90	1238660	198698	1.2	4821

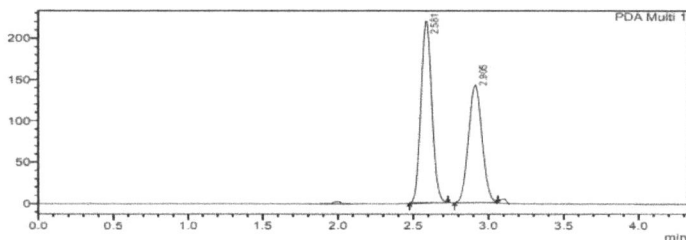

Figure 20.9. Photolytic degradation chromatogram.

Source: Author.

Entry. No.	Analyte.	RT index	Peak Area Value	Peak Length	T.F. (USP)	Plate. count
1	Valsartan	2.58	1487373	186954	1.0	4857
2	Sacubitril	2.90	1238660	169587	1.2	3635

HO. Two milliliters of 3% hydrogen peroxide (HO) were added separately to two milliliters of stock solution. The solutions were stored for a full day (Figure 20.10).

6.6. *Neutral condition*

In order to investigate stress testing in neutral medium, the drug was refluxed in water for six hours at a temperature of 60°C (Figure 20.11).

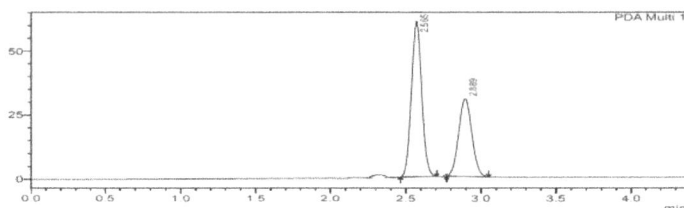

Figure 20.10. Oxidative degradation chromatogram.

Source: Author.

Entry. No.	Analyte.	RT index	Peak Area Value	Peak Length	T.F. (USP)	Plate. count
1	Valsartan	2.56	1427579	198595	1.19	4634
2	Sacubitril	2.88	1156663.	177854	1.50	3458

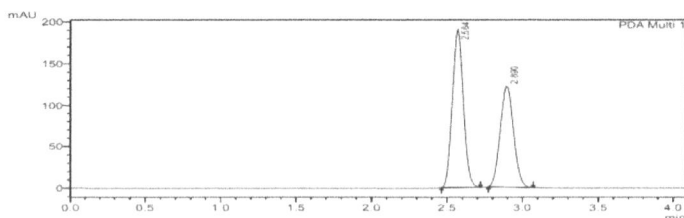

Figure 20.11. Chromatogram shows Degradation in Neutral condition at room temperature for 7 days.

Source: Author.

Entry. No.	Analyte.	RT index	Peak Area Value	Peak Length	T.F. (USP)	Plate. count
1	Valsartan	2.56	1492007	198596	1.2	4635
2	Sacubitril	2.89	1239903	177854	1.5	3458

6.7. *Results of forced degradation studies*

The developed method's specificity was demonstrated by the stress study findings. Under photolytic, thermal and basic stress conditions, sacubitril and valsartan remained stable. Tables 20.11 and 20.12 presents the findings from studies on forced deterioration.

Table 20.11. Stability studies of Sacubitril

Conditions of Stress Testing	Duration	Decomposition product Analysis.	Potency of Active compound	Recover balance (%)
AcidHydrolysis. (1 N HCl)	6Hrs.	8.1	91.9	100
Basic. Hydrolysis. (0.I M NaOH)	48Hrs.	5.6	94.4	100
Thermal Degradation (50C)	24Hrs.	1.01	98.9	100
UV (254nm)	7 days.	0.3	99.7	100
3 % Hydrogen peroxide	6Hrs.	3.9	96.1	100
Neutral Condition	24Hrs.	0.2	99.8	

Source: Author.

Table 20.12. Stability studies of Valsartan

Conditions of Stress Testing	Duration	Decomposition product Analysis	Potency of Active compound	Recover balance (%)
AcidHydrolysis. (1 N HCl)	6Hrs.	6.2	93.8	100
Basic. Hydrolysis. (0.I M NaOH)	48Hrs.	3.1	96.9	100
Thermal Degradation (50C)	24Hrs.	1.9	98.1	100
UV (254 nm)	7 days.	0.51	99.49	100
3 % Hydrogen peroxide	6Hrs.	4.8	95.2	100
Neutral Condition	7 days.	0,19	99.81	

Source: Author.

7. Conclusion

It was discovered that the recommended method for identifying the pure and dosage forms of sacubitril and valsartan was rapid, simple, accurate and precise. The mobile phase is inexpensive and easy to set up. The recoveries of samples in each formulation demonstrated good agreement within the limit. For routine testing of sacubitril and valsartan in both dose and pure form, this method is therefore easy to use and practical.

References

[1] https://www.webmd.com/drugs/2/drug-5891/amlodipine-oral/details

[2] https://www.webmd.com/drugs/2/drug-63172/olmesartan-oral/details

[3] Draft ICH Guidelines on Validation of Analytical Procedures Definitions and terminology. Federal Register, volume 60. IFPMA, Switzerland, PP 1126 (1995).

[4] Patel, K. H., Luhar, S. V., & Narkhede, S. B. (2016). Simultaneous estimation of sacubitril and valsartan in synthetic mixture by RP-HPLC method. *J Pharm Sci Bioscientific Res, 6*(3), 262–269.

[5] Pachauri, S., Paliwal, S., Srinivas, K. S., Singh, Y., & Jain, V. (2010). Development & validation of HPLC method for analysis of some antihypertensive agents in their pharmaceutical dosage forms. *Journal of Pharmaceutical sciences and Research, 2*(8), 459.

[6] Jignesh, S., & Kokila, P. (2014). Development and validation of HPLC method for analysis of some anti hypertensive agents in their pharmaceutical dosage forms. *J Drug Delivery Ther*, *4*, 12–15.

[7] Naazneen, S., & Sridevi, A. (2017). Development of assay method and forced degradation study of valsartan and sacubitril by RP-HPLC in tablet formulation. *Int J App Pharm*, *9*(1), 9–15.

[8] Manoranjani, M., & Bhagyakumar, T. (2017). RP-HPLC method for the estimation of valsartan in pharmaceutical dosage forms. *Int J Sci Innov Dis*, *1*, 101–108,

[9] El-Gizawy, S. M., Abdelmageed, O. H., Omar, M. A., Deryea, S. M., & Abdel-Megied, A. M. (2012). Development and validation of HPLC method for simultaneous determination of amlodipine, valsartan, hydrochlorothiazide in dosage form and spiked human plasma. *American Journal of Analytical Chemistry*, *3*(6), 422–430.

[10] Vaka, S., & Parthiban, P. (2017). New method development and validation for the simultaneous estimation of sacubitril and valsartan in a bulk and pharmaceutical dosage forms. *Int J Res*, *4*(1), 17–24.

[11] Singh, T., Kumar, D., Mishra, A., Arora, S., Kumar, D. V., Pamu, S., ... & Nishanth, V. (2017). Development of validated RP-HPLC method formultanious estimation of cefadroxil and ambroxol in pure and pharmaceutical dosage form. *British Journal of Pharmaceutical Research*, *15*(6), 1–7.

21 Development of sophisticated RP-HPLC approach for quantifying lornoxicam in API and pharmaceutical formulation

Biswa Mohan Sahoo[1,a], Chandan Kumar Brahma[1],
Bikash Ranjan Jena[1], Kirtimaya Mishra[1], Debabrata
Panda[1], Satya Narayan Tripathy[1], and Dipika Rani Sahu[2]

[1]School of Pharmacy and Life Sciences, Centurion University of Technology and Management, Jatni, Khurda, Odisha, India
[2]Pandaveswar School of Pharmacy, Pandaveswar, Paschim Bardhaman, West Bengal, India

Abstract: The present experiment is designed to develop an appropriate, accurate, precise and reproducible RP-HPLC technique for estimating Lornoxicam in API and pharmaceutical dosage form. The HPLC instrument equipped with a pump, injector and PDA detector is used for conducting assay of the samples. Under isocratic conditions, the chromatogram was run through an inertsil ODS C-18 column (250 × 4.6 mm, 5 μ). The mobile phase was prepared using methanol and distilled water (70/30, v/v) with ammonium dihydrogen phosphate buffer of pH 7.3 adjusted with triethylamine. The solution of sample was pumped across the column at a flow rate of 1 mL/min. The eluent was analyzed at 290 nm for the estimation of Lornoxicam. For Lornoxicam, the retention time was found to be 4.293 min. Similarly, the % RSD of the Lornoxicam was found to be 0.63. Whereas the percent recovery was obtained between 98 to 102. For lornoxicam, the linearity range was observed from 57.59 to 106.94 μg/ml. The developed technique was validated by assessing linearity, system suitability, sensitivity, selectivity, precision, accuracy, robustness and stability following ICH requirements.

Keywords: Lornoxicam, estimation, RP-HPLC, development, mobile phase, formulations

1. Introduction

Lornoxicam is considered as a NSAID of the oxicam class with the molecular formula of CHClNOS [1] (Figure 21.1). It exerts antipyretic, analgesic and anti-inflammatory properties because it inhibits the production of prostaglandin (PG) and thromboxane (Tx) through the inhibitory action on both cyclooxygenase enzymes [2, 3]. It received patent approval in 1977 and medical use approval in 1997 [4]. It is recommended for the management of arthritis. It is available in both oral and parenteral dosage form [5–8]. The recommended dosage of lornoxicam is 8–16mg/day divided into 2–3 times [9–12]. The maximum dose of Lornoxicam is 16mg/day [13–18]. Lornoxicam is categorized as a drug with low solubility and high permeability through biological membranes under the BCS class [19–22]. It is a yellow crystalline powder. Lornoxicam can form salts because of its acidic nature with a pKa value of 4.7 [23].

[a]drbiswamohansahoo@gmail.com

DOI: 10.1201/9781003672869-21

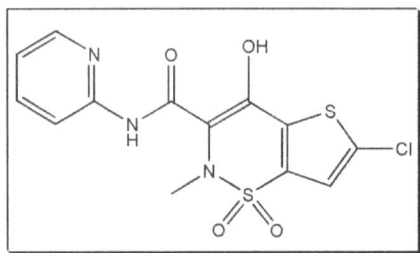

Figure 21.1. Molecular structure of lornoxicam.

Source: Author.

Several spectrophotometric and chromatographic methods are reported to estimate Lornoxicam in the pharmaceutical preparation [24]. So, the purpose of the current study is to develop a simple, specific and precise RP-HPLC technique for the assessment of lornoxicam in pharmaceutical formulation. The suggested approach has been verified according to the guidelines mentioned in ICH [25].

2. Material and Method

2.1. Chemicals and reagents

Lornoxicam samples were collected from Hetero Drugs in Hyderabad, India. Lornoxicam tablets were procured from the local medicine store. Water from Milliq (Millipore) was used for the entire experiment. HPLC grade solvents were procured from Sigma Aldrich. The analytical grade chemicals, solvents and reagents were utilized in this investigation.

2.2. Equipment

The components of an Agilent LC system included a UV/VIS photodiode array (PDA) detector, a degasification system, an auto-injector with a 100 μL syringe and a solvent delivery system. Chem station software (Agilent) was used to monitor and integrate the output signal.

2.3. Chromatographic condition

The study was conducted using an Agilent 1200 HPLC apparatus from Germany with an inertsil ODS C-18 analytical column (250 × 4.6 mm, 5 μ). The mobile phase was prepared using methanol-water with ammonium dihydrogen phosphate buffer of pH 7.3 adjusted with Triethylamine. A freshly prepared mobile phase was allowed to pass through a Millipore nylon filter with a pore size of 0.45 μm. The sample solution was injected through the column at a flow rate of 1 mL/min. The injection volume was set at 20 μl. To measure Lornoxicam, the wavelength was selected at 290 nm in UV detection.

2.4. Preparation of standard and stock solutions

10 mg of lornoxicam was dissolved in methanol and fill a volumetric flask (100 mL) to the desired level to prepare a stock solution of concentration 100 μg/ml. Aliquots were made from the stock solution ranging from 0.5 to 20 μg/ml [26].

2.5. Preparation of sample solutions

Weighed accurately 20 tablets and estimated their average weight. After crushing the tablets into a homogeneous powder, 8 mg of the drug was placed to a volumetric flask of capacity 50 mL, diluted in methanol and filtered using 0.45 μm Millipore nylon filters [27, 28].

2.6. Validation of the HPLC method

The validation of the RP-HPLC method was conducted by the ICH procedures and the parameters for validation were selected based on the availability of resources and limitations. The suggested method was validated by analyzing several validation parameters such as linearity, accuracy, precision, specificity, LOD and LOQ and system suitability [29].

2.7. Linearity

The linearity for Lornoxicam was determined in the range of 57.59 mcg/ml to 106.94 mcg/ml. For the Lornoxicam peak, the correlation coefficient was calculated by plotting a graph with the conc on the X-axis and the mean areas on the Y-axis [30].

2.8. Precision

Through the daily preparation and analysis of three distinct sample solutions at low, medium and high conc, the repeatability and intermediate precision of the proposed approach were evaluated. These experiments were repeated over 2 days to estimate intermediate precision [31].

2.9. Specificity

The capacity of the procedure to precisely quantify the analyte response in the presence of every possible sample component is known as specificity. The results were compared to standard drug (Lornoxicam) and its tablet dosage form. The excipients of the solid dosage form had no effect on the analyte [32].

2.10. Accuracy

The accuracy was measured as the % recovery of the known amount of API in the presence of excipients. The known amount of placebo spiked with the known amount of Lornoxicam at 70%, 100% and 130% of test concentration as Lornoxicam. The quantity of Lornoxicam in dosage form was measured following the test protocol. The % recovery was calculated using the quantity present and the actual amount added [33].

2.11. Robustness

The robustness of the HPLC technique was determined by the fact that minor changes in the extraction duration, mobile phase composition, flow rate and wavelength did not significantly impact the results [34].

2.12. System suitability

Six replicate injections from newly prepared standard solutions were used to examine the system applicability of the HPLC procedure. Each solute was analyzed for its peak area, theoretical plates (N), resolution (R) and tailing factors (T) [35].

2.13. Degradation study

The sample was subjected to a forced deterioration investigation by treating the sample under several conditions (acid, alkali, peroxide, photolytic, thermal) [36].

Acid degradation: The sample solution was exposed to 5 ml of 0.1N HCl for 24.5 hours. The treated sample solution was investigated as per the test process.

Alkali degradation: 5ml of 0.1N NaOH solution was added to the sample solution and left for 25 hours. The treated sample solution was analyzed in accordance with the test methodology.

Peroxide degradation: The sample solution was treated with 5ml of 3% v/v solution of hydrogen peroxide for 25.5 hours. The treated sample solution was assessed as per the test process.

Photolytic degradation: The sample was exposed to UV light under 254 nm for about 7 days. The treated sample was examined as per the test procedure.

Thermal degradation: The sample was kept in an oven at 80°C for about 26 hours. The test protocol was followed in the analysis of the treated sample.

3. Results and Discussion

Several mobile phases and columns were used to accomplish effective resolution to create a reliable and appropriate HPLC technique for the estimation of lornoxicam. Both the cost and the time needed for the analysis were considered while selecting the suitable mobile phase for analyzing the drug. The suggested approach was able to separate lornoxicam selectively from the sample in a short chromatographic run time (4.5 min). A retention time of 4.293 minutes was observed. The retention time was found to be 4.293min and the chromatogram was displayed in Figure 21.2.

3.1. Linearity

The evaluation of linearity was performed by assessing lornoxicam working standard solutions at 6 several concentrations. Linearity ranged from 0.5 to 20 µg/ml. Regression analysis was applied to determine the calibration equations and correlation coefficients for each drug's peak area and concentration. Table 21.1 shows the regression data

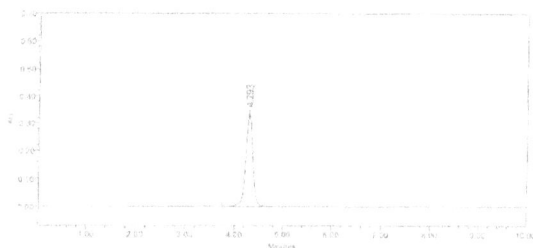

Figure 21.2. Chromatogram of Lornoxicam at optimized condition.

Source: Author.

Table 21.1. Linearity study of Lornoxicam

Spike level (%)	Concentration of Lornoxicam (mcg/ml)	Mean peak areas
70	57.59	2312197
80	65.81	2610075
90	74.04	2993666
100	82.26	3286386
110	90.49	3597780
120	98.72	3945727
130	106.94	4313224
Slope	40282	
Intercept	−19617	
CC	0.99949	
RSS	3132624702	

Source: Author.

that were acquired. The experiment's results show that there was a good relationship between peak area and conc of drug within the range as described. The correlation coefficient (CC) value must meet a minimum value of 0.99 as the acceptance criterion (Figure 21.3).

3.2. Precision

3.2.1. System precision

The HPLC apparatus received six replicate injections of standard solution. The mean, SD and percentage RSD for Lornoxicam peak regions were computed. The acceptance criteria for the % RSD of Lornoxicam peak areas should not be more than 2. The study reports are displayed in Table 21.2.

3.2.2. Method precision

A single batch of six samples was examined using the test procedure. The percentage assay for Lornoxicam in six samples was computed and the findings are reported in Table 21.3. The acceptance criteria for the % RSD for the assay of six samples of Lornoxicam should not be more than 2.0. The acceptability criterion for the percentage RSD for the assay of six Lornoxicam samples should not exceed 2.0 (Figures 21.4 and 21.5).

3.2.3. Intermediate precision (Ruggedness)

The six samples from the same batch that were utilized to test the method's robustness were analyzed by several analysts

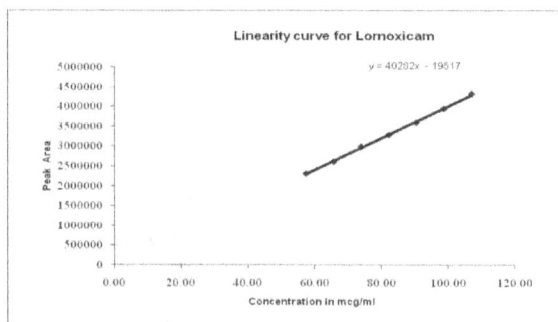

Figure 21.3. Calibration curve for Lornoxicam.

Source: Author.

Table 21.2. System precision analysis

Sl. No.	Peak areas of Lornoxicam
1	3314316
2	3372238
3	3315465
4	3332533
5	3335937
6	3330321
Mean	3333468
SD	21028.05
RSD (%)	0.63

Source: Author.

Table 21.3. Method precision study

Sample No.	% Assay of Lornoxicam
1	101.17
2	101.39
3	101.42
4	102.49
5	101.76
6	101.99
Mean	101.70
SD	0.483
RSD (%)	0.47

Source: Author.

Figure 21.4. Standard chromatogram (Retention time versus peak height).

Source: Author.

Figure 21.5. Chromatogram of method precision.

Source Author.

on different days, using various columns and instruments, following the test protocol. The % assay was determined and represented in Table 21.4.

3.2.4. Accuracy

The percentage of analytes retrieved by the assay was used to assess accuracy. The drug recoveries from a range of spiked concentrations are listed in Table 21.5. The findings show that the procedure is quite accurate in determining the lornoxicam. Using the standard addition approach, a recovery experiment was conducted to examine the method's accuracy. For each drug, a known amount equivalent to 70%, 100% and 130% of the label claim was added. This was done to see if there were any

Table 21.4. Results of precision study (Intermediate)

Sample No.	% Assay for Lornoxicam	
	Analyst-I	Analyst-II
1	101.17	100.54
2	101.39	100.44
3	101.42	102.52
4	102.49	99.90
5	101.76	100.39
6	101.99	101.25
Mean	101.70	100.84
SD	0.483	0.930
%RSD	0.47	0.92
Over all Mean	101.27	
Over all SD	0.838	
Over all %RSD	0.83	

Source: Author.

Table 21.5. Recovery data of lornoxicam

Level No/ Spike level (%)	Amount of Lornoxicam added (mg)	Amount of Lornoxicam found (mg)	Recovery (%)	Mean	SD	RSD (%)
Level-1 (70%)	57.34	57.25	99.84	100.54	1.108	1.10
	57.25	57.23	99.97			
	57.20	58.82	101.82			
Level-2 (100%)	81.09	80.31	99.04	98.98	0.095	0.10
	81.16	80.24	98.87			
	81.11	80.32	99.03			
Level-3 (130%)	104.93	105.33	100.38	100.30	0.071	0.07
	104.99	105.24	100.24			
	105.25	105.55	100.29			
	Over all Mean		99.94			
	Over all SD		0.971			
	Over all % RSD		0.97			

Source: Author.

+ve or -ve interactions with the excipients in the pharmaceutical dosage form.

3.2.5. Robustness

The samples were examined three times and the method's modest changes were performed on purpose in order to assess the robustness.

- Flow rate (±0.1ml)
- Column Oven Temperature. (± 5°C)
- Wavelength (± 2 nm)
- Mobile phase (± 2%) organic
- Buffer pH (±0.1units)

Every condition's system suitability was assessed and the outcomes were compared to the precision of the approach. The results are represented in Tables 21.6a and 21.6b. The acceptance criteria for the overall % RSD should not exceed 2 using the precision data approach for individual tests.

3.2.6. Specificity

Blank, placebo and sample solutions were all injected into the HPLC apparatus. At the analyte peak retention time, neither the blank nor the placebo caused any interference. Peak purity data shows that there was no interference during the Lornoxicam peak's retention period and that the peak was homogeneous. The acceptance criteria state that no peaks will elute at the Lornoxicam peak retention time in the placebo and blank and that the sample must pass the Lornoxicam peak purity test (Table 21.7).

3.3 Degradation

This technique was proved to be specific for lornoxicam since the drug peaks were not affected by the peaks of the degradation products (Table 21.8).

Table 21.6(a). Robustness study of Lornoxicam

Sl. No.	A	B	C	D	E	F	G	H	I	J	K
1	101.17	99.32	101.44	97.38	100.19	104.09	104.80	102.90	103.60	105.79	104.23
2	101.39	100.32	101.16	98.04	100.97	103.32	104.11	99.89	100.35	102.47	103.63
3	101.42	100.56	101.23	98.35	101.46	106.82	107.07	100.16	101.24	103.61	104.64
4	102.49	-	-	-	-	-	-	-	-	-	-
5	101.76	-	-	-	-	-	-	-	-	-	-
6	101.99	-	-	-	-	-	-	-	-	-	-
Over all mean		101.16	101.56	100.44	101.43	102.72	102.91	101.46	101.71	102.45	102.52
Over all SD		0.961	0.444	1.944	0.649	1.817	2.007	0.984	0.923	1.458	1.314
Over all %RSD		0.95	0.44	1.94	0.64	1.77	1.95	0.97	0.91	1.42	1.28

Source: Author.

Table 21.6(b). Experimental data

Sl. No.	Experimental value
A	Method precision data
B	Column oven temperature (35°C)
C	Column oven temperature (25°C)
D	Flow rate (1.1 mL/min)
E	Flow rate (0.9 mL/min)
F	Wavelength (292nm)
G	Wavelength (288nm)
H	Mobile phase (+2%)organic
I	Mobile phase (-2%)organic
J	Buffer pH(+0.1)
K	Buffer pH(-0.1)

Source: Author.

4. Conclusion

The suggested technique is sensitive and effective for quantifying lornoxicam in pharmaceutical formulations. The analytical process was not affected by the excipients present in the commercial sample under investigation, indicating the method's specificity for this particular formulation. The HPLC method

Table 21.7. Specificity study of Lornoxicam

Sample type	Peak name	Match factor
Control sample	Lornoxicam	999.989

Source: Author.

Table 21.8. Degradation study of Lornoxicam

Sl. No.	Condition	Assay (%)	Degradation w.r.t. control sample (%)	Match Factor
1	Acid degradation	97.32	4.3	999.789
2	Base degradation	97.24	4.4	999.775
3	Peroxide degradation	65.71	35.4	999.100
4	Thermal degradation	94.07	7.5	999.677

Source: Author.

proved simple, easy to use, quick, accurate and sensitive. Its benefits over other approaches include being less expensive, requiring fewer buffers and less time-consuming. The routine analysis of lornoxicam in commercial samples can be accomplished with this procedure.

References

[1] Jeffery, G. H., Bassett, J., Mendham, J., & Denney, R. C. (2001). *Vogel's Textbook of Quantitative Chemical Analysis*, 5th Edition (pp. 3–5). Addison Wesley Longman Inc, Singapore.

[2] Bramley, R. K., Bullock, D. G., & Garein, J. R. (2004). Quality control and Assessment. In Moffat, A. C., Osselton, M. D., & Widdop, B. *Clarke's Analysis of Drugs and Poisons*, 3rd Edition (Vol I, pp. 161–171). Pharma Press, London.

[3] Moores, C. J. (2003). Quality Control and Regulation. In: Lee, D. C., & Webb, M. *Pharmaceutical Analysis*, First Edition (pp. 9–12). Blackwell Publishing CRC Press, USA..

[4] Garatt, D. C. (2005). *The Quantitative Analysis of Drugs*, 3rd Edition (pp. 876–877). CBS Publishers and Distributors, New Delhi.

[5] Connors, K. A. (2004). *A Textbook of Pharmaceutical Analysis*, 3rd Edition (pp. 373–385). Wiley Interscience Publication, New York.

[6] Markert, B. (1996). *Instrumental Element and Multi-element Analysis of plant samples*, 1st Edition (pp. 3–5). John Wiley and Sons, NY.

[7] Gorog, S. (2007). *Steroid Analysis in the Pharmaceutical Industry*, 1st Edition (pp. 5–6). Pharma Book Syndicate, Hyderabad.

[8] Sachdeva, M. S., & Babu, R. J. (2007). Chromatography. In: *Remington: The Science and Practice of Pharmacy*, 21st Edition (Vol I, pp. 599–615). Lippincott Williams and Wilkins, published by Wolters Kluwer Health (India) Pvt. Ltd., New Delhi.

[9] Sharma, B. K. (2005). *Instrumental Methods of Chemical Analysis*, 24th Edition (p. C-3-8). Goel Publishing House, Meerut.

[10] Settle, F. A. (2004). *Instrumental Techniques for Analytical Chemistry*, 1st Edition (pp. 147–152). Pearson Education Publication, New Delhi.

[11] Braun, R. D. (2006). *Introduction to Instrumental Analysis*, 1st Edition (pp. 839–865). Pharma Book Syndicate, Hyderabad.

[12] Stahl, E. (2005). *Thin Layer Chromatography*, 2nd Edition (pp. 1–5). Springer International Edition, New Delhi.

[13] Sethi, P. D., & Charegaonkar, D. (2005). *Identification of Drugs in Pharmaceutical Formulation by Thin Layer Chromatography*, 2nd Edition (pp. 8–10). CBS Publishers and Distributors, New Delhi.

[14] Feigl, F., Anger, V., & Oesper, R. (2005). *Spot tests in Organic Analysis*, 1st Edition (pp. 76–89). Elsevier, a division of Reed Elsevier India Pvt. Ltd., New Delhi.

[15] Sethi, P. D. (2001). *High Performance Liquid Chromatography*, 1st Edition (pp. 3–16). CBS Publishers and Distributors, New Delhi.

[16] Christian, G. D. (2003). *Analytical Chemistry*, 6th Edition (pp. 604–620).

John Wiley and Sons (Asia) Pte. Ltd., Singapore.

[17] Staut, T. H., & Dorsay, J. G. (2005). High Performance Liquid Chromatography. In: Ohnnesian, L., Streeter, A. J. *Handbook of Pharmaceutical Analysis*, 1st Edition (Vol 117, pp. 87–90). Marcel Dekker, Inc., New York.

[18] Pavia, D. L., Lampman, G. M., & Kriz, G. S. (2001). *Introduction to Spectroscopy*, 3rd Edition (pp. 13–82). Thomson Brooks/Cole, Chennai.

[19] Kemp, W. (2004). *Organic Spectroscopy*, 3rd Edition (pp. 14–15). Palgrave Publication, New York.

[20] Parimoo, P. (2005). *Pharmaceutical Analysis*, 1st Edition (pp. 294–299). CBS Publishers and Distributors, New Delhi.

[21] Snyder, L. R., Glajch, J. L., & Kirkland, J. J. (1997). *Practical HPLC Method Development*, 2nd Edition (pp. 2–9). John Wiley & Sons, Inc., A Wiley-interscience Publication, USA.

[22] Chatwal, G. R., & Anand, S. K. (2007). *Instrumental Methods of Chemical Analysis*, 5th Edition (pp. 2.624–2.629). Himalaya Publishing House, New Delhi.

[23] Kalsi, P. S. (2006). *Spectroscopy of Organic Compounds*, 6th Edition (pp. 9–16). New age International Publishers, New Delhi..

[24] Verma, R. M. (2002). *Analytical Chemistry: Theory and Practice*, 3rd Edition (pp. 291–293). CBS Publishers and Distributors, New Delhi.

[25] Jag, M. (2008). *Organic Analytical Chemistry: Theory and Practice*, 1st Edition (pp. 347–353). Narosa Publishing House, New Delhi.

[26] Mahesh, A. (2010). Rapid Rp HPLC Method for Quantitative Determination of Lornoxicam in Tablets *Journal of Basic and Clinical Pharmacy*, *1*(2), 115–118.

[27] Patil, K. R., Rane, V. P., Sangshetti, J. N., & Shinde, D. B. (2009).

Stability-indicating LC method for analysis of lornoxicam in the dosage form. *Chromatographia*, *69*, 1001–1005.

[28] Bhavsar, S. M., Patel, D. M., Khandhar, A. P., & Patel, C. N. (2010). Validated RP-HPLC method for simultaneous estimation of lornoxicam and thiocolchicoside in solid dosage form. *J.Chem Pharm.Res*, *2*(2), 563–572.

[29] Nemutlu, E., & Kịr, S. (2005). Determination of lornoxicam in pharmaceutical preparations by zero and first order derivative UV spectrophotometric methods. *Die Pharmazie-An International Journal of Pharmaceutical Sciences*, *60*(6), 421–425.

[30] Suwa, T., Urano, H., Shinohara, Y., & Kokatsu, J. (1993). Simultaneous high-performance liquid chromatographic determination of lornoxicam and its 5'-hydroxy metabolite in human plasma using electrochemical detection. *Journal of Chromatography B: Biomedical Sciences and Applications*, *617*(1), 105–110.

[31] Bae, J. W., Kim, M. J., Jang, C. G., & Lee, S. Y. (2007). Determination of meloxicam in human plasma using a HPLC method with UV detection and its application to a pharmacokinetic study. *Journal of chromatography B*, *859*(1), 69–73.

[32] Taha, E. A., Salama, N. N., & Fattah, L. E. S. A. (2004). Stability-indicating chromatographic methods for the determination of some oxicams. *Journal of AOAC International*, *87*(2), 366–373.

[33] Radhofer-Welte, S., & Dittrich, P. (1998). Determination of the novel non-steroidal anti-inflammatory drug Lornoxicam and its main metabolic in Plasma and Synovial fluid. *J Chromatogr B Biomed Sci Appl*, *707*(1–2), 151–159.

[34] Hasan, N. Y., Elkawy, M. A., & Elzeany, B. E. (2010). Rapid RP-HPLC method for quantitative determination of Lornoxicam in Tablets. *Journal of Basic and Clinical Pharmacy JBCP*.

[35] Bendale, A. R., Narkhede, S. P., Palande, S., Nagar, A. A., & Vidyasagar, G. L. (2013). Validated RP-HPLC method as a tool for estimation of lornoxicam in pharmaceutical dosage form. *Indo Am J Pharm Res*, *3*(7), 5491–5498.

[36] Amin, N. M., Sen, D. B., Khandhar, A. P., & Seth, A. K. (2012). Development and validation of stability indicating assay method for Lornoxicam & Tramadol in tablet dosage form by RP-HPLC. *Pharma Science Monitor*, *3*(2), 11–29.

22 COVID-19: Immergence, molecular mechanism of transmission and pathogenesis

Chinmaya Chidananda Behera[1], Sagar K Mishra[1], Biswajit Mishra[2], Gurudutta Pattnaik[1], Bikash Ranjan Jena[2], and Suman K Mekap[2,a]

[1]Pharmacognosy and Phytochemistry Division, University Department of Pharmaceutical Sciences, Utkal University, Vani Vihar, Bhubaneswar, Odisha, India

[2]Centurion University of Technology and Management, Ramchandrapus, Odisha, India

Abstract: Since December 2019, a novel form of pneumonia began spreading globally, raising widespread concern due to its link to a newly identified coronavirus, now commonly referred to as COVID-19. The virus has been officially named SARS-CoV-2 (Severe Acute Respiratory Syndrome Coronavirus 2) because of its genetic resemblance to the earlier SARS virus. In India, the situation became significantly more severe starting on March 22, 2020. By May 30, 2021, the World Health Organization (WHO) had documented around 169,118,995 confirmed infections and 3,519,175 mortality reports globally, indicating a mortality rate of approximately 2%. To combat the virus, many existing medications have been repurposed, and a wide range of compounds are being evaluated through both clinical and preclinical studies. However, a conclusive cure or treatment has yet to be found. This study aims to investigate the replication process of SARS-CoV-2 to identify potential therapeutic targets and to assess the challenges and possibilities in developing effective treatments for both current and future reports of the infection.

Keywords: COVID-19, pathogenesis, antiviral targets, *in-silico*, preclinical-clinical cases

1. Introduction

COVID-19 is a Inter-species transmissible infection caused by a newly discovered coronavirus, first referred to as 2019-nCoV. Given the widespread severity of the outbreak, the International Committee on Taxonomy of Viruses (ICTV) officially designated the virus as Severe Acute Respiratory Syndrome Coronavirus 2 (SARS-CoV-2) on February 11, 2020. The name 'SARS' was included due to the virus's genetic similarity to the one that caused the SARS outbreak in 2002 [1, 2]. Coronaviruses are part of the Coronaviridae family within the Nidovirales order. They are primarily classified into four main types: alpha (B.1.1.7), beta (B.1.351), gamma (P.1), and delta (B.1.617.2). The variant linked to the COVID-19 outbreak in 2019 belongs to the beta group and is recognized as the seventh known strain in the coronavirus family. On a genetic level, it shares approximately 70% similarity with the SARS-CoV virus that

[a]suman.mekap@cutm.ac.in

DOI: 10.1201/9781003672869-22

led to the 2002 epidemic [3, 4]. By the end of 2020 and the beginning of 2021, several new coronavirus variants had emerged worldwide. These included the D614G and B.1.1.7 strains identified in the United Kingdom, along with the B.1.617, B.1.617.1, and B.1.617.2 variants first reported in India [5–7]. By 30th May 2021, WHO recorded 169,118,994 confirmed infections of COVID-19 and 3,529,175 mortality reports globally, indicating an estimated case fatality rate of around 2%. Structurally, the coronavirus measures between 70 and 100 nanometers in size and features a protein envelope surrounding a large, positive-sense RNA genome [8] (Figure 22.1).

The membrane (M) protein is a triple transmembrane glycoprotein with an N-terminal exposed outside the virus that undergoes N-glycosylation. It facilitates host entry, contributes to pathogenesis, and is responsible for RNA replication. The C-terminal, located inside the virus and made up of 123 amino acids, binds to viral RNA and nucleocapsid (N) proteins to support RNA packaging and replication. The envelope (E) protein forms a pentameric ion channel, crucial for the virus assembly, release from the host cell, and disease progression. The viral genetic material is about 30–32 kilobases long and consists of a positive-sense, single-stranded RNA enclosed in a lipid bilayer. The nucleocapsid (N) protein associates with viral RNA in a helical arrangement to form the nucleocapsid complex, whose RNA-binding and oligomerization capabilities

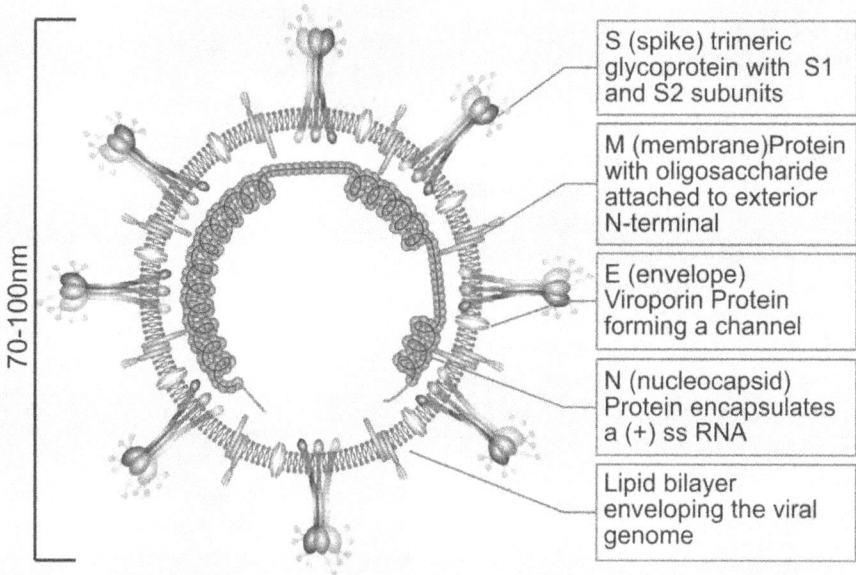

Figure 22.1. The structural characteristics of SARS-CoV-2 include the spike (S) protein, a homotrimeric glycoprotein that extends roughly 20 nanometers from the viral envelope with S1 and S2subunits, which are vital for facilitating membrane fusion by binding to the host cell receptor.

Source: Author.

are vital for the reproduction of the virus. Additionally, the RNA contains a 5' cap structure and a 3' polyadenylated (poly-A) tail [8].

2. History and Epidemiology

Prior to the COVID-19 pandemic, the world witnessed two significant coronavirus outbreaks. The first was Severe Acute Respiratory Syndrome (SARS) outbreak during 2002 and 2003 [9] in China's Guangdong province. This virus, a beta-coronavirus, was transmitted from bats to civets and then to humans [10], resulting in a global outbreak that caused 916 deaths and had an estimated case fatality rate of around 11%. Another epidemic, referred to as Middle East Respiratory Syndrome (MERS) outbreak of Saudi Arabia during 2012 and rapidly spread to multiple other nations. Also caused by a beta-coronavirus, MERS was transmitted from camels to humans and, by 2019, had resulted in about 2,494 confirmed cases and 858 deaths – equating to a case fatality rate of roughly 34%. The most recent coronavirus outbreak, caused by the novel virus initially labeled 2019-nCoV, appeared in Wuhan, China, in December 2019 or earlier. This is the seventh strain of coronavirus, which is identified to infect humans, following the patterns of SARS and MERS. MERS-CoV, the virus behind the MERS outbreak, became endemic about ten years after the SARS epidemic [11–13].

3. Transmission pattern

Coronaviruses are capable of infecting both birds and mammals, but bats are believed to carry the greatest variety of coronavirus strains. While they serve as natural reservoirs for many of these viruses, bats typically do not exhibit symptoms due to the unique strength and adaptability of their immune systems [14]. These viruses are capable of mutating their envelope proteins, enhancing their ability to bind to host cells. This adaptation increases their potential to cross species barriers and infect new hosts more easily [15, 16]. In humans, coronaviruses can lead to both respiratory and gastrointestinal infections. It is estimated that they are responsible for about 5–10% of upper respiratory tract illnesses in both adults and children, typically starting with symptoms of the common cold and potentially progressing to pneumonia or acute respiratory distress syndrome (ARDS). Viral mutations observed after each epidemic or endemic may be linked to shifts in host immune responses. In recent outbreaks, cross-species transmission has played a significant role in the emergence of new coronavirus strains [17].

According to the most recent guidelines from the WHO-China Joint Mission on Coronavirus Disease 2019 [18] and the International Pharmaceutical Federation (FIP) [19] health advisory, 2019-nCoV primarily spreads through three modes of transmission: direct contact, aerosol transmission, and contact via surfaces. In direct transmission, inhalation or ingestion of respiratory droplets can occur from infected person by a healthy individual, typically within a 6-foot range. Aerosol transmission occurs when respiratory droplets become airborne, forming aerosols that can be inhaled by others. In contact transmission, proper sanitation is crucial to prevent the virus's spread through surfaces, as the virus can be transferred to a host and replicate. Cases have also been observed where patients with digestive issues, such as diarrhea and abdominal discomfort, show single-cell transcriptomes of the virus in their digestive systems, suggesting potential contact transmission [20].

4. The Reproduction Cycle

Upon reaching the alveoli, the viral spike (S) glycoprotein occupies the host cell ACE2 receptor facilitating cellular entry [21]. These receptors are abundantly found on alveolar type II cells, which generate a surfactant – a vital lipoprotein that reduces surface tension in the alveoli and helps maintain their structure, ensuring efficient breathing [22]. After binding to ACE2, the spike protein facilitates cellular entry through two primary mechanisms: direct fusion with the host cell (early pathway), or through endocytosis (late pathway).

Several forms of endocytosis have been proposed for this process, including macropinocytosis, clathrin- and caveolae-dependent pathways, as well as clathrin- and caveolae-independent pathways [23, 24]. After occupying thr ACE2 receptor the spike protein undergoes a structural change, making itself vulnerable to cleavage by host proteases. This cleavage separates the spike into two subunits: S1 and S2, responsible for membrane fusion [25].

The cleavage can occur either at the host cell membrane or within endosomes. Different proteases are involved in these processes: TMPRSS2, a transmembrane serine protease, facilitates early entry by enabling direct release of the viral genome into the cytoplasm [26]. Alternatively, cathepsin B and L – cysteine proteases activated in mildly acidic conditions (~pH 6) – are involved in the endosomal route of entry [27, 28].

Cleavage of the spike protein at the S1/S2 junction initiates the insertion of the S2′fusion peptide into the host's cell membrane. This action exposes a previously hidden trimeric heptad repeat regions HR1 and HR2, which interact through hydrophobic forces to form a six-helix bundle commonly known as the fusion core. This structure facilitates Viral-host membrane fusion, permitting the viral nucleocapsid entry into the host cytoplasm [27, 28].

Upon cytoplasmic entry, the viral RNA is decoated & translated by host ribosomes into replicase polyproteins. The open reading frames (ORFs) being crucial at this point, produce polyprotein 1a (pp1a) upon translation of ORF1a and a larger polyprotein 1ab (pp1ab) upon translation of ORF1b. Undergoing an autoproteolytic fragmentation the polyprotein pp1a yields 11 non-structural proteins (nsp1 to nsp11), while pp1ab is processed into 15 non-structural proteins [29] (nsp1–nsp10 and nsp12–nsp16). Two viral proteases: nsp3 and nsp5 are responsible for this cleavage [30].

Coronaviruses have evolved a range of non-structural proteins with specialized roles in replication and transcription. These proteins form the replicase-transcriptase complex (RTC), which is anchored to the endoplasmic reticulum through the transmembrane domains of nsp3, nsp4, and nsp6. The double-membrane vesicles or DMVs which evolve due to this anchoring are crucial for viral replication.

Coronavirus replication begins with the (+)ssRNA genome, which acts as a template for the production of a complementary (−)ssRNA intermediate by the viral RNA-dependent RNA polymerase (RdRp). This (−)ssRNA strand subsequently serves as a template for generating new (+)ssRNA genomes, essential for viral genome propagation and packaging [31]. The virus can effectively hide its replication activities by temporarily shifting to negative-sense RNA.

At the same time, the replicase enzyme performs discontinuous transcription of the genome, producing a series of subgenomic mRNAs [32]. These subgenomic RNAs are translated into structural proteins – such as spike (S), membrane (M), envelope (E), and accessory proteins, which are folded and processed in the endoplasmic reticulum, then moved to the Endoplasmic Reticulum-Golgi intermediate compartment (ERGIC) via Golgi apparatus [33].

Meanwhile, nucleocapsid (N) proteins bind to the nascent viral RNA genome in the cytoplasm to form nucleocapsids. Assembly of new virions takes place in the ERGIC, where the M protein organizes the viral structure by interacting with N proteins. Although the specific role is not fully understood, furin protease is thought to assist in processing the spike protein during this stage. Once fully assembled, mature virions are packaged into smooth-walled vesicles and released from the host cell through exocytosis [29].

5. Pathogenesis During COVID-19

Viral replication within alveolar cells induces cellular damage, thereby initiating an inflammatory response. Damaged cells release inflammatory mediators such as interferons, cytokines, and signaling molecules that alert neighbouring cells and activate the body's defense mechanisms [34]. Alveolar macrophages

sense the cell damage and begin secreting pro-inflammatory cytokines, interleukins and chemokines.

TNF-α and IL-1β enhance vascular permeability and encourages the migration of immune cells such as neutrophils and monocytes to site of infection. IL-8 specifically draws neutrophils, while other chemokines guide monocytes to the area [35, 36]. This increased permeability leads to fluid accumulation in the alveoli, causing pulmonary edema, breathing difficulty (dyspnea), and reduced oxygen exchange, resulting in low blood oxygen levels (hypoxemia) [37].

Neutrophils begin clearing viruses and cellular debris through phagocytosis, but their toxic secretions can also harm surrounding lung tissue, disrupt surfactant production, and contribute to alveolar collapse [38]. In addition, immune cells may damage the lung's endothelial lining and release arachidonic acid metabolites, such as leukotrienes and prostaglandins, which cause bronchial constriction and reduce airflow [39].

Alveolar macrophages recognize the virus using toll-like receptors, engulf it via phagocytosis, process it, and present viral antigens to T and B cells – initiating an adaptive immune response [40]. Chronic inflammatory activity in the lung parenchyma sensitizes afferent nerve fibers, resulting in the initiation of a reflexive cough response. Inflammatory molecules like prostaglandins, interferon-alpha (INF-α), IL-1, and IL-6 contribute to fever symptoms during infection [41–43].

As blood oxygen levels drop, the brain's cardiopulmonary center is activated, increasing both respiratory and heart rates – resulting in tachypnea and tachycardia [44]. IL-6 induces the hepatic synthesis of C-reactive protein (CRP), a widely recognized biomarker employed in the diagnosis and monitoring of COVID-19 infection [45].

5.1. Other immunological pathways

The attachment of the viral spike protein to the host's ACE2 receptor, causes the receptor phosphorylation, a process regulated by casein kinase II (CK II). This event initiates a chain reaction within the cell that activates the ERK1/2 and AP-1 signaling pathways, leading to increased production of the chemokine CCL2, which is associated with lung fibrosis [46]. Moreover, virus-infected cells activate the PI3K/AKT/mTOR signaling cascade, which is responsible for cellular apoptosis and proliferation – a condition that the virus takes advantage of to enhance its own replication [47].

German research has also highlighted the importance of the calcineurin/NF-AT signaling pathway in coronavirus infection and immune activation. They discovered a strong link between the SARS-CoV non-structural protein Nsp1 and host proteins involved in NFAT signaling, suggesting that the virus manipulates this pathway to stimulate the immune system [43]. Calcineurin, also known as protein phosphatase 2B (PP2B), is a calcium-dependent enzyme that activates the NF-AT transcription factor by removing phosphate groups. This transcription factor regulates genes responsible for activating T cells (Table 22.1).

Table 22.1. Biochemical role of Coronavirus Nonstructural proteins (Nsps)

Nsps (1–16)	Biochemical role of Nonstructural Proteins (nsps)	Refs
nsp1	Binds to 40S ribosomes to degrade cellular mRNA, suppressing host translation and gene expression, while also blocking type I interferon (IFN) signaling and halting the cell cycle.	49
nsp2	It is proposed to modulate host cell survival signaling pathways through interaction with host proteins Prohibitin (PHB and PHB2), which are essential for maintaining mitochondrial integrity and cellular stress resistance. This interaction may also facilitate increased viral transmissibility.	50, 51
nsp3	Papain-like proteases (PL1pro and PL2pro), which participate in the processing of viral polyproteins, also modulate host immune responses by enhancing cytokine expression and suppressing innate immunity through deubiquitination. Localized on double-membrane vesicles (DMVs), these proteases antagonize type I interferon (IFN) responses and exhibit binding affinity for ADP-ribose and nucleic acids.	33, 51, 52
nsp4	Double-membrane vesicle (DMV) formation is facilitated by multiple nonstructural proteins, including nsp3 and nsp6 and serve as viral reproduction sites.	33, 53, 54
nsp5	Chymotrypsin-like protease (3CLpro) is involved in the processing of viral polyproteins and acts as a potent antagonist of IFN-β (type I interferon) by cleaving the NEMO (NF-κB essential modulator) complex.	33, 55

(continue)

Table 22.1. (continued)

Nsps (1–16)	Biochemical role of Nonstructural Proteins (nsps)	Refs
nsp6	In conjunction with nsp3, it is present over the surface of DMVs and aids in their formation. As coronavirus infection progresses, autophagy expansion are regulated, with DMVs exhibiting characteristics similar to autophagosomes of smaller diameter.	33, 56
nsp7	Nsp7, in association with nsp8, forms a hexadecameric complex that exhibits RNA-binding activity, which is crucial for both genomic and subgenomic RNA replication and transcription. Additionally, it serves as a cofactor for nsp8 and nsp12.	57
nsp8	Serves as a cofactor for nsp7; experiments have demonstrated that both the nsp7-nsp8 complex and nsp8 alone are capable of extending RNA primers beyond the length of the template, thereby functioning as a primase in viral replication.	57, 58
nsp9	Nsp9 dimerizes to facilitate effective RNA binding activity, with a conserved GXXXG sequence playing a critical role in maintaining the dimerization between its N- and C-terminal regions.	59, 60
nsp10	To evade host immunity, coronaviruses cap the 5' position of their RNA through methylation. nsp16 is the catalytic subunit, while nsp10 serves as the stimulatory subunit. Thus, nsp10 functions as a scaffold protein, coordinating the activity of both nsp16 and nsp14.	61
nsp11	Nsp11 functions as an endoribonuclease (NendoU) in members of the Arterivirus family; however, its precise role in coronaviruses remains unclear.	62
nsp12	Nsp12, also known as RNA-dependent RNA polymerase, along with the nsp7-nsp8 complex (a hexadecameric structure), is capable of initiating RNA synthesis via the de novo pathway.	57
nsp13	In conjunction with nsp12, nsp13 enhances helicase activity by up to twofold, increasing the step size of nucleic acid unwinding, whether for RNA-RNA or DNA-DNA strands. Nsp13 is a member of the SF1 helicase family and exhibits NTPase activity. Additionally, it possesses RNA 5'-triphosphatase activity.	63
nsp14	A distinctive characteristic of the Coronaviridae family is its proofreading activity, mediated by a 3'→5' exoribonuclease. This enzyme functions as an S-adenosylmethionine (SAM)-dependent guanine-N7 methyltransferase (N7-MTase) during RNA cap formation in viral RNA synthesis. Additionally, it forms a complex with nsp10.	64
nsp15	Nsp15 exhibits endonuclease activity (NendoU) with specificity for uridylate, a characteristic feature shared uniquely by nidoviruses when compared to other RNA viruses. It is also involved in the evasion of double-stranded RNA (dsRNA) sensors, and nsp15 partially restricts apoptosis in macrophages.	65
nsp16	In addition to nsp10 and nsp14, nsp16 is involved in RNA cap formation as a 2'-O-methyltransferase (2'-O-MTase). In Porcine epidemic diarrhea virus (PEDV), nsp10 enhances the activity of nsp16, negatively regulating innate immunity, particularly by inhibiting IFN-β production.	66

IFN, Interferon; ADP, Adenosine diphosphate
Source: Author.

By inhibiting calcineurin, T-cell activation can be suppressed, which in turn may reduce inflammation. Notably, the immunosuppressive drug Cyclosporin A has been found to inhibit coronavirus genome replication and/or transcription in laboratory studies [48] (Table 22.2).

6. Prospects of COVID-19 Drug discovery

Several obstacles complicate the development of anti-COVID drugs. One major challenge is that coronaviruses are among the most diverse and rapidly mutating viruses, and new strains continue to emerge unpredictably. Due to their zoonotic origin and frequent mutations in structural proteins, these viruses adapt easily across different geographic regions and human populations. As a result, drugs targeting viral entry have shown limited success in clinical trials.

However, certain conserved regions within viral proteins present promising opportunities for drug development, either through novel compounds or drug repurposing. Additionally, structural proteins involved in viral exit may also serve as viable drug targets. While drug repurposing [74–78] has opened new avenues for identifying potential treatments, gaps in understanding the virus's replication process and mechanisms continue to hinder the discovery of

Table 22.2. Results of some preclinical/clinical studies for identification of drug candidates

Results of Preclinical/Clinical study	Refs
The antiviral efficacy of chloroquine and hydroxychloroquine was assessed in multiple cell lines, including Vero E6, Vero 76, human epithelial lung cells, and Huh7 (human liver cell line). A total of 20 clinical studies were conducted in various hospitals across China.	67
Clinical trials were performed with 42 participants, comprising of 26 individuals in the treatment group and 16 in the control group. SARS-CoV-2 RNA levels were quantified by RT-PCR, and the serum concentrations of hydroxychloroquine and azithromycin were quantified by UHPLC.	68
The cytotoxicity and antiviral efficacy of the candidate drugs were evaluated in vitro using the CCK-8 assay in Vero E6 cells (ATCC-1586 for ribavirin, penciclovir, itazoxanide, nafamostat, chloroquine, remdesivir and favipiravir, against the clinical isolate of 2019-nCoV.	69
Teicoplanin, a glycopeptide antibiotic used primarily for Gram-positive bacterial infections, has shown antiviral activity against viruses including Ebola, influenza, flavivirus, hepatitis C, HIV, and MERS-CoV. Teicoplanin inhibited SARS-CoV with an IC50 of 1.66 µM.	70
A meta-analysis of data from 15 studies, encompassing 5,270 patients with coronavirus pneumonia, revealed an association between corticosteroid treatment and an elevated mortality rate.	71
According to the COVID-19 treatment guidelines in China, a combination of 5 million units of inhaled interferon alpha (IFNα) and ribavirin is recommended.	72
A clinical evaluation was conducted to assess the impact of protease inhibitors like E-64d (25 mM) and camostat mesylate (1–500 mM), on the host cell entry.	73

PCR, Polymerase chain reaction; mM, millimoles
Source: Author.

more effective molecules. Nonetheless, non-structural proteins remain critical targets in the search for therapeutic agents [78]. Fragment-based de novo drug design is also being explored as a promising strategy to develop compounds capable of targeting multiple viral components [79].

Another challenge lies in the availability of appropriate animal models for testing anti-SARS-CoV-2 therapies. Both small animal and non-human primate models are in short supply, and there are only a limited number of biosafety level 3 (BSL-3) labs equipped to carry out such research. Technical limitations further complicate these efforts [80].

To address these challenges, a deeper and more comprehensive understanding of the virus's transmission, replication, and disease-causing mechanisms is essential for developing effective broad-spectrum antiviral agents against SARS-CoV-2.

7. Acknowledgements

The authors are thankful to Centurion University of Technology and Management, Bhubaneswar, Odisha, India; School of Pharmaceutical Sciences, and University Department of Pharmaceutical Sciences, Utkal University, Vani Vihar, Bhubaneswar, Odisha, India for encouragement and support to carry out this study.

References

[1] Shereen, M. A., Khan, S., Kazmi, A., Bashir, N., & Siddique, R. (2020). COVID-19 infection: Origin, transmission, and characteristics of human coronaviruses. *Journal of Advanced Research, 24*, 91–98.

[2] Harapan, H., *et al.* (2020). Coronavirus disease 2019 (COVID-19): A literature review. *Journal of Infection and Public Health, 13*, 667–673.

[3] Rabi, F. A., Al Zoubi, M. S., Al-Nasser, A. D., Kasasbeh, G. A., & Salameh, D. M. (2020). Sars-cov-2 and coronavirus disease 2019: What we know so far. *Pathogens, 9*, 231.

[4] WHO. (2020). WHO | Novel Coronavirus – China. *World Health Organization.* https://www.who.int/csr/don/12-january-2020-novel-coronavirus-china/en/

[5] Schoeman, D., & Fielding, B. C. (2019). Coronavirus envelope protein: Current knowledge. *Virology Journal, 16*, 1–22.

[6] Wang, L. F., *et al.* (2006). Review of bats and SARS. *Emerging Infectious Diseases, 12*, 1834–1840.

[7] Ji, W., Wang, W., Zhao, X., Zai, J., & Li, X. (2020). Cross-species transmission of the newly identified coronavirus 2019-nCoV. *J Med Virol, 92*, 433–440.

[8] Fung, W. K., & Yu, P. L. H. (2003). SARS case-fatality rates [2]. *CMAJ, 169*, 277–278.

[9] Alsolamy, S., & Arabi, Y. M. (2015). Infection with Middle East respiratory syndrome coronavirus. *Can J Respir Ther CJRT = Rev Can la Ther Respir RCTR, 51*, 102.

[10] Mahase, E. (2020). Coronavirus covid-19 has killed more people than SARS and MERS combined, despite lower case fatality rate. *BMJ, 368*, m641.

[11] World Health Organization (WHO). (2021). Case and death count data (by WHO Region). *Data as received by WHO through official communications under the International Health Regulations (IHR, 2005).* https://covid19.who.int/info, retrieved on 30th may 2021 at 8:15pm.

[12] da Costa, V. G., Moreli, M. L., & Saivish, M. V. (2020). The emergence of SARS, MERS and novel SARS-2

coronaviruses in the 21st century. *Archives of Virology*, *1*, 3.

[13] Liu, Y., Gayle, A. A., Wilder-Smith, A., & Rocklöv, J. (2020). The reproductive number of COVID-19 is higher compared to SARS coronavirus. *J Travel Med*, *27*, 1–4.

[14] Brook, C. E., *et al.* (2020). Accelerated viral dynamics in bat cell lines, with implications for zoonotic emergence. *Elife*, *9*.

[15] Belouzard, S., Millet, J. K., Licitra, B. N., & Whittaker, G. R. (2012). Mechanisms of coronavirus cell entry mediated by the viral spike protein. *Viruses*, *4*, 1011–1033.

[16] Menachery, V. D., Graham, R. L., & Baric, R. S. (2017). Jumping species—a mechanism for coronavirus persistence and survival. *Current Opinion in Virology*, *23*, 1–7.

[17] MacIntyre, C. R., *et al.* (2017). Viral and bacterial upper respiratory tract infection in hospital health care workers over time and association with symptoms. *BMC Infect Dis*, *17*, 553.

[18] World Health Organization (WHO). (2020). *Report of the WHO-China Joint Mission on Coronavirus Disease 2019 (COVID-19). The WHO-China Joint Mission on Coronavirus Disease 2019*, *1*. https://www.who.int/docs/default-source/coronaviruse/who-china-joint-mission-on-covid-19-final-report.pdf.

[19] International Pharmaceutical Federation (FIP). (2020). Coronavirus SARS-CoV-2 / COVID-19 pandemic: Information and Guidelines for Pharmacists and the Pharmacy Workforce, *48*.

[20] Li, L. Y., *et al.* (2020). Digestive system involvement of novel coronavirus infection: Prevention and control infection from a gastroenterology perspective. *Journal of Digestive Diseases*, *21*, 199–204.

[21] Wan, Y., Shang, J., Graham, R., Baric, R. S., & Li, F. (2020). Receptor recognition by the novel coronavirus from

Wuhan: An analysis based on decade-long structural studies of SARS coronavirus. *J Virol*, *94*.

[22] Nkadi, P. O., Merritt, T. A., & Pillers, D. A. M. (2009). An overview of pulmonary surfactant in the neonate: Genetics, metabolism, and the role of surfactant in health and disease. *Molecular Genetics and Metabolism*, *97*, 95–101.

[23] Yang, N., & Shen, H. M. (2020). Targeting the endocytic pathway and autophagy process as a novel therapeutic strategy in COVID-19. *International Journal of Biological Sciences*, *16*, 1724–1731.

[24] Wang, H., *et al.* (2008). SARS coronavirus entry into host cells through a novel clathrin- and caveolae-independent endocytic pathway. *Cell Res*, *18*, 290–301.

[25] Stebbing, J., *et al.* (2020). COVID-19: Combining antiviral and anti-inflammatory treatments. *The Lancet Infectious Diseases*, *20*, 400–402.

[26] McKee, D. L., Sternberg, A., Stange, U., Laufer, S., & Naujokat, C. (2020). Candidate drugs against SARS-CoV-2 and COVID-19. *Pharmacological Research*, *157*.

[27] Xia, S., *et al.* (2020). Inhibition of SARS-CoV-2 (previously 2019-nCoV) infection by a highly potent pan-coronavirus fusion inhibitor targeting its spike protein that harbors a high capacity to mediate membrane fusion. *Cell Res*, *30*, 343–355.

[28] Xia, S., *et al.* (2019). A pan-coronavirus fusion inhibitor targeting the HR1 domain of human coronavirus spike. *Sci Adv*, *5*.

[29] Tang, T., Bidon, M., Jaimes, J. A., Whittaker, G. R., & Daniel, S. (2020). Coronavirus membrane fusion mechanism offers a potential target for antiviral development. *Antiviral Research*, *178*, 104792.

[30] Krichel, B., Falke, S., Hilgenfeld, R., Redecke, L., & Uetrecht, C. (2020).

Processing of the SARS-CoV pp1a/ab nsp7–10 region. *Biochem J, 477,* 1009–1019.

[31] Yin, C. (2020). Genotyping coronavirus SARS-CoV-2: Methods and implications. *Genomics.* doi:10.1016/j.ygeno.2020.04.016.

[32] Sola, I., Almazán, F., Zúñiga, S., & Enjuanes, L. (2015). Continuous and discontinuous RNA synthesis in coronaviruses. *Annu Rev Virol, 2,* 265–288.

[33] Chen, Y., Liu, Q., & Guo, D. (2020). Emerging coronaviruses: Genome structure, replication, and pathogenesis. *Journal of Medical Virology, 92,* 418–423.

[34] Newton, A. H., Cardani, A., & Braciale, T. J. (2016). The host immune response in respiratory virus infection: Balancing virus clearance and immunopathology. *Seminars in Immunopathology, 38,* 471–482.

[35] Moore, B. B., & Kunkel, S. L. (2019). Attracting attention: Discovery of IL-8/CXCL8 and the Birth of the Chemokine Field. *J Immunol, 202,* 3–4.

[36] Duque, G. A., & Descoteaux, A. (2014). Macrophage cytokines: Involvement in immunity and infectious diseases. *Frontiers in Immunology, 5.*

[37] Elbehairy, A. F., *et al.* (2015). Pulmonary gas exchange abnormalities in mild chronic obstructive pulmonary disease: Implications for dyspnea and exercise intolerance. *Am J Respir Crit Care Med, 191,* 1384–1394.

[38] Cheng, O. Z., & Palaniyar, N. (2013). NET balancing: A problem in inflammatory lung diseases. *Front Immunol, 4.*

[39] Robb, C. T., Regan, K. H., Dorward, D. A., & Rossi, A. G. (2016). Key mechanisms governing resolution of lung inflammation. *Seminars in Immunopathology, 38,* 425–448.

[40] Lester, S. N., & Li, K. (2014). Toll-like receptors in antiviral innate immunity. *Journal of Molecular Biology, 426,* 1246–1264.

[41] Alberts, B., *et al.* (2008). T Cells and MHC Proteins. In *Molecular Biology of the Cell,* 1569–1588 (Garland Science).

[42] Polverino, M., *et al.* (2012). Anatomy and neuro-pathophysiology of the cough reflex arc. *Multidisciplinary Respiratory Medicine, 7,* 5.

[43] Pfefferle, S., *et al.* (2011). The SARS-Coronavirus-host interactome: Identification of cyclophilins as target for pan-Coronavirus inhibitors. *PLoS Pathog, 7.*

[44] Çimen, T. *et al.* (2015). Hypotension, tachycardia, and tachypnea in a patient with coronary artery disease. *Anadolu Kardiyoloji Dergisi, 15,* 430.

[45] Wang, L. (2020). C-reactive protein levels in the early stage of COVID-19. *Med Mal Infect, 50,* 332–334.

[46] Chen, I.-Y., *et al.* (2010). Upregulation of the Chemokine (C-C Motif) Ligand 2 via a Severe Acute Respiratory Syndrome Coronavirus Spike-ACE2 Signaling Pathway. *J Virol, 84,* 7703–7712.

[47] Diehl, N., & Schaal, H. (2013). Make yourself at home: Viral hijacking of the PI3K/Akt signaling pathway. *Viruses, 5,* 3192–3212.

[48] Willicombe, M., Thomas, D., & McAdoo, S. (2020). COVID-19 and Calcineurin Inhibitors: Should They Get Left Out in the Storm? *J Am Soc Nephrol,* ASN.2020030348. doi:10.1681/ASN.2020030348.

[49] Huang, C., *et al.* (2011). SARS coronavirus nsp1 protein induces template-dependent endonucleolytic cleavage of mRNAs: Viral mRNAs are resistant to nsp1-induced RNA cleavage. *PLoS Pathog, 7.*

[50] Graham, R. L., Sims, A. C., Brockway, S. M., Baric, R. S., & Denison, M. R. (2005). The nsp2 replicase proteins of murine hepatitis virus and severe acute respiratory syndrome coronavirus are dispensable for viral replication. *J Virol, 79,* 13399–13411.

[51] Angeletti, S., *et al.* (2020). COVID-2019: The role of the nsp2 and nsp3 in its pathogenesis. *J Med Virol, 92*, 584–588.

[52] Lei, J., Kusov, Y., & Hilgenfeld, R. (2018). Nsp3 of coronaviruses: Structures and functions of a large multi-domain protein. *Antiviral Research, 149*, 58–74.

[53] Sakai, Y., *et al.* (2017). Two-amino acids change in the nsp4 of SARS coronavirus abolishes viral replication. *Virology, 510*, 165–174.

[54] J Alsaadi, E. A., & Jones, I. M. (2019). Membrane binding proteins of coronaviruses. *Future Virology, 14*, 275–286.

[55] Tahir ul Qamar, M., Alqahtani, S. M., Alamri, M. A., & Chen, L. L. (2020). Structural basis of SARS-CoV-2 3CLpro and anti-COVID-19 drug discovery from medicinal plants. *J Pharm Anal.* doi:10.1016/j.jpha.2020.03.009.

[56] Cottam, E. M., Whelband, M. C., & Wileman, T. (2014). Coronavirus NSP6 restricts autophagosome expansion. *Autophagy, 10*, 1426–1441.

[57] Kirchdoerfer, R. N., & Ward, A. B. (2019). Structure of the SARS-CoV nsp12 polymerase bound to nsp7 and nsp8 co-factors. *Nat Commun, 10*, 1–9.

[58] Zhai, Y., *et al.* (2005). Insights into SARS-CoV transcription and replication from the structure of the nsp7-nsp8 hexadecamer. *Nat Struct Mol Biol, 12*, 980–986.

[59] Zeng, Z. *et al.* (2018). Dimerization of coronavirus nsp9 with diverse modes enhances its nucleic acid binding affinity. *J Virol, 92.*

[60] Prajapat, M., *et al.* (2020). Drug targets for corona virus: A systematic review. *Indian Journal of Pharmacology, 52*, 56–65.

[61] Wang, Y., *et al.* (2015). Coronavirus nsp10/nsp16 methyltransferase can be targeted by nsp10-derived peptide in vitro and in vivo to reduce replication and pathogenesis. *J Virol, 89*, 8416–8427.

[62] Zhang, M. *et al.* (2017). Structural biology of the arterivirus nsp11 endoribonucleases. *J Virol, 91.*

[63] Jang, K. J., *et al.* (2020). A high ATP concentration enhances the cooperative translocation of the SARS coronavirus helicase nsP13 in the unwinding of duplex RNA. *Sci Rep, 10*, 1–13.

[64] Bouvet, M., *et al.* (2012). RNA 3'-end mismatch excision by the severe acute respiratory syndrome coronavirus non-structural protein nsp10/nsp14 exoribonuclease complex. *Proc Natl Acad Sci U S A, 109*, 9372–9377.

[65] Kim, Y. *et al.* (2020). Crystal structure of Nsp15 endoribonuclease NendoU from SARS-CoV-2. *Protein Sci,* pro.3873. doi:10.1002/pro.3873.

[66] Menachery, V. D., Debbink, K., & Baric, R. S. (2014). Coronavirus non-structural protein 16: Evasion, attenuation, and possible treatments. *Virus Research, 194*, 191–199.

[67] Colson, P., Rolain, J. M., Lagier, J. C., Brouqui, P., & Raoult, D. (2020). Chloroquine and hydroxychloroquine as available weapons to fight COVID-19. *International Journal of Antimicrobial Agents, 55.*

[68] Gautret, P., *et al.* (2020). Hydroxychloroquine and azithromycin as a treatment of COVID-19: Results of an open-label non-randomized clinical trial. *Int J Antimicrob Agents*, 105949. doi:10.1016/j.ijantimicag.2020.105949.

[69] Wang, M., *et al.* (2020). Remdesivir and chloroquine effectively inhibit the recently emerged novel coronavirus (2019-nCoV) in vitro. *Cell Research, 30*, 269–271.

[70] Baron, S. A., Devaux, C., Colson, P., Raoult, D., & Rolain, J. M. (2020). Teicoplanin: An alternative drug for the treatment of COVID-19? *International Journal of Antimicrobial Agents, 55.*

[71] Yang, Z., *et al.* (2020). The effect of corticosteroid treatment on patients with coronavirus infection: A systematic review

and meta-analysis. *Journal of Infection.* doi:10.1016/j.jinf.2020.03.062.

[72] Dong, L., Hu, S., & Gao, J. (2020). Discovering drugs to treat coronavirus disease 2019 (COVID-19). *Drug Discov Ther, 14*, 58–60.

[73] Hoffmann, M., *et al.* (2020). SARS-CoV-2 cell entry depends on ACE2 and TMPRSS2 and is blocked by a clinically proven protease inhibitor. *Cell, 181*, 271–280.e8.

[74] Yuan, M., *et al.* (2020). A highly conserved cryptic epitope in the receptor binding domains of SARS-CoV-2 and SARS-CoV. *Science, 368*, 630–633.

[75] Robson, B. (2020). Computers and viral diseases. Preliminary bioinformatics studies on the design of a synthetic vaccine and a preventative peptidomimetic antagonist against the SARS-CoV-2 (2019-nCoV, COVID-19) coronavirus. *Comput Biol Med, 119*, 103670.

[76] Lan, J., *et al.* (2020). Structure of the SARS-CoV-2 spike receptor-binding domain bound to the ACE2 receptor. *Nature, 581*, 215–220.

[77] Wu, Y., *et al.* (2020). A noncompeting pair of human neutralizing antibodies block COVID-19 virus binding to its receptor ACE2. *Science, 368*(6496), eabc2241. doi:10.1126/science.abc2241.

[78] Li, G., & De Clercq, E. (2020). Therapeutic options for the 2019 novel coronavirus (2019-nCoV). *Nature Reviews. Drug Discovery, 19*, 149–150.

[79] Choudhury, C. (2020). Fragment tailoring strategy to design novel chemical entities as potential binders of novel corona virus main protease. *J Biomol Struct Dyn,* 1–15. doi:10.1080/073911 02.2020.1771424.

[80] Zumla, A., Chan, J. F. W., Azhar, E. I., Hui, D. S. C., & Yuen, K. Y. (2016). Coronaviruses-drug discovery and therapeutic options. *Nature Reviews Drug Discovery, 15*, 327–347.

For Product Safety Concerns and Information please contact our EU
representative GPSR@taylorandfrancis.com
Taylor & Francis Verlag GmbH, Kaufingerstraße 24, 80331 München, Germany

www.ingramcontent.com/pod-product-compliance
Lightning Source LLC
Chambersburg PA
CBHW060358220326
41598CB00023B/2961

* 9 7 8 1 0 4 1 1 4 0 8 1 8 *